Lecture Notes in Computer Science 13388

More information about this series at https://link.springer.com/bookseries/558

Ivan Varzinczak (Ed.)

Foundations of Information and Knowledge Systems

12th International Symposium, FoIKS 2022
Helsinki, Finland, June 20–23, 2022
Proceedings

Springer

Editor
Ivan Varzinczak
Université d'Artois and CNRS
Lens, France

ISSN 0302-9743 ISSN 1611-3349 (electronic)
Lecture Notes in Computer Science
ISBN 978-3-031-11320-8 ISBN 978-3-031-11321-5 (eBook)
https://doi.org/10.1007/978-3-031-11321-5

This Springer imprint is published by the registered company Springer Nature Switzerland AG
The registered company address is: Gewerbestrasse 11, 6330 Cham, Switzerland

Preface

These proceedings contain the papers that have been selected for presentation at the 12th International Symposium on Foundations of Information and Knowledge Systems, FoIKS 2022. The symposium was held during June 20–23, 2022, at the University of Helsinki in Finland.

The FoIKS symposia provide a biennial forum for presenting and discussing theoretical and applied research on information and knowledge systems. The goal is to bring together researchers with an interest in this subject, share research experiences, promote collaboration, and identify new issues and directions for future research. Previous FoIKS meetings were held in Schloss Salzau (Germany, 2002), Vienna (Austria, 2004), Budapest (Hungary, 2006), Pisa (Italy, 2008), Sofia (Bulgaria, 2010), Kiel (Germany, 2012), Bordeaux (France, 2014), Linz (Austria, 2016), Budapest (Hungary, 2018), and Dortmund (Germany, 2020).

The call for papers solicited original contributions dealing with any foundational aspect of information and knowledge systems, including submissions that apply ideas, theories, or methods from specific disciplines to information and knowledge systems. Examples of such disciplines are discrete mathematics, logic and algebra, model theory, databases, information theory, complexity theory, algorithmics and computation, statistics, and optimization.

Traditionally, the FoIKS symposia are a forum for intensive discussion where speakers are given sufficient time to present their ideas and results within the larger context of their research. Furthermore, participants are asked to prepare a first response to another contribution in order to initiate discussion.

FoIKS 2022 received 21 submissions, which were evaluated by the Program Committee on the basis of their significance, novelty, technical soundness, and appropriateness for the FoIKS audience. Each paper was subjected to three reviews (two reviews in only one case). In the end, 13 papers were selected for oral presentation at the symposium and publication in the archival proceedings.

We were delighted to have five outstanding keynote speakers. The abstracts of their talks were included in this volume:

- Patricia Bouyer-Decitre (CNRS, France): *Memory complexity for winning games on graphs*
- Gabriele Kern-Isberner (Technische Universität Dortmund, Germany): *The Relevance of Formal Logics for Cognitive Logics, and Vice Versa*
- Jussi Rintanen (Aalto University, Finland): *More automation to software engineering*
- Jouko Väänänen (University of Helsinki, Finland): *Dependence logic: Some recent developments*
- Zeev Volkovich (ORT Braude College of Engineering, Israel): *Text classification using "imposter" projections method*

We want to thank all the people who contributed to making FoIKS 2022 a successful meeting. In particular, we thank the invited speakers for their inspirational talks, the authors for providing their high-quality submissions and revising and presenting their work, and all the attendees for contributing to the discussions at the symposium. We thank the Program Committee and the external reviewers for their prompt and careful reviews.

We extend our thanks to the Local Organizing Committee, chaired by Matti Järvisalo and Juha Kontinen, who have always been proactive and responsive throughout the whole organization process, and to the FoIKS Steering Committee members for their trust and support. We gratefully acknowledge the financial support of FoIKS 2022 by the University of Helsinki, the Finnish Academy of Science and Letters, and the Artificial Intelligence Journal.

June 2022 Ivan Varzinczak

Organization

Program Chair

Ivan Varzinczak	CRIL, Université d'Artois, and CNRS, France

Program Committee

Yamine Ait Ameur	IRIT/INPT-ENSEEIHT, France
Kim Bauters	University of Bristol, UK
Christoph Beierle	FernUniversität in Hagen, Germany
Leopoldo Bertossi	Universidad Adolfo Ibanez, Chile
Meghyn Bienvenu	CNRS and University of Bordeaux, France
Joachim Biskup	Technische Universität Dortmund, Germany
Elena Botoeva	University of Kent, UK
Arina Britz	CAIR, Stellenbosch University, South Africa
Giovanni Casini	ISTI-CNR, Italy
Fabio Cozman	University of São Paulo, Brazil
Dragan Doder	Utrecht University, The Netherlands
Thomas Eiter	Vienna University of Technology, Austria
Christian Fermüller	Vienna University of Technology, Austria
Flavio Ferrarotti	Software Competence Center Hagenberg, Austria
Laura Giordano	Università del Piemonte Orientale, Italy
Dirk Van Gucht	Indiana University Bloomington, USA
Marc Gyssens	Universiteit Hasselt, Belgium
Andreas Herzig	IRIT, CNRS, and Université de Toulouse, France
Tomi Janhunen	Tampere University, Finland
Matti Järvisalo	University of Helsinki, Finland
Gabriele Kern-Isberner	Technische Universität Dortmund, Germany
Elena Kleiman	Braude College Karmiel, Israel
Sébastien Konieczny	CRIL, CNRS, and Université d'Artois, France
Juha Kontinen	University of Helsinki, Finland
Antti Kuusisto	Tampere University, Finland
Sebastian Link	University of Auckland, New Zealand
Thomas Lukasiewicz	University of Oxford, UK
Pierre Marquis	CRIL, Université d'Artois, and CNRS, France
Thomas Meyer	CAIR, University of Cape Town, South Africa
Nicola Olivetti	LSIS, Aix-Marseille University, France
Iliana Petrova	Inria - Sophia Antipolis, France

Ramon Pino Perez	CRIL, Université d'Artois, and CNRS, France
Gian Luca Pozzato	Università di Torino, Italy
Sebastian Rudolph	TU Dresden, Germany
Attila Sali	Alfréd Rényi Institute, Hungary
Katsuhiko Sano	Hokkaido University, Japan
Kostyantyn Shchekotykhin	Alpen-Adria University Klagenfurt, Austria
Klaus-Dieter Schewe	Zhejiang University, China
Steven Schockaert	Cardiff University, UK
Guillermo Simari	Universidad del Sur in Bahia Blanca, Argentina
Mantas Simkus	Vienna University of Technology, Austria
Michael Sioutis	Otto-Friedrich-University Bamberg, Germany
Umberto Straccia	ISTI-CNR, Italy
Karim Tabia	CRIL, Université d'Artois, and CNRS, France
Bernhard Thalheim	University of Kiel, Germany
Alex Thomo	University of Victoria, Canada
David Toman	University of Waterloo, Canada
Qing Wang	Australian National University, Australia
Renata Wassermann	University of São Paulo, Brazil
Stefan Woltran	Vienna University of Technology, Austria

Local Organizing Chairs

Matti Järvisalo	University of Helsinki, Finland
Juha Kontinen	University of Helsinki, Finland

Local Organizing Committee

Jeremias Berg	University of Helsinki, Finland
Miika Hannula	University of Helsinki, Finland
Minna Hirvonen	University of Helsinki, Finland
Tuomo Lehtonen	University of Helsinki, Finland
Andreas Niskanen	University of Helsinki, Finland
Max Sandström	University of Helsinki, Finland

Additional Reviewers

Tamara Drucks
Anna Rapberger

Abstracts of Invited Talks

Memory Complexity for Winning Games on Graphs

Patricia Bouyer-Decitre

CNRS, France

Abstract. Two-player games are relevant models for reactive synthesis, with application to the verification of systems. Winning strategies are then viewed as controllers, which guarantee that the system under control will satisfy its specification. In this context, simpler controllers will be easier to implement. Simplicity will refer to the memory complexity, that is, how much memory is needed to win the games. We will in particular discuss cases where the required memory is finite.

This talk will be based on recent works made with my colleagues Mickael Randour and Pierre Vandenhove.

The Relevance of Formal Logics for Cognitive Logics, and Vice Versa

Gabriele Kern-Isberner

Technische Universität Dortmund, Germany

Abstract. Classical logics like propositional or predicate logic have been considered as the gold standard for rational human reasoning, and hence as a solid, desirable norm on which all human knowledge and decision making should be based, ideally. For instance, Boolean logic was set up as kind of an arithmetic framework that should help make rational reasoning computable in an objective way, similar to the arithmetics of numbers. Computer scientists adopted this view to (literally) implement objective knowledge and rational deduction, in particular for AI applications. Psychologists have used classical logics as norms to assess the rationality of human commonsense reasoning. However, both disciplines could not ignore the severe limitations of classical logics, e.g., computational complexity and undecidedness, failures of logic-based AI systems in practice, and lots of psychological paradoxes. Many of these problems are caused by the inability of classical logics to deal with uncertainty in an adequate way. Both disciplines have used probabilities as a way out of this dilemma, hoping that numbers and the Kolmogoroff axioms can do the job (somehow). However, psychologists have been observing also lots of paradoxes here (maybe even more).

So then, are humans hopelessly irrational? Is human reasoning incompatible with formal, axiomatic logics? In the end, should computer-based knowledge and information processing be considered as superior in terms of objectivity and rationality?

Cognitive logics aim at overcoming the limitations of classical logics and resolving the observed paradoxes by proposing logic-based approaches that can model human reasoning consistently and coherently in benchmark examples. The basic idea is to reverse the normative way of assessing human reasoning in terms of logics resp. probabilities, and to use typical human reasoning patterns as norms for assessing the cognitive quality of logics. Cognitive logics explore the broad field of logic-based approaches between the extreme points marked by classical logics and probability theory with the goal to find more suitable logics for AI applications, on the one hand, and to gain more insights into the rational structures of human reasoning, on the other.

More Automation to Software Engineering

Jussi Rintanen

Aalto University, Finland

Abstract. My background in AI is in knowledge representation and later in automated planning and other problems that can be solved with automated reasoning methods. When challenged about the applications of my research, I identified software production as a promising area to look at, differently from many in automated planning who have considered robotics or generic problem-solving as the potential future applications of their work. In this talk, I will describe my work since in 2016 on producing software fully automatically from high-level specifications of its functionalities. I will compare my approach to existing commercial tools and to works in academic AI research, especially to models used in automated planning and knowledge representation, and discuss my experiences in moving research from academia to real world use.

Dependence Logic: Some Recent Developments

Jouko Väänänen

University of Helsinki, Finland

Abstract. In the traditional so-called Tarski's Truth Definition the semantics of first-order logic is defined with respect to an assignment of values to the free variables. A richer family of semantic concepts can be modelled if semantics is defined with respect to a set (a "team") of such assignments. This is called team semantics. Examples of semantic concepts available in team semantics but not in traditional Tarskian semantics are the concepts of dependence and independence. Dependence logic is an extension of first-order logic based on team semantics. It has emerged that teams appear naturally in several areas of sciences and humanities, which has made it possible to apply dependence logic and its variants to these areas. In my talk, I will give a quick introduction to the basic ideas of team semantics and dependence logic as well as an overview of some new developments, such as an analysis of dependence and independence in terms of diversity, a quantitative analysis of team properties in terms of dimension, and a probabilistic independence logic inspired by the foundations of quantum mechanics.

Text Classification Using "Imposter" Projections Method

Zeev Volkovich

Ort Braude College of Engineering, Israel

Abstract. The paper presents a novel approach to text classification. The method comprehends the considered documents as random samples generated from distinct probability sources and tries to estimate a difference between them through random projections. A deep learning mechanism composed, for example, of a Convolutional Neural Network (CNN) and a Bi-Directional Long Short-Term Memory Network (Bi-LSTM) deep networks together with an appropriate word embedding is applied to estimate the "imposter" projections using imposters data collection. As a result, each studied document is transformed into a digital signal aiming to exploit signal classification methods. Examples of the application of the suggested method include studies of the Shakespeare Apocrypha, the New and Old Testament, and the Abu Hamid Al Ghazali creations particularly.

Contents

On Sampling Representatives of Relational Schemas with a Functional Dependency

Maximilian Berens and Joachim Biskup(✉)

Fakultät für Informatik, Technische Universität Dortmund, 44227 Dortmund, Germany
{maximilian.berens,joachim.biskup}@cs.tu-dortmund.de

Abstract. We further contribute to numerous efforts to provide tools for generating sample database instances, complement a recent approach to achieve a uniform probability distribution over all samples of a specified size, and add new insight to the impact of schema normalization for a relational schema with one functional dependency and size restrictions on the attribute domains. These achievements result from studying the problem how to probabilistically generate a relation instance that is a representative of a class of equivalent or similar instances, respectively. An instance is equivalent to another instance if there are bijective domain mappings under which the former one is mapped on the other one. An instance is similar to another instance if they share the same combinatorial counting properties that can be understood as a solution to a layered system of equalities and lessthan-relationships among (non-negative) integer variables and some (non-negative) integer constants. For a normalized schema, the two notions turn out to coincide. Based on this result, we conceptually design and formally verify a probabilistic generation procedure that provides a random representative of a randomly selected class, i.e., each class is represented with the same probability or, alternatively, with the probability reflecting the number of its members. We also discuss the performance of a prototype implementation and further optimizations. For a non-normalized schema, however, the coincidence of the respective notions does not hold. So we only present some basic features of these notions, including a relationship to set unification.

Keywords: Combinatorial analysis · Domain cardinality · Duplicate-freeness · Functional dependency · Instance equivalence · Instance similarity · Integer partition · Normalized schema · Non-normalized schema · Random representative · Relational schema · Set unification · System of relationships among integer variables · Uniform probability distribution

1 Introduction

Since many years, there have been numerous efforts to provide tools for generating sample database instances under various aspects of software engineering, see

I. Varzinczak (Ed.): FoIKS 2022, LNCS 13388, pp. 1–19, 2022.
https://doi.org/10.1007/978-3-031-11321-5_1

Table 1. Array representations $r_1[]$, $r_2[]$ and $r_3[]$, also indicating colored domain renamings to show the equivalence of r_1 with r_2.

$r_1[]$	A	B	C
	1	1	1
	2	2	1
	3	2	1
	4	2	1
	4	2	2

$r_2[]$	A / A	B / B	C
	1 / 2	1 / 2	1
	2 / 3	1 / 2	1
	3 / 1	2 / 1	1
	4	1 / 2	1
	4	1 / 2	2

$r_3[]$	A	B	C
	1	1	1
	2	2	1
	3	2	2
	4	2	1
	4	2	2

[2,14] just to mention two of them, or inspect the section on related work in our previous articles [4,5]. To perform reasonably unbiased experiments, a software engineer will expect the samples generated to be random in some situation-specific sense. As a default, an engineer might want to achieve a *uniform* probability distribution over *all* possible samples. We followed such a wish in our previous work [4,5] for the special case that relation instances of a relational schema with just one functional dependency should be generated.

As already discussed but left open in [4], more sophisticated experimental situations might demand for achieving probabilistic uniformity over some more specific range of possibilities. For our case, dealing with both normalized and non-normalized relational schemas, in this article we investigate how to achieve uniformity over *representatives* of suitably defined *classes* of relation instances. Intuitively, two relation instances are seen to belong to a same class if they *share* some *structural properties* of interest and might only *differ* in the *actually occurring values* taking from the declared domains. We will consider two approaches to formalize this intuition, called *equivalence* and *similarity*, respectively.

The notion of equivalence focuses on the abstraction from concrete values: formally, a relation instance r_1 is defined to be *equivalent* to a relation instance r_2 if there are *bijective domain renamings* under which r_1 is mapped on r_2. Notably, such mappings *preserve all equalities and inequalities* among values occurring under the same attribute.

Example 1. We assume a relational schema with an attribute A from which the attribute B functionally depends, but not the additional attribute C. Table 1 shows three relation instances r_i in *array representations* $r_i[]$, for which the order of tuples (rows) is irrelevant since relation instances are defined as (unordered, duplicate-free) *sets* of tuples. The instances r_1 and r_2 are equivalent: leaving the attribute C untouched, the B-domain renaming exchanges the values 1 and 2, and a cyclic (+1 mod 3) A-domain renaming of the values 1, 2 and 3 maps the instance r_2 on r_1. The instances r_1 and r_3 are *not* equivalent: there is *no way to unify* the first three identical entries in the C-column of r_1 with the corresponding two identical and one different entries in the C-column of r_3.

The notion of similarity focuses on the sharing of structural properties, fully abstracting from concrete and identifiable values: still roughly outlined, a relation instance r_1 is defined to be (combinatorially) *similar* to a relation instance

r_2 if the following basic *counting properties* are the same: (i) the *number of different values* occurring under an attribute and (ii) the *multiplicities* of such existing values, as well as (iii) the *multiplicities* of *combinations* and *associations* between such existing values under different attributes. Notably, the considered counting properties of a relation instance can be understood as a solution to a layered system of equalities and lessthan-relationships among (non-negative) integer variables and some (non-negative) integer constants.

Example 2. Resuming Example 1, the relation instance r_1 satisfies the following combinatorial properties, which are intended to describe the internal structure of the instance merely in terms of pure *existences*. Under attribute A, there are $ak_A = 4$ different values; $m_1 = 3$ of them occur only $s_1 = 1$ often; $m_2 = 1$ of them occurs $s_2 = 2$ often. Under attribute B, there are $ak_B = 2$ different values; one of them occurs $\overline{m}_1^1 = 1$ often combined with an A-value that appears only once; the other one appears $\overline{m}_1^2 = 2$ often combined with an A-value that appears only once and $\overline{m}_2^2 = 2$ often together with an A-value that appears twice. Under attribute C, there are $ak_C = 2$ different values; one of them together with that B-value that appears only once occurs $\widehat{m}_1^{1,1} = 1$ often in combination/association with an A-value with multiplicity 1; moreover, that C-value together the other B-value occurs $\widehat{m}_1^{2,1} = 2$ often in combination/association with an A-value with multiplicity 1, and $\widehat{m}_2^{2,1} = 1$ often in combination/association with an A-value with multiplicity 2; the other C-value together with the latter B-value occurs $\widehat{m}_2^{2,2} = 1$ often in combination/association with an A-value with multiplicity 2. The reader might want to explicitly verify that the same combinatorial properties hold for the relation instance r_2, but this fact is an easy consequence of the equivalence. In contrast, the combinatorial properties of the relation instance r_3 differ, since we have $\widehat{m}_1^{1,1} = 1$, $\widehat{m}_1^{2,1} = 1$, $\widehat{m}_2^{2,1} = 1$, $\widehat{m}_1^{2,2} = 1$ and $\widehat{m}_2^{2,2} = 1$.

The examples deal with a non-normalized schema, since the left-hand side A of the assumed functional dependency does not form a primary key. For a normalized schema the situation is essentially less complicated.

Example 3. Let r_1' and r_2' be the relation instances that would result from storing only the first four subtuples over the attributes A and B of the arrays $r_1[]$ and $r[]_2$, respectively. For A, we still have $ak_A = 4$, such that there are $m_1 = 4$ many values with multiplicity $s_1 = 1$; and for B, we still have $ak_B = 2$, one value with $\overline{m}_1^1 = 1$ and another one with $\overline{m}_1^2 = 3$.

For either kind of classes, our task is to exhibit a *probabilistic generation procedure* for relation instances of a given relational schema and having a required size (number of tuples), such that each class is represented "uniformly". A straightforward interpretation of "uniformity" just means that each class is represented with *exactly the same probability.* An alternative interpretation means that each class is represented with the probability that is *proportional to the contribution of the class* to the set of all relation instances, which is the cardinality of the class divided by the cardinality of the set of all instances. A prerequisite for the task outlined so far is to characterize classes in more operational terms such that the required counting and sampling can effectively be computed.

Section 2 specifies the problem more formally. Section 3 deals with normalized schemas, laying the combinatorial foundations for a probabilistic generation procedure. Section 4 describes our prototype implementation and indicates possible optimizations. Section 5 discusses partial results and open issues for non-normalized schemas. Section 6 summarizes and suggests future research.

2 Formal Problem Specification

Throughout this work, we are mostly employing a kind of standard terminology, see, e.g., [1]. We will consider *relational schemas* over a set of *attributes* $\mathcal{U}$ with just one *functional dependency* $fd : \mathcal{A} \rightarrow \mathcal{B}$. This semantic constraint requires that values occurring under the left-hand side $\mathcal{A}$ uniquely determine the values under the disjoint right-hand side $\mathcal{B}$, not affecting the possibly existing remaining attributes in $\mathcal{C} = \mathcal{U} \setminus (\mathcal{A} \cup \mathcal{B})$. If $\mathcal{C} = \emptyset$, then $\mathcal{A}$ forms a *primary key* and, thus, such a schema is *normalized*. Otherwise, if $\mathcal{C} \neq \emptyset$, then $\mathcal{A} \cup \mathcal{C}$ forms a *unique primary key* and, thus, such a schema in *non-normalized*, satisfying the *multivalued dependency* $mvd : \mathcal{A} \twoheadrightarrow \mathcal{B}|\mathcal{C}$. Without loss of generality, we can assume that the left-hand side, the right-hand side and, if existing, the set of remaining attributes contain just one attribute, denoted by A, B and C, respectively.

For each attribute, the relational schemas considered will also specify a *finite domain*, i.e., a set of values that might occur under the attribute. Notably, theoretical work on relational databases often assumes an infinite global domain for all attributes. In contrast: to come up with discrete probability spaces, we assume finite domains; to emphasize the specific roles of the attributes (in applications presumably stemming from some real-world observations) we intuitively think of assigning a different domain as some kind of a "semantic type" to each of the attributes. This assumption implies that a value occurring under some attribute cannot occur under any of the other attributes and, thus, our notion of equivalence will be based on domain-specific renamings.

Formally summarizing these settings, we consider either a *normalized relational schema* $\langle R(\{A : dom_A, B : dom_B\}, \{fd : A \rightarrow B\}\rangle$ or a *non-normalized relational schema* $\langle R(\{A : dom_A, B : dom_B, C : dom_C\}, \{fd : A \rightarrow B\}\rangle$. Each *attribute att* has a finite *domain* dom_{att} of cardinality $k_{att} \geq 2$. A *relation instance* r is a (duplicate-free and unordered) set of tuples over the pertinent set of attributes, complying with the declared domains and satisfying the functional dependency. Further requiring a *size* n that is compatible with the declared domains, i.e., with $1 \leq n \leq k_A$ or $1 \leq n \leq k_A \cdot k_C$, respectively, let *Ins* be the set of all relation instances that contain exactly n many (different) tuples[1].

In [4,5] we were primarily interested in uniformly generating each of the relation instances $r \in Ins$ with the same probability $1/\|Ins\|$. Now we aim to probabilistically draw samples that are seen as a *representative* of a *class* of relation instances. Such classes are induced by some suitably defined reflexive, sym-

[1] To keep the notations simple, we deliberately refrain from explicitly denoting the dependence of *Ins*—and related items—from the size n, as well as from the relational schema, most relevantly including the attributes and their cardinalities.

metric and transitive relationship on *Ins*, namely either an equality/inequality-preserving *equivalence* relationship $\equiv$ or a combinatorial *similarity* relationship $\approx$, denoting such classes by $[r]_\equiv \in Ins/\equiv$ or $[r]_\approx \in Ins/\approx$, respectively.

The equivalence relationship $\equiv \subseteq Ins \times Ins$ can uniformly be defined for both kinds of relational schemas, as indicated below. However, the concrete definition of the similarity relationship will depend on the normalization status of the relational schema and will later be specified in the pertinent sections.

Definition 1 (Equivalence by equality/inequality preservation).
1. *A* domain renaming *for an attribute att is a function* $ren_{att} : dom_{att} \longrightarrow dom_{att}$ *that is both injective and surjective.*
2. *An* instance renaming, *based on some domain renamings* ren_{att}*, is a function* $ren : Ins \longrightarrow Ins$ *such that* $ren(r) :=$ $\{\, \nu \mid$ *there exists* $\mu \in r$ *such that for all att* : $\nu(att) = ren_{att}(\mu(att))\,\}$.
3. *The relation instances* r_1 *and* r_2 *of Ins are* equivalent, $r_1 \equiv r_2$, *iff there exists an instance renaming ren such that* $ren(r_1) = r_2$.

The task of *sampling representative instances* is then to design, formally verify and implement a *probabilistic generation procedure* with the following properties. On input of a relational schema together with a size n, the procedure returns three outputs: a relation instance $r \in Ins$ of size n and a description and the cardinality of its class $[r]$, i.e., r is a representative of $[r]$, defined as $[r]_\equiv \in Ins/\equiv$ or $[r]_\approx \in Ins/\approx$, achieving a probability distribution over the pertinent quotient *Cla* defined as $Ins/\equiv$ or $Ins/\approx$, respectively, such that

- either each class is represented with the same probability $1/\|Cla\|$
- or, alternatively, a class $[r]$ is represented with probability $\|[r]\|/\|Ins\|$, i.e., by its relative contribution to the set *Ins* of all instances.

Basically, solving this task requires us to determine the cardinalities involved, i.e., the cardinality $\|Ins\|$ of the set of all relational instances, the cardinality $\|Cla\|$ of the quotient and the cardinalities $\|[r]\|$ of each of its classes. While the first cardinality has already been given in our previous work [4,5], the other ones are central issues in this article.

3 Equivalence and Similarity for Normalized Schemas

In this section we treat the *normalized* case of a single relational schema with one functional dependency in the simplified form, i.e., we consider a schema $\langle R(\{A : dom_A, B : dom_B\}, \{fd : A \rightarrow B\}\rangle$ with finite *domains* dom_{att} of cardinality $k_{att} \geq 2$. In this case, the left-hand side A of the functional dependency alone forms a key. Hence, a relation instance of size n contains exactly n different key values taken from the domain dom_A of the attribute A, each of which is independently combined with some value of the domain dom_B of the attribute B, in particular allowing repeated occurrences of a value under the attribute B. Accordingly, as already reported in [4,5], we can immediately count the possible relation instances as expressed by the following proposition.

Proposition 1 (Instance count for normalized schemas). *For a normalized relational schema, the* number of relation instances *of size* n *with* $1 \leq n \leq k_A$*, satisfying the given functional dependency and complying with the given domains, equals*

$$\binom{k_A}{n} \cdot k_B^n. \tag{1}$$

While the notion of (equality/inequality-preserving) *equivalence* has generically been specified by Definition 1, we define the notion of (combinatorial) *similarity* for normalized schemas as a specialization of the more general notion for non-normalized schemas, to be later discussed in Sect. 5 below.

Definition 2 (Combinatorial similarity for normalized schemas). *For a normalized relational schema and an instance size* n *with* $1 \leq n \leq k_A$*:*[2]

1. *A relation instance* r satisfies a specialized A-structure *of the (uniquely determined) kind* $S_A = (s_1 = 1, m_1 = n)$ *if there are exactly* $ak_A := n$ *many different values that form the active domain* $act_A \subseteq dom_A$ *of the attribute* A*, each of them occurring exactly* $s_1 = 1$ *often, i.e., exactly once (as required by the key property implied by the functional dependency).*
2. *A relation instance* r satisfies *a* specialized B-structure *of the kind* $S_B = (\overline{m}_1^j)_{j=1,\dots,ak_B}$ *with* $\overline{m}_1^j \geq 1$, $1 \leq ak_B \leq \min(k_B, n)$ *and* $\sum_{j=1,\dots,ak_B} \overline{m}_1^j = n$*, if there are exactly* ak_B *many different values that form the active domain* $act_B \subseteq dom_B$ *of the attribute* B*, such that the multiplicities of their occurrences are given by the numbers* $\overline{m}_1^j$ *(in any order).*
3. *The set* Str *of all* specialized structures *is defined as* $\{\langle S_A, S_B\rangle \mid S_A$ *is specialized* A*-structure and* S_B *is specialized* B*-structure*$\}$.
4. *The relation instances* r_1 *and* r_2 *of* Ins *are* similar, $r_1 \approx r_2$*, iff they satisfy the same specialized structure* $str = \langle S_A, S_B\rangle \in Str$.

To prove the coincidence of equivalence and similarity of relation instances (seen as unordered sets of tuples) for normalized schemas, in the following we will establish a one-to-one correspondence between the purely arithmetic counting properties of specialized structures and a dedicated kind of suitably normalized abstract arrays of value variables, instantiations of which will lead to representations of relation instances as (ordered) arrays. Figure 1 visualizes our concepts.

More specifically, for each size n of relation instances, we choose a fixed sequence of A-value variables $\alpha_1, \dots, \alpha_n$ and a fixed sequence of B-value variables $\beta_1, \dots, \beta_n$. For each specialized structure $str = \langle S_A, S_B\rangle$ such that $S_A = (s_1 = 1, m_1 = n)$ and $S_B = (\overline{m}_1^j)_{j=1,\dots,ak_B}$ with $\overline{m}_1^j \geq 1$, $1 \leq ak_B \leq \min(k_B, n)$ and $\sum_{j=1,\dots,ak_B} \overline{m}_1^j = n$, we create an *abstract array* $vr[]$ with n rows for tuples/pairs of value variables and two columns for the attributes A and B and uniquely *populate* its $n \cdot 2$ many entries with the chosen value variables as follows:

[2] For the sake of brevity, in the following we combine the definition of a "structure" with the definition of "satisfaction" of a structure by a relation instance; we will proceed similarly in Sect. 5.

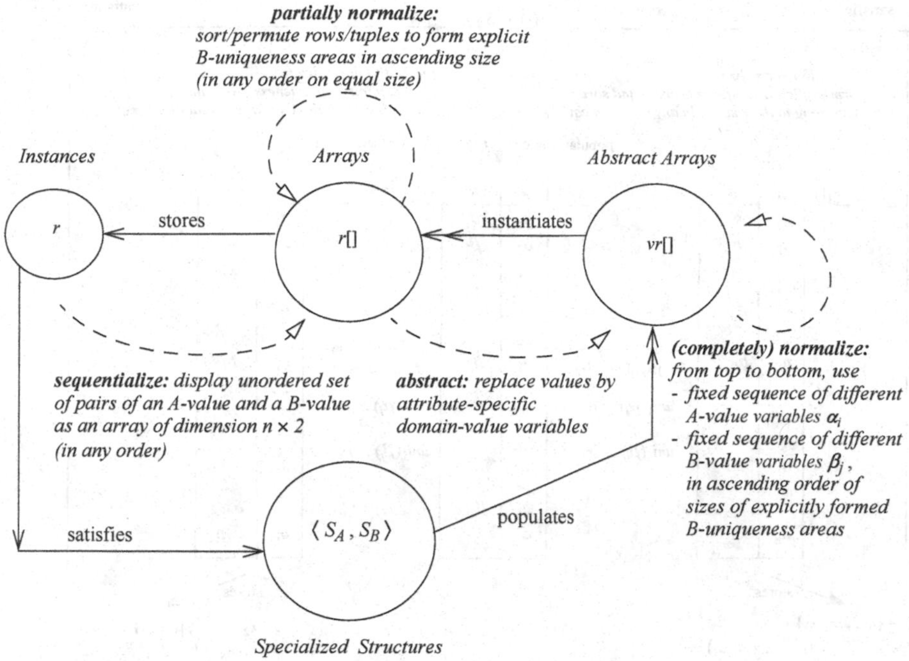

Fig. 1. Instances, arrays, abstract arrays and specialized structures.

1. $vr[i, A] = \alpha_i$ for $i = 1, \ldots, n$.
2. Assuming w.l.o.g. the numbers of $S_B = (\overline{m}_1^j)_{j=1,\ldots,ak_B}$ being ascendingly sorted according to numerical values,
$vr[i, B] = \beta_j$ for $i = \sum_{0 \leq j' < j} \overline{m}_1^{j'} + 1, \ldots, \sum_{0 \leq j' \leq j} \overline{m}_1^{j'}$ and $j = 1, \ldots, ak_B$,
i.e., intuitively, we arrange the multiplicities of the existing B-values from top to bottom and insert the B-value variables accordingly.

By abuse of language, we denote this normalized abstract array by $vr[] = [\alpha_i, \beta_j]_{i=1,\ldots,n;\ S_B:j=1,\ldots,ak_B}$. Using this notation, we aim to capture the somehow complex conditions for S_B in a short way by just annotating the range "$j = 1, \ldots, ak_B$" by the prefix "S_B". In the remainder, we will introduce further similarly normalized arrays and then employ a similar denotation.

Evidently, we can *store* a relation instance r as an *array* $r[]$, also of dimension $n \times 2$, filling the n rows with the tuples in any order. Following the same approach as described above, we can *partially normalize* such an array by (i) permuting rows, to form explicit B-uniqueness areas and (ii) sorting these areas according to their sizes, in any order on equal size. In general, however, the result is not unique, since there might be B-uniqueness areas of equal size.

An *instantiation* of an abstract array $vr[]$, by assigning different concrete values to different value variables in a domain-complying way, leads to a partially normalized array $r[]$. *Abstracting* a partially normalized array $r[]$, by replacing values by value variables in a domain-complying way and according to the chosen fixed sequences from top to bottom, results in a unique abstract array $vr[]$.

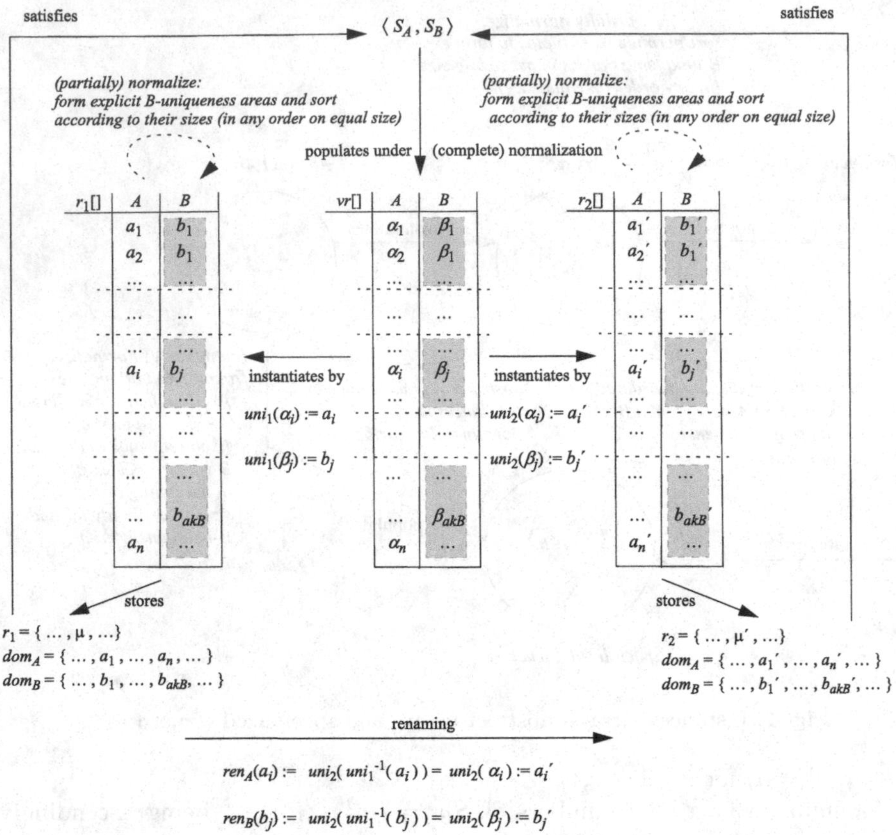

Fig. 2. Visualization of the basic features of the proof of Theorem 1, with explicit B-uniqueness areas marked in green. (Color figure online)

Theorem 1 (Coincidence of classes for normalized schemas). *For a normalized schema, the following assertions about relation instances r_1 and r_2 mutually imply each other:*

1. *$r_1 \equiv r_2$, as witnessed by some domain renamings.*
2. *$r_1 \approx r_2$, satisfying the same specialized structure $\langle S_A, S_B \rangle$.*
3. *r_1 and r_2 can be stored as arrays that are instantiations of the normalized abstract array $vr[]$ corresponding to the shared specialized structure $\langle S_A, S_B \rangle$.*

Proof. The basic features of the proof are visualized by Fig. 2.
"1 ⇒ 2": For any relation instance, its specialized structure only refers to properties that are invariant under equality/inequality-preserving domain renamings.
"2 ⇒ 3": First, we store the relation instances r_1 and r_2 as partially normalized arrays $r_1[] = [a_i, b_j]_{i=1,\ldots,n;\ S_B:j=1,\ldots,ak_B}$ and $r_2[] = [a'_i, b'_j]_{i=1,\ldots,n;\ S_B:j=1,\ldots,ak_B}$, respectively, both satisfying the given $\langle S_A, S_B \rangle$. Second, we generate the completely normalized abstract array $vr[] = [\alpha_i, \beta_j]_{i=1,\ldots,n;\ S_B:j=1,\ldots,ak_B}$, uniquely

corresponding to $\langle S_A, S_B \rangle$. Finally, the following instantiations witness the claim: $uni_1(\alpha_i) := a_i$ and $uni_1(\beta_j) := b_j$ as well as $uni_2(\alpha_i) := a'_i$ and $uni_2(\beta_j) := b'_j$.
"3 ⇒ 1": The claimed equivalence of r_1 and r_2 is witnessed as follows:
$ren_A(a_i) := uni_2(uni_1^{-1}(a_i)) = uni_2(\alpha_i) := a'_i$;
$ren_B(b_j) := uni_2(uni_1^{-1}(b_j)) = uni_2(\beta_j) := b'_j$. □

Given a specific size n and the cardinality k_B, the preceding theorem immediately implies that the pertinent number of classes is exactly determined by the number of pertinent integer partitions. Various kinds of such partitions have extensively been studied in mathematical and algorithmic number theory, see, e.g., [7,11,13]. So our task of identifying the number of classes is closely related to *combinatorial counting problems* for finite structures that are known to have (in the required size n) exponentially growing solution spaces.

Corollary 1 (Class count for normalized schemas). *For a normalized schema, for each size n of relation instances, the* number of classes *equals the number of integer partitions of n with up to* $\min(k_B, n)$ *parts, denoted by*

$$p_{\leq}(n, \min(k_B, n)). \tag{2}$$

Proof. By Theorem 1, each class is uniquely determined by a specialized structure $\langle S_A, S_B \rangle$. There is only one choice for S_A, namely $(s_1 = 1, m_1 = n)$, and each $S_B = (\overline{m}_1^j)_{j=1,\ldots,ak_B}$ defines an integer partition of n of the specified kind. □

Now, already knowing the number of instances by Proposition 1 and the number of classes (for both equivalence and similarity) by Corollary 1 based on Theorem 1, we are left with investigating the sizes of the individual classes.

Theorem 2 (Member count for normalized schemas). *For a normalized schema, for the size n of relation instances, for the class* $[r]$ *uniquely determined by the specialized structure* $\langle S_A, S_B \rangle$ *with* $S_B = (\overline{m}_1^j)_{j=1,\ldots,ak_B}$*, as an integer partition having an (ascendingly ordered) multiplicity representation with different summands* $\overline{m}_1^{j_1}, \ldots, \overline{m}_1^{j_{\bar{k}}}$ *and multiplicities* $\bar{j}_1, \ldots, \bar{j}_{\bar{k}}$ *such that* $\sum_{l=1,\ldots,\bar{k}} \overline{m}_1^{j_l} \cdot \bar{j}_l = n$ *and* $\sum_{l=1,\ldots,\bar{k}} \bar{j}_l = ak_B$*, the* number of members *equals*

$$\frac{\binom{k_A}{n} \cdot n! \cdot \binom{k_B}{ak_B} \cdot ak_B!}{\prod_{l=1}^{\bar{k}} \; (\overline{m}_1^{j_l}!)^{\bar{j}_l} \cdot \bar{j}_l!}. \tag{3}$$

Proof. By Theorem 1, the class $[r]$ is uniquely determined by the specialized structure $\langle S_A, S_B \rangle$, which in turn uniquely determines a normalized abstract array $vr[] = [\alpha_i, \beta_j]_{i=1,\ldots,n;\; S_B: j=1,\ldots,ak_B}$. Moreover, each relation instance in $[r]$ can be obtained as an instantiation of that $vr[]$.

According to S_B and its multiplicity representation, there are ak_B many *B-uniqueness sections* [range/set of (indices for) rows] (as determined by B-uniqueness areas, of respective sizes $\overline{m}_1^j$): for $j = 1, \ldots, ak_B$, define $I_j :=$

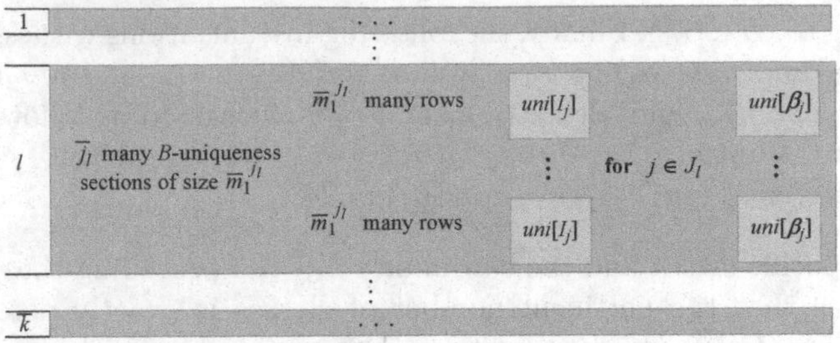

Fig. 3. Visualization of instantiated B-multiplicity sections.

$\{ i \mid \sum_{j'<j} \overline{m}_1^{j'} < i \le \sum_{j' \le j} \overline{m}_1^{j'} \}$. Moreover, there are $\bar{k}$ many *B-multiplicity sections* [range/set of indices for B-uniqueness sections] (as determined by B-uniqueness areas of the same size, of respective size $\bar{j}_l$): for $l = 1, \ldots, \bar{k}$, define $J_l := \{ j \mid \sum_{l'<l} \bar{j}_{l'} < j \le \sum_{l' \le l} \bar{j}_{l'} \}$.

Based on these definitions, we investigate the properties of any instantiation *uni* of the normalized abstract array $vr[]$. First, for $j = 1, \ldots, ak_B$, let $uni[I_j] := \{ uni(\alpha_i) \mid i \in I_j \}$ be the set of A-values combined with the B-value $uni(\beta_j)$. Second, for $l = 1, \ldots, \bar{k}$, let $uni[J_l] := \{ uni(\beta_j) \mid j \in J_l \}$ be the set of B-values within the B-multiplicity section determined by the B-uniqueness areas of the same size $\bar{j}_l$. The preceding settings are partially visualized by Fig. 3.

Finally, the transformation of the completely normalized abstract array into a relation instance as a set of tuples can be described as follows:

- *normalized abstract array*: $vr[] = [\alpha_i, \beta_j]_{i=1,\ldots,n;\ S_B:j=1,\ldots,ak_B}$;
- *instantiation*: $uni(\alpha_i) \in dom_A$ and $uni(\beta_j) \in dom_B$;
- *partially normalized array*: $uni(vr[]) = uni([\alpha_i, \beta_j]_{i=1,\ldots,n;\ S_B:j=1,\ldots,ak_B})$
 $:= [uni(\alpha_i), uni(\beta_j)]_{i=1,\ldots,n;\ S_B:j=1,\ldots,ak_B}$;
- *relation instance*: $stores(\, uni(vr[])\,)$
 $= stores(\, [uni(\alpha_i), uni(\beta_j)]_{i=1,\ldots,n;\ S_B:j=1,\ldots,ak_B})$
 $:= \{ (a, b) \mid$ exists $i \in \{1, \ldots, n\}$, exists $j \in \{1, \ldots, ak_B\}$ such that
 $i \in I_j$ and $(a, b) = (uni(\alpha_i), uni(\beta_j)) \}$.

Accordingly, the function *stores* induces a set partition $Uni/stores$ on the set Uni of all pertinent instantiations, the blocks of which are defined by $[uni_1]_{stores} := \{ uni_2 \mid stores(\, uni_1(vr[])\,) = stores(\, uni_2(vr[])\,) \}$. The following lemma completely characterizes these blocks.

Lemma 1. *For instantiations uni_1 and uni_2 of the abstract array $vr[]$: $stores(\, uni_1(vr[])\,) = stores(\, uni_2(vr[])\,)$ [same relation instance generated] iff for all $1 \le l \le \bar{k}$ [within each B-*multiplicity section*]:*

1. $\{ uni_1[I_j] \mid j \in J_l \} = \{ uni_2[I_j] \mid j \in J_l \}$
 [same *sets* of A-values within a B-uniqueness area];
2. $\{ uni_1(\beta_j) \mid j \in J_l \} = \{ uni_2(\beta_j) \mid j \in J_l \}$
 [same *set* of B-values];

3. for all $j, j' \in J_l$: $uni_1(\beta_j) = uni_2(\beta_{j'})$ iff $uni_1[I_j] = uni_2[I_{j'}]$
   [same *combinations* of a B-value with a set of A-values within a B-uniqueness area].

Proof (Lemma). "$\Rightarrow$": Consequences of the assumed set equality regarding the definitions involved.
"$\Leftarrow$": Let $(a, b) \in stores(uni_1(vr[]))$. Then, as explained above, there exists $i \in \{1, \ldots, n\}$ and there exists $j \in \{1, \ldots, ak_B\}$ such that $i \in I_j$ and $(a, b) = (uni_1(\alpha_i), uni_1(\beta_j))$. Moreover, by the definition of the B-multiplicity sections, there exists $l \in \{1, \ldots, \bar{k}\}$ such that $j \in J_l$. By the assumption, namely the validity of the assertions 1, 2 and 3, there exist $i' \in I_j$ and $j' \in J_l$ such that $(uni_1(\alpha_i), uni_1(\beta_j)) = (uni_2(\alpha_{i'}), uni_2(\beta_{j'}))$. In turn, again as explained above, this implies that $(uni_2(\alpha_{i'}), uni_2(\beta_{j'})) \in stores(uni_2(vr[]))$. □

To finalize the proof of the theorem, we justify the numerator and the denominator of the quotient (3) in turn.

The *numerator* counts the number $\|Uni\|$ of pertinent instantiations. In fact, there are $\binom{k_A}{n}$ possibilities to select an active domain for the attribute A, the elements of which can be assigned to the A-value variables occurring in $vr[]$ in $n!$ many ways and, independently, there are $\binom{k_B}{ak_B}$ possibilities to select an active domain for the attribute B, the elements of which can be assigned to the B-value variables occurring in $vr[]$ in $ak_B!$ many ways.

The *denominator* counts the number $\|[uni]_{stores}\|$ of pertinent instantiations in each of the blocks $[uni]_{stores}$ of the set partition $Uni/stores$. In fact, if we fix a particular instantiation uni of the abstract array $vr[]$, then the lemma tells us precisely which other instantiations yield the same relation instance, implying that this common relation instance remains invariant exactly under the following mutually independent permutations:

- For the $\bar{k}$ many B-multiplicity sections numbered by $1 \leq l \leq \bar{k}$, for the included $\bar{j}_l$ many B-uniqueness areas denoted by $j \in J_l$, permute the A-values within $uni[I_j]$, which has size $\overline{m}_1^{j_l}$.
- For the $\bar{k}$ many B-multiplicity sections numbered by $1 \leq l \leq \bar{k}$, permute the included $\bar{j}_l$ many B-uniqueness areas as a whole.

The denominator just counts these permutations.

The *quotient* is then justified by the mutual disjointness of the blocks of the set partition $Uni/stores$. □

4 Probabilistic Generation for Normalized Schemas

Depending on the variant of uniformity, we may choose different probabilistic generation procedures. In the following, we describe our implementations of both versions and discuss respective optimizations. Further, we briefly assess the runtime performance of our prototype implementation of the second and more critical case of uniformity.

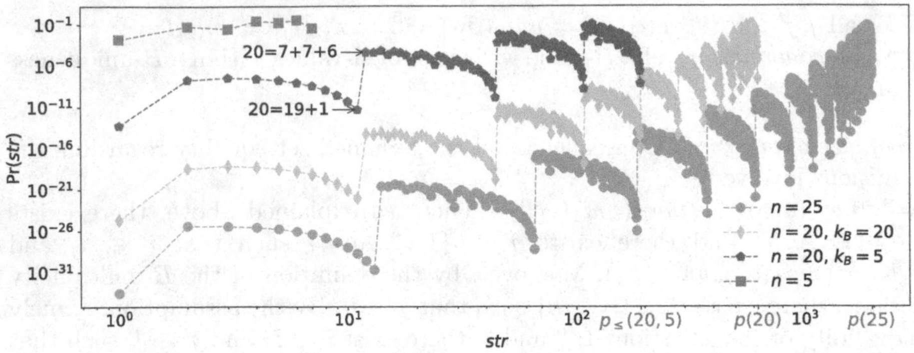

Fig. 4. The discrete probability distribution $Pr(str)$ for the parameter configurations $n = k_A = k_B = 5$ (red squares) and $n = k_A = k_B = 20$ (yellow diamonds) and $n = k_A = 20, k_B = 5$ (blue pentagons) and $n = k_A = k_B = 25$ (green dots). $p(n)$ denotes the (unrestricted) partition function. Both axes are log scaled. (Color figure online)

4.1 Drawing a Class $[r]$ with Proportional Probability

If our goal is to sample relation instances with uniform probability $1/\|[r]\|$ from the respective class $[r]$, where the class was selected with a probability proportional to the cardinality $\|[r]\|/\|Ins\|$ of the class, a simple generation procedure can be used. Instead of determining the class first, we may generate the relation instance directly from the set of all instances with uniform probability $1/\|Ins\|$. This can be achieved by first drawing a random active domain act_A from dom_A, with $\|act_A\| = n$. Many algorithms for the generation of random subsets exist and may perform differently, depending on the ratio of $\|act_A\|/\|dom_A\|$. To illustrate runtimes, drawing a subset of 500000 values from a domain with cardinality 1000000 takes about 5.2 ms. See [4] for other examples and a more detailed discussion on this topic.

After selecting the active domain act_A, we form tuples by combining every value of act_A with a B-value, drawn uniformly at random from dom_B. This procedure can generate any relation instance r with probability $1/\|Ins\|$. Further, the specialized structure *str* of the produced instance r, which can be derived from its value distributions, is selected with probability $Pr(str) = \|[r]\|/\|Ins\|$, since $\|[r]\|$ many instances share the same structure and all those instances are equally likely.

Figure 4 shows the discrete probability distribution of the specialized structures for the following parameter configurations: $n = k_A = k_B = 5$ and $n = k_A = 20, k_B = 5$ and $n = k_A = k_B = 20$ and $n = k_A = k_B = 25$. Dashed lines between the probabilities are added for visual clarity. The specialized structures $str \in Str$ are given in lexicographically ascending order along the x-axis, as they are produced by Gupta's partition generation algorithm [9]. For example, the respectively *smallest* partition/structure, n itself, is located to the far-left side, and the partitions $20 = (19, 1), (1, 1)$ and $25 = (22, 1), (3, 1)$ are shown at location 11 on the x-axis (just right of position 10^1). The plot indicates that a class $[r]$ contains more instances (has a higher probability), if ak_B is large and if the B-uniqueness areas are similar in size. For example, the class with $S_B = (19,1),(1,1)$ (small ak_B; non-similar sizes) has cardinality 400, whereas the class with $S_B = (7,2),(6,1)$ (larger ak_B; similar sizes) already contains about $400 \cdot 10^7$ relation instances. In particular, this number—as most of the other ones depicted in Fig. 4—indicates that in general we are dealing with pretty large probability spaces.

4.2 Drawing a Class $[r]$ with Uniform Probability

If we are instead interested in generating relation instances from a class $[r]$ selected with *uniform* probability $1/\|Cla\| = 1/p_{\leq}(n, \min(k_B, n))$, we require the explicit generation of the representing structure *str*.

Conceptually, but not regarding the computational complexity, this approach is comparable to the one used for random relation instances of non-normalized schemas, presented in our previous work [4,5]. It has been based on the counting formula (4–5) for relation instances of such schemas recalled and explained in Sect. 5 below. Roughly outlined, the procedure explicitly generates the desired probability distribution by computing all relevant structures first and then determining their probabilities from the respective instance counts. Due to the exponentially increasing number of structures, generating the distribution for $n > 90$ quickly becomes infeasible. See [4] for details.

In our case, however, the desired distribution is uniform (i.e., the instance counts do not matter), which enables efficient procedures for the random generation of a specialized structure $str \in Str$. Since S_A is trivial in the normalized case, we get structures by drawing a random integer partition as S_B. Although a thorough treatment of the more general field of efficient random integer partition generation is beyond the scope of this article, we will now describe one simple example and its adoption to our context.

4.3 Efficient Random Generation of a Specialized B-Structure S_B

One algorithm is given by Fristedt [8]: Treating the integer n itself as a random variable, the iterative algorithm generates partitions by independently drawing the number of occurrences of every possible part. Specifically, the number of occurrences of a part $\overline{m}_1^j$ is drawn from a geometric distribution with parameter $1-\exp(-\overline{m}_1^j\pi/(\sqrt{6n}))$. If the sum of all drawn parts equals n, an *unbiased* integer partition of n was found. If the drawn parts do not sum up to n, the algorithm is repeated. The procedure is expected to make $2\sqrt[4]{6}n^{3/4}$ proposals until a valid partition is found [3].

If $k_B < n$, however, not all partitions qualify. This constraint can be implemented by restricting the number of parts in a partition, because the number of parts in S_B corresponds to the active domain size $ak_B = \|act_B\|$. In general, there are $p_{\leq}(n, \min(k_B, n))$ valid integer partitions (see Corollary 1). Various approaches generate restricted partitions with at most $\min(k_B, n)$ parts. One way is to adopt Fristedt's algorithm by simply ignoring the multiplicities of the invalid parts of size larger then ak_B. Since the *conjugate* of an integer partition [12] with largest part ak_B corresponds to a partition with exactly ak_B parts, the adopted algorithm produces the desired partitions uniformly at random.

On a Intel Xeon Gold 6226 CPU with 2.7 GHz, generating relation instances with $n = 1000$ and $k_A = k_B = 100000$ with our (single-core) Python- and Numpy-based [10] implementation takes about 366.517 ms (average over 200 runs). Notably, for $n = 1000$, the step of drawing a valid structure S_B alone took 366.390 ms. The runtime for increasing n progresses in a slightly below quadratic fashion, as indicated by the expected number of proposals, and suggests that it is generally possible to generate even very large instances. For a faster implementation, more efficient random partition generation algorithms are necessary. For example, Arratia and DaSalvo [3] introduce a more involved, probabilistic divide-and-conquer-based approach, which requires only about $2\pi\sqrt[4]{6}n^{1/4}$ proposals. Further, they describe an implementation that generates unrestricted partitions up to $n = 2^{58}$, before running out of memory.

5 Results and Issues for Non-normalized Schemas

In this section we discuss partial results and open issues for the *non-normalized* case of a single relational schema with one functional dependency in the simplified form, i.e., we consider a schema $\langle R(\{A : dom_A, B : dom_B, C : dom_C\}, \{fd : A \to B\}\rangle$ with finite *domains* dom_{att} of cardinality $k_{att} \geq 2$. For this case, we tacitly adapt the previously introduced notations. Moreover, the cardinality $\|Ins\|$ has already been determined in [4,5], as shown next.

Proposition 2 (Instance count for non-normalized schemas). *For a non-normalized relational schema, the* number of relation instances *of size* n *with* $1 \leq n \leq k_A \cdot k_C$ *that satisfy the given functional dependency and comply with the given domains equals*

$$\sum_{\lceil \frac{n}{k_C} \rceil \leq ak_A \leq \min(k_A, n)} \quad \sum_{\substack{1 \leq k \leq n, \\ (s_1,m_1),\ldots,(s_i,m_i),\ldots,(s_k,m_k): \\ 1 \leq s_i < s_{i+1} \leq k_C; \\ 1 \leq m_i \leq n; \\ n = \sum_{i=1}^{k} s_i \cdot m_i; \\ ak_A = \sum_{i=1}^{k} m_i}} \tag{4}$$

$$\left[\prod_{i=1}^{k} \binom{k_A - \sum_{1 \leq j < i} m_j}{m_i} \quad \cdot \quad k_B^{ak_A} \quad \cdot \quad \prod_{i=1}^{k} \binom{k_C}{s_i}^{m_i} \right]. \tag{5}$$

In the counting formula (4–5), in the first summation of (4) the term ak_A ranges over the possible cardinalities of the *active domain* act_A of A, and in the second summation of (4) the suitably normalized *size–multiplicity-list* $(s_i, m_i)_{i=1,\ldots,k}$ (for A-uniqueness areas) ranges over all possibilities to let m_i elements of the active domain of A appear exactly s_i often. Each accepted pair of ak_A and $(s_i, m_i)_{i=1,\ldots,k}$ together (i.e., jointly satisfying the conditions in the two summations of (4)) constitutes a *solution* to a *restricted integer partition problem* for the number n, and also vice versa. We shortly refer to such a pair as an *n-partition* or an *A-structure*, denoted as $S_A = (s_i, m_i)_{i=1,\dots,k}$ leaving ak_A implicit.

To prepare for a general definition of (combinatorial) *similarity*, let us inspect a relation instance r, the active domains of which are enumerated (in some fixed order) as $(a_e)_{e=1,\dots,ak_A}$, $(b_j)_{j=1,\dots,ak_B}$ and $(c_d)_{d=1,\dots,ak_C}$, respectively. Where convenient, we also employ alternative indices for A-values and enumerate the active domain of A as $a_{1,1}, \dots a_{1,m_1}, \dots, a_{i,1}, \dots a_{i,m_i}, \dots, a_{k,1}, \dots a_{k,m_k}$.

Regarding the attribute A, let m_i be the multiplicity of A-uniqueness *areas* of size s_i, containing some different values $a_{i,1}, \dots, a_{i,m_i}$, respectively, each of them occurring s_i often. Thus[3], the inspected relation instance r *satisfies*

the *A-structure* $S_A = (s_i, m_i)_{i=1,\dots,k}$ (with implicit size ak_A of act_A) for which, as already stated in (4), the following properties hold:

$$\lceil \frac{n}{k_C} \rceil \leq ak_A \leq \min(k_A, n) \quad \text{and} \quad \begin{array}{c} 1 \leq k \leq n\,; \\ 1 \leq s_i < s_{i+1} \leq k_C\,; \\ 1 \leq m_i \leq n\,; \\ n = \sum_{i=1}^{k} s_i \cdot m_i\,; \\ ak_A = \sum_{i=1}^{k} m_i\,. \end{array} \tag{6}$$

Regarding the attribute B, already given the A-structure of r, let $\overline{m}_i^j$ be the number of combinations of partial B-uniqueness *areas* (defined by A-uniqueness areas) containing the value b_j with an A-uniqueness area of size s_i (for short, A-B-combinations). Thus, the relation instance r inspected *satisfies*

the *B-structure* $S_B = (\overline{m}_i^j)_{i=1,\dots,k;\ j=1,\dots,ak_B}$ (with implicit size ak_B of act_B), for which the following properties hold:

$$1 \leq ak_B \leq \min(k_B, ak_A) \quad \text{and} \quad \begin{array}{c} 0 \leq \overline{m}_i^j \leq m_i\,; \\ \sum_{j=1}^{ak_B} \overline{m}_i^j = m_i\,; \\ 1 \leq \sum_{i=1}^{k} \overline{m}_i^j\,. \end{array} \tag{7}$$

Regarding the attributes B and C jointly, already given the A-structure and the B-structure of r, let $\widehat{m}_i^{j,d}$ be the number of occurrences of the subtuple (b_j, c_d) where the second component c_d occurs in a C-diversity area associated with different A-uniqueness *areas* (defining partial B-uniqueness areas) of size s_i (for short: A-(Bv, Cv)-multiplicities), equivalently the number of combinations/associations of the (B, C)-uniqueness area containing (b_j, c_d) with different A-uniqueness areas of size s_i. Thus, the relation instance r inspected *satisfies*

the *BC-structure* $S_{BC} = (\widehat{m}_i^{j,d})_{i=1,\dots,k;\ j=1,\dots,ak_B;\ d=1,\dots,ak_C}$ (with implicit size ak_C of act_C), for which the following properties hold:

$$s_k \leq ak_C \leq \min(k_C, n) \quad \text{and} \quad \begin{array}{c} 0 \leq \widehat{m}_i^{j,d} \leq \overline{m}_i^j \quad (\leq m_i \ \textit{redundant!})\,; \\ (\sum_{1 \leq j \leq ak_B} \widehat{m}_i^{j,d} \leq m_i \ \textit{redundant!})\,; \\ \sum_{1 \leq d \leq ak_C} \widehat{m}_i^{j,d} = s_i \cdot \overline{m}_i^j\,; \\ (\sum_{1 \leq j \leq ak_B} \sum_{1 \leq d \leq ak_C} \widehat{m}_i^{j,d} = s_i \cdot m_i \ \textit{redundant!})\,; \\ (\sum_{1 \leq i \leq k} \sum_{1 \leq j \leq ak_B} \sum_{1 \leq d \leq ak_C} \widehat{m}_i^{j,d} = n \ \textit{redundant!})\,; \\ 1 \leq \sum_{1 \leq i \leq k} \sum_{1 \leq j \leq ak_B} \widehat{m}_i^{j,d} \quad (\leq ak_A \ \textit{redundant!})\,. \end{array} \tag{8}$$

[3] As in Definition 2, here and in the following we integrate the definition of a "structure" with the definition of "satisfaction".

Conversely, we can treat (6), (7) and (8) as a layered system *ELTS* of equalities and less–than-relationships among *integer variables* and some constants and additive operations on such terms. Accordingly, the combinatorial structure $\langle S_A, S_B, S_{BC}\rangle$ of the relation instance r inspected constitutes a specific *solution* to *ELTS*. Further on, we will consider *each* solution to *ELTS* as a structure, where the sequencing of the lists for B, C and BC regarding the respective index does not matter[4] (all sequences are seen to be equivalent).

Definition 3 (Combinatorial similarity for non-normalized schemas). *For a non-normalized relational schema and an instance size n with $1 \le n \le k_A \cdot k_C$:*

1. *The set Str_{SS} of* structures as system solutions *is defined as* $\{\, \langle S_A, S_B, S_{BC}\rangle \mid S_X$ *solves stepwise* $(6),(7),(8)$, *respectively*$\}$.
2. *A relation instance $r \in Ins$ satisfies $\langle S_A, S_B, S_{BC}\rangle \in Str_{SS}$ if there are exactly ak_A, ak_B and ak_C many different values that form the active domains $act_A \subseteq dom_A$ of the attribute A, $act_B \subseteq dom_B$ of the attribute B and $act_C \subseteq dom_C$ of the attribute C, respectively, such that (6), (7), and (8) stepwise hold for suitable enumerations of the active domains of r.*
3. *The relation instances r_1 and r_2 of Ins are* similar, $r_1 \approx r_2$, *if they satisfy the same structure as system solution $str = \langle S_A, S_B, S_{BC}\rangle \in Str_{SS}$.*

The following theorem confirms that each solution to *ELTS* can indeed be seen as the description of the structure of some relation instances.

Theorem 3 (Satisfiability of system solutions for non-normalized schemas). *For each structure as system solution $str = \langle S_A, S_B, S_{BC}\rangle \in Str_{SS}$ there exists a relation instance $r \in Ins$ that satisfies str.*

Proof. (*sketch*) We construct satisfying relation instances in two steps: (i) for a given $\langle S_A, S_B, S_{BC}\rangle$, we *populate* an abstract array $vr[]$ with rows $i = 1, \ldots, n$ and columns $att = A, B, C$ by *domain-value variables*; (ii) we *instantiate* the domain-value variables with different *values* of the respective domains, getting an array $r[]$ that *stores* a relation instance r of the required kind.
ad (i): The definition of $vr[]$ is given attribute-wise from A over B to C according to the given A-structure, B-structure and BC-structure in turn, as roughly outlined in the following.
According to S_A, the A-column of $vr[]$ is populated from top to bottom by the A-value variables $\alpha_{1,1}, \ldots \alpha_{1,m_1}, \ldots, \alpha_{i,1}, \ldots \alpha_{i,m_i}, \ldots, \alpha_{k,1}, \ldots \alpha_{k,m_k}$, such that an A-value variable of the form $\alpha_{i,\cdot}$ is placed s_i often in sequence in the pertinent A-uniqueness area. According to S_B, the B-column of $vr[]$ is populated by the B-value variables $\beta_1, \ldots, \beta_{ak_B}$, such that, for $i = 1, \ldots, k$, the B-value variable β_j is placed in combination with $\overline{m}_i^j$ many A-uniqueness areas of size i, for each of them s_i often in sequence, thus forming the pertinent combined *partial* B-uniqueness area (for the sake of satisfying the functional dependency). According

[4] However, since such a list might contain duplicates, a usual set notation would not be appropriate; instead we could use multi-sets.

Table 2. Two arrays $r_1[]$ and $r_2[]$ storing relation instances r_1 and r_2 that share the combinatorial structure as system solution but are *not equivalent*.

$r_1[]$	A	B	C
	1	1	1
	1	1	2
	2	1	1
	2	1	2
	3	1	3
	3	1	4

$r_2[]$	A	B	C
	1	1	1
	1	1	4
	2	1	1
	2	1	2
	3	1	2
	3	1	3

to S_{BC}, the C-column of $vr[]$ is populated by the C-value variables $\gamma_1, \ldots, \gamma_{ak_C}$, such that in total $1 \leq \sum_{1 \leq i \leq k} \sum_{1 \leq j \leq ak_B} \widehat{m}_i^{j,d}$ copies of γ_d are placed, following a circularly sequential choice/selection strategy: While repeatedly [in total s_i often] circling through the $\overline{m}_i^j$ many determined combinations of an A-uniqueness area of size s_i with a partial B-uniqueness area also of size s_i, as w.l.o.g. numbered $1, \ldots, \overline{m}_i^j$, in the round h visiting the h-th entries (rows) of the areas.
ad (ii): We just *instantiate* each of the pairwise different A-value variables $\alpha_{i,e}$ with a different value $a_{i,e} \in dom_A$, each of the pairwise different B-value variables β_j with a different value $b_j \in dom_B$, and each of the pairwise different C-value variables γ_d with a different value $c_d \in dom_C$. By the construction, the resulting array $r[]$ stores a relation instance $r \in Ins$. □

Proposition 3 (Similarity strictly generalizes equivalence for non-normalized schemas). *For a non-normalized schema, if two relation instances are equivalent then they are similar. Moreover, there exist relation instances r_1 and r_2 that are similar but not equivalent.*

Proof (*sketch*). The first claim follows from the invariance of the combinatorial properties of structures under equality/inequality-preserving renamings. The second claim is witnessed by the example shown in Table 2. The *similarity* can straightforwardly be verified. The *non-equivalence* can be formally proved by some subtle case considerations that—though the *purely combinatorial* requirements of the *BC-structure* $S_{BC} = (2, 2, 1, 1)$ are satisfied by both relation instances— the respective sets of associations, each formed by an A-value with two C-values, cannot faithfully be mapped on each other. □

Accordingly, the cardinalities $\| Cla \|$ and $\| [r] \|$ for the classes differ for the two notions of classes, and their determination for equivalence cannot directly be based on the determination for similarity, as successfully done for normalized schemas. Moreover, even for similarity an explicit determination would be much more involved, since the corresponding integer partitions are much more involved: they have to be layered like *ELTS*, possibly contain $0's$ as summands, and on each layer forming a suitable collection of somehow "compatible" ones.

Considering an imagined relation instance $r \in Ins$, we can *abstract* from an *array representation* $r[]$ of the *set* r by replacing concrete values by attribute-specific domain-value variables, resulting in an *abstract array* $vr[]$. Conversely, we can again *directly instantiate* $vr[]$ by assigning some different domain-complying values to the domain-value variables in $vr[]$, getting an array representation $\tilde{r}[]$ of a relation instance $\tilde{r}$ that is obviously equivalent with r. Furthermore, defining vr as the *set* of rows (tuples of variables) of the abstract *array* $vr[]$, by the same assignment we directly get the relation instance $\tilde{r}$. We extend such consideration by identifying a relationship to the notion of *set unification* [6].

Theorem 4 (Equivalence, unification and matching for non-normalized schemas). *Let r_1 and r_2 be relation instances with some array representations $r_1[]$ and $r_2[]$ and corresponding abstract arrays $vr_1[]$ and $vr_2[]$, defining sets vr_1 and vr_2 of rows (tuples of variables), respectively. Then the following assertions mutually imply each other:*

1. $r_1 \equiv r_2$, *by a (bijective and thus injective)* renaming *ren with* $ren(r_1) = r_2$ *(here: set-equality).*
2. $ren(r_1[]) = r_2[]_{per}$, *with an (injective)* renaming *ren and a* permutation *per of the array rows (tuples) (here: array-equality).*
3. $vren(vr_1[]) = vr_2[]_{per}$, *with an injective* variable renaming *vren and a* permutation *per of the array rows (tuples of variables) (here: array-equality).*
4. vr_1 *and* vr_2 *are injectively set-unifiable, by an injective* set-unifier *uni with* $uni(vr_1) = vr_2$ *(here: set-equality).*
5. vr_1 *and* r_2 *are injectively set-matching, by an injective* set-matcher *mat with* $mat(vr_1) = r_2$ *(here: set-equality).*

Proof. (*sketch*) Basically, the implications can be circularly verified by suitably applying the pertinent definitions. □

6 Conclusions

We aimed at probabilistically generating *representatives* of *classes* of relation instances of a relational schema with one functional dependency and specified attribute domains. Such classes are defined by means of either invariance under bijective *functions* on the domains or sharing of combinatorial counting properties related to the existence and multiplicities of *values* and their combinations and associations.

For *normalized* schemas, the two notions of classes coincide, the cardinalities of the set of all instances, the set of all classes and the single classes, respectively, can effectively be determined, and the designed generation procedure has formally been verified. The availability of efficient random integer partition generation algorithms indicates general feasibility, up to suitable optimizations. The main optimization issues to be further explored regard options for selecting active domains and integer partitions uniformly at random.

In contrast, for *non-normalized* schemas the two notions differ and, as a consequence, the further features (as stated above) remain mostly open. We suggest future research to understand the deeper reasons for the observed differences, to identify further manageable cases, and to improve the combinatorial analysis.

As already remarked in our previous work [4,5], further research should also treat more advanced relational schemas including, e.g., any number of suitable equality-generating and tuple-generating dependencies, possibly spanning over more than one relation scheme. For such situations, but also for the currently considered quite simple case, we could refine the investigations by studying meaningful variants of the notion of equivalence.

Acknowledgements. We sincerely thank the anonymous reviewers for their careful evaluations and constructive remarks.

References

1. Abiteboul, S., Hull, R., Vianu, V.: Foundations of Databases. Addison-Wesley, Reading (1995)
2. Arasu, A., Kaushik, R., Li, J.: Data generation using declarative constraints. In: Sellis, T.K., Miller, R.J., Kementsietsidis, A., Velegrakis, Y. (eds.) SIGMOD 2011, pp. 685–696. ACM (2011)
3. Arratia, R., DeSalvo, S.: Probabilistic divide-and-conquer: a new exact simulation method, with integer partitions as an example. Comb. Probab. Comput. **25**(3), 324–351 (2016)
4. Berens, M., Biskup, J., Preuß, M.: Uniform probabilistic generation of relation instances satisfying a functional dependency. Inform. Syst. **103**, 101848 (2021)
5. Biskup, J., Preuß, M.: Can we probabilistically generate uniformly distributed relation instances efficiently? In: Darmont, J., Novikov, B., Wrembel, R. (eds.) ADBIS 2020. LNCS, vol. 12245, pp. 75–89. Springer, Cham (2020). https://doi.org/10.1007/978-3-030-54832-2_8
6. Dovier, A., Pontelli, E., Rossi, G.: Set unification. Theory Pract. Log. Program. **6**(6), 645–701 (2006)
7. Flajolet, P., Sedgewick, R.: Analytic Combinatorics. Cambridge University Press, Cambridge (2009)
8. Fristedt, B.: The structure of random partitions of large integers. Trans. Am. Math. Soc. **337**(2), 703–735 (1993)
9. Gupta, U.I., Lee, D.T., Wong, C.K.: Ranking and unranking of B-trees. J. Algorithms **4**(1), 51–60 (1983)
10. Harris, C.R., Millman, K.J., van der Walt, S.J., et al.: Array programming with numpy. Nature **585**(7825), 357–362 (2020)
11. Nijenhuis, A., Wilf, H.S.: A method and two algorithms on the theory of partitions. J. Comb. Theory Ser. A **18**(2), 219–222 (1975)
12. Stanley, R.P.: Enumerative Combinatorics, vol. 1, 2nd edn. Cambridge University Press, Cambridge (2012)
13. Stojmenovic, I., Zoghbi, A.: Fast algorithms for generating integer partitions. Int. J. Comput. Math. **70**(2), 319–332 (1998)
14. Transaction Processing Performance Council, TPC: TCP Benchmarks & Benchmark Results. http://www.tpc.org

On the Expressive Power of Message-Passing Neural Networks as Global Feature Map Transformers

Floris Geerts[1], Jasper Steegmans[2(✉)], and Jan Van den Bussche[2]

[1] University of Antwerp, Antwerp, Belgium
[2] Hasselt University, Hasselt, Belgium
jasper.steegmans@uhasselt.be

Abstract. We investigate the power of message-passing neural networks (MPNNs) in their capacity to transform the numerical features stored in the nodes of their input graphs. Our focus is on global expressive power, uniformly over all input graphs, or over graphs of bounded degree with features from a bounded domain. Accordingly, we introduce the notion of a global feature map transformer (GFMT). As a yardstick for expressiveness, we use a basic language for GFMTs, which we call MPLang. Every MPNN can be expressed in MPLang, and our results clarify to which extent the converse inclusion holds. We consider exact versus approximate expressiveness; the use of arbitrary activation functions; and the case where only the ReLU activation function is allowed.

Keywords: Closure under concatenation · Semiring provenance semantics for modal logic · Query languages for numerical data

1 Introduction

An important issue in machine learning is the choice of formalism to represent the functions to be learned [24,25]. For example, feedforward neural networks with hidden layers are a popular formalism for representing functions from $\mathbb{R}^n$ to $\mathbb{R}^p$. When considering functions over graphs, graph neural networks (GNNs) have come to the fore [18]. GNNs come in many variants; in this paper, specifically, we will work with the variant known as message-passing neural networks (MPNNs) [12].

MPNNs compute numerical values on the nodes of an input graph, where, initially, the nodes already store vectors of numerical values, known as *features*. Such an assignment of features to nodes may be referred to as a *feature map* on the graph [15]. We can thus view an MPNN as representing a function that maps a graph, together with a feature map, to a new feature map on that graph. We refer to such functions as *global feature map transformers (GFMTs)*.

Of course, MPNNs are not intended to be directly specified by human designers, but rather to be learned automatically from input–output examples. Still,

I. Varzinczak (Ed.): FoIKS 2022, LNCS 13388, pp. 20–34, 2022.
https://doi.org/10.1007/978-3-031-11321-5_2

MPNNs do form a language for GFMTs. Thus the question naturally arises: what is the expressive power of this language?

We believe GFMTs provide a suitable basis for investigating this question rigorously. The G for 'global' here is borrowed from the terminology of *global function* introduced by Gurevich [16,17]. Gurevich was interested in defining functions in structures (over some fixed vocabulary) *uniformly*, over all input structures. Likewise, here we are interested in expressing GFMTs uniformly over *all* input graphs. We also consider infinite subclasses of all graphs, notably, the class of all graphs with a fixed bound on the degree.

As a concrete handle on our question about the expressive power of MPNNs, in this paper we define the language MPLang. This language serves as a yardstick for expressing GFMTs, in analogy to the way Codd's relational algebra serves as a yardstick for relational database queries [2]. Expressions in MPLang can define features built arbitrarily from the input features using three basic operations also found in MPNNs:

1. Summing a feature over all neighbors in the graph, which provides the message-passing aspect;
2. Applying an activation function, which can be an arbitrary continuous function;
3. Performing arbitrary affine transformations (built using constants, addition, and scalar multiplication).

The difference between MPLang-expressions and MPNNs is that the latter must apply the above three operations in a rigid order, whereas the operations can be combined arbitrarily in MPLang. In particular, every MPNN is readily expressible in MPLang.

Our research question can now be made concrete: is, conversely, every GFMT expressible in MPLang also expressible by an MPNN? We offer the following answers.

1. We begin by considering the case of the popular activation function ReLU [3,13]. In this case, we show that every MPLang expression can indeed be converted into an MPNN (Theorem 1).
2. When arbitrary activation functions are allowed, we show that Theorem 1 still holds in restriction to any class of graphs of bounded degree, equipped with features taken from a bounded domain (Theorem 2).
3. Finally, when the MPNN is required to use the ReLU activation function, we show that every MPLang expression can still be *approximated* by an MPNN; for this result we again restrict to graphs of bounded degree, and moreover to features taken from a compact domain (Theorem 3).

This paper is organized as follows. Section 2 discusses related work. Section 3 defines GFMTs, MPNNs and MPLang formally. Sections 4, 5 and 6 develop our Theorems 1, 2 and 3, respectively. We conclude in Sect. 7.

In this paper, proofs of some lemmas and theorems are only sketched. Detailed proofs will be given in the journal version of this paper. Certain concepts and arguments assume some familiarity with real analysis [23].

2 Related Work

The expressive power of GNNs has received a great deal of attention in recent years. A very nice introduction, highlighting the connections with finite model theory and database theory, has been given by Grohe [15].

One important line of research is focused on characterizing the distinguishing power (also called separating power) of GNNs, in their many variants. There, one is interested in the question: given two graphs, when can they be distinguished by a GNN? This question is closely related to strong methods for graph isomorphism checking, and more specifically, the Weisfeiler-Leman algorithm. A recent overview has been given by Morris et al. [21].

Another line of research has as goal to extend classical results on the "universality" of neural networks [22] to graphs [1,4]. (There are close connections between this line of research and the one just mentioned on distinguishing power [11].) These results consider graphs with a fixed number n of nodes; functions on graphs are shown to be approximable by appropriate variants of GNNs, which, however, may depend on n.

A notable exception is the work by Barceló et al. [6,7], which inspired our present work. Barceló et al. were the first to consider expressiveness of GNNs uniformly over all graphs (note, however, the earlier work of Hella et al. [19] on similar message-passing distributed computation models). Barceló et al. focus on MPNNs, which they fit in a more general framework named AC-GNNs, and they also consider extensions of MPNNs. They further focus on *node classifiers*, which, in our terminology, are GFMTs where the input and output features are boolean values. Using the truncated ReLU activation function, they show that MPNNs can express every node classifiers expressible in *graded modal logic* (the converse inclusion holds as well).

In a way, our work can be viewed as generalizing the boolean setting considered by Barceló et al. to the numerical setting. Indeed, the language MPLang can be viewed as giving a numerical semantics to positive modal logic without conjunction, following the established methodology of semiring provenance semantics for query languages [9,14], and extending the logic with application of arbitrary activation functions. By focusing on boolean inputs and outputs, Barceló et al. are able to capture a stronger logic than our positive modal logic, notably, by expressing negation and counting.

We note that MPLang is a sublanguage of the Tensor Language defined recently by one of us and Reutter [11]. That language serves to unify several GNN variants and clarify their separating power and universality (cf. the first two lines of research on GNN expressiveness mentioned above).

Finally, one can also take a matrix computation perspective, and view a graph on n nodes, together with a d-dimensional feature map, as an $n \times n$ adjacency matrix, together with d column vectors of dimension n. To express GFMTs, one may then simply use a general matrix query language such as MATLANG [8]. Indeed, results on the distinguishing power of MATLANG fragments [10] have been applied to analyze the distinguishing power of GNN variants [5]. Of course, the specific message-passing nature of computation with MPNNs is largely lost when performing general computations with the adjacency and feature matrices.

3 Models and Languages

In this section, we recall preliminaries on graphs; introduce the notion of global feature map transformer (GFMT); formally recall message-passing neural networks and define their semantics in terms of GFMTs; and define the language MPLang.

3.1 Graphs and Feature Maps

We define a *graph* as a pair $G = (V, E)$, where V is the set of nodes and $E \subseteq V \times V$ is the edge relation. We denote V and E of a particular graph G as $V(G)$ and $E(G)$ respectively. By default, we assume graphs to be finite, undirected, and without loops, so E is symmetric and antireflexive. If $(v, u) \in E(G)$ then we call u a neighbor of v in G. We denote the set of neighbors of v in G by $N(G)(v)$. The number of neighbors of a node is called the degree of that node, and the degree of a graph is the maximum degree of its nodes. We use $\mathbb{G}$ to denote the set of all graphs, and $\mathbb{G}_p$, for a natural number p, to denote the set of all graphs with degree at most p.

For a natural number d, a *d-dimensional feature map* on a graph G is a function $\chi : V(G) \to \mathbb{R}^d$, mapping the nodes to *feature vectors*. We use $\mathit{Feat}(G, d)$ to denote the set of all possible d-dimensional feature maps on G. Similarly, for a subset X of $\mathbb{R}^d$, we write $\mathit{Feat}(G, d, X)$ for the set of all feature maps from $\mathit{Feat}(G, d)$ whose image is contained in X.

3.2 Global Feature Map Transformers

Let d and r be natural numbers. We define a *global feature map transformer (GFMT) of type $d \to r$*, to be a function $T : \mathbb{G} \to (\mathit{Feat}(G, d) \to \mathit{Feat}(G, r))$, where $G \in \mathbb{G}$ is the input of T. Thus, if G is a graph and χ is a d-dimensional feature map on G, then $T(G)(\chi)$ is an r-dimensional feature map on G. We call d and r the *input* and *output arity* of T, respectively.

Example 1. We give a few simple examples, just to fix the notion, all with output arity 1. (GFMTs with higher output arities, after all, are just tuples of GFMTs with output arity 1.)

1. The GFMT T_1 of type $2 \to 1$ that assigns to every node the average of its two feature values. Formally, $T_1(G)(\chi)(v) = (x + y)/2$, where $\chi(v) = (x, y)$.
2. The GFMT T_2 defined like T_1, but taking the maximum instead of the average.
3. The GFMT T_3 of type $1 \to 1$ that assigns to every node the maximum of the features of its neighbors. Formally, $T_3(G)(\chi)(v) = \max\{\chi(u) \mid u \in N(G)(v)\}$.

4. The GFMT T_4 of type $1 \to 1$ that assigns to every node v the sum, over all paths of length two from v, of the feature values of the end nodes of the paths. Formally,

$$T_4(G)(\chi)(v) = \sum_{(v,u)\in E(G)} \sum_{(u,w)\in E(G)} \chi(w).$$

3.3 Operations on GFMTs

If $T_1, \ldots, T_r$ are GFMTs of type $d \to 1$, then the tuple $(T_1, \ldots, T_r)$ defines a GFMT T of type $d \to r$ in the obvious manner:

$$T(G)(\chi)(v) := (T_1(G)(\chi)(v), \ldots, T_r(G)(\chi)(v)) \tag{1}$$

Conversely, it is clear that any T of type $d \to r$ can be expressed as a tuple $(T_1, \ldots, T_r)$ as above, where $T_i(G)(\chi)(v)$ equals the i-th component in the tuple $T(G)(\chi)(v)$.

Related to the above tupling operation is concatenation. Let T_1 and T_2 be GFMTs of type $d \to r_1$ and $d \to r_2$, respectively. Their *concatenation* $T_1 \mid T_2$ is the GFMT T of type $d \to r_1 + r_2$ defined by $T(G)(\chi)(v) = T_1(G)(\chi)(v) \mid T_2(G)(\chi)(v))$, where $\mid$ denotes concatenation of vectors. Concatenation is associative. Thus, we could write the previously defined $(T_1, \ldots, T_r)$ also as $T_1 \mid \cdots \mid T_r$.

We also define the *parallel composition* $T_1 \parallel T_2$ of two GFMTs T_1 and T_2, of type $d_1 \to r_1$ and $d_2 \to r_2$, respectively. It is the GFMT T of type $(d_1 + d_2) \to (r_1 + r_2)$ defined by $T(G)(\chi)(v) = T_1(G)(\chi_1)(v) \mid T_2(G)(\chi_2)(v)$, where χ_1 (χ_2) is the feature map that assigns to any node w the projection of $\chi(w)$ to its first (last) d_1 (d_2) components.

In contrast, the *sequential composition* $T_1; T_2$ of two GFMTs T_1 and T_2, of type $d_1 \to d_2$ and $d_2 \to d_3$ respectively, is the GFMT T of type $d_1 \to d_3$ that maps every graph G to $T_2(G) \circ T_1(G)$. In other words, $(T_1; T_2)(G)(\chi)(v) = T_2(G)(T_1(G)(\chi))(v)$.

Finally, for two GFMTS T_1 and T_2 of type $d \to r$, we naturally define their sum $T_1 + T_2$ by $(T_1 + T_2)(G)(\chi)(v) := T_1(G)(\chi)(v) + T_2(G)(\chi)(v)$ (addition of r-dimensional vectors). The difference $T_1 - T_2$ is defined similarly.

Example 2. Recall T_1 and T_4 from Example 1, and consider the following simple GFMTs:

- For $j = 1, 2$, the GFMT P_j of type $2 \to 1$ defined by $P_j(G)(\chi)(v) = x_j$, where $\chi(v) = (x_1, x_2)$.
- The GFMT T_{half} of type $1 \to 1$ defined by $T_{\text{half}}(G)(\chi)(v) = \chi(v)/2$.
- The GFMT T_{sum} of type $1 \to 1$ defined by $T_{\text{sum}}(G)(\chi)(v) = \sum_{u\in N(G)(v)} \chi(u)$.

Then T_1 equals $(P_1 + P_2); T_{\text{half}}$, and T_4 equals $T_{\text{sum}}; T_{\text{sum}}$.

3.4 Message-passing Neural Networks

A *message-passing neural network (MPNN)* consists of layers. Formally, let d and r be natural numbers. An *MPNN layer of type* $d \to r$ is a 4-tuple $L = (W_1, W_2, b, \sigma)$, where $\sigma : \mathbb{R} \to \mathbb{R}$ is a continuous function, and W_1, W_2 and b are real matrices of dimensions $r \times d$, $r \times d$ and $r \times 1$, respectively. We call σ the *activation function* of the layer; we also refer to L as a *σ-layer*.

An MPNN layer L as above defines a GFMT of type $d \to r$ as follows:

$$L(G)(\chi)(v) := \sigma\big(W_1\chi(v) + W_2 \sum_{u \in N(G)(v)} \chi(u) + b\big). \tag{2}$$

In the above formula, feature vectors are used as *column vectors*, i.e., $d \times 1$ matrices. The matrix multiplications involving W_1 and W_2 then produce $r \times 1$ matrices, i.e., r-dimensional feature vectors as desired. We see that matrix W_1 transforms the feature vector of the current node from a d-dimensional vector to an r-dimensional vector. Matrix W_2 does a similar transformation but for the sum of the feature vectors of the neighbors. Vector b serves as a *bias*. The application of σ is performed component-wise on the resulting vector.

We now define an MPNN as a finite, nonempty sequence $L_1, \ldots, L_p$ of MPNN layers, such that the input arity of each layer, except the first, equals the output arity of the previous layer. Such an MPNN naturally defines a GFMT that is simply the sequential composition $L_1; \ldots; L_p$ of its layers. Thus, the input arity of the first layer serves as the input arity, and the output arity of the last layer serves as the output arity. Next we shall give examples of MPNNs that express commonly known functions.

Example 3. Recall the "rectified linear unit" function $\mathrm{ReLU} : \mathbb{R} \to \mathbb{R} : z \mapsto \max(0, z)$. Observe that $\max(x, y) = \mathrm{ReLU}(y - x) + x$, and also that $x = \mathrm{ReLU}(x) - \mathrm{ReLU}(-x)$. Hence, T_2 from Example 1 can be expressed by a two-layer MPNN, where the first layer L_1 transforms input feature vectors (x, y) to feature vectors $(y - x, x, -x)$ and then applies ReLU, and the second layer L_2 transforms the feature vector (a, b, c) produced by L_1 to the final result $a + b - c$. Formally, $L_1 = (A, 0^{3\times2}, 0^{3\times1}, \mathrm{ReLU})$, with

$$A = \begin{pmatrix} -1 & 1 \\ 1 & 0 \\ -1 & 0 \end{pmatrix},$$

and $L_2 = ((1, 1, -1), (0, 0, 0), 0, \mathrm{id})$, with id the identity function.

For another, simple, example, T_{sum} from Example 2 is expressed by the single layer $(0, 1, 0, \mathrm{id})$.

Same activation function If, for a particular MPNN, and an activation function σ, all layers except the last one are σ-layers, and the last layer is either also a σ-layer, or has the identity function as activation function, we refer to the MPNN as a σ-MPNN. Thus, the two MPNNs in the above example are ReLU-MPNNs.

3.5 MPLang

We introduce a basic language for expressing GFMTs. The syntax of expressions e in MPLang is given by the following grammar:

$$e ::= 1 \mid P_i \mid ae \mid e + e \mid f(e) \mid \Diamond e$$

where i is a non-zero natural number, $a \in \mathbb{R}$ is a constant, and $f : \mathbb{R} \to \mathbb{R}$ is continuous.

An expression e is called *appropriate for input arity* d if all subexpressions of e of the form P_i satisfy $1 \leq i \leq d$. In this case, e defines a GFMT of type $d \to 1$, as follows:

- if $e = 1$, then $e(G)(\chi)(v) := 1$
- if $e = P_i$, then $e(G)(\chi)(v) :=$ the i-th component of $\chi(v)$
- if $e = ae_1$, then $e(G)(\chi)(v) := ae_1(G)(\chi)(v)$
- if $e = e_1 + e_2$, then $e(G)(\chi)(v) := e_1(G)(\chi)(v) + e_2(G)(\chi)(v)$
- if $e = f(e_1)$, then $e(G)(\chi)(v) := f(e_1(G)(\chi)(v))$
- if $e = \Diamond e_1$, then $e(G)(\chi)(v) := \sum_{u \in N(G)(v)} e_1(G)(\chi)(u)$

Notice how there is no concatenation operator since the output arity of an expression is always 1. To express higher output arities, we agree that a GFMT T of type $d \to r$ is expressible in MPLang if there exists a tuple $(e_1, \ldots, e_r)$ of expressions that defines T in the sense of Eq. 1. We further agree:

- The constant a will be used as a shorthand for the expression $a1$, i.e., the scalar multiplication of expression 1 by a.
- For any fixed function f, we denote by f-MPLang the language fragment of MPLang where all function applications apply f.

Example 4. Continuing Example 3, we can also express T_2 and T_{sum} in MPLang, namely, T_2 as $\text{ReLU}(P_2 - P_1) + P_1$, and T_{sum} as $\Diamond P_1$.

3.6 Equivalence

Let T_1 and T_2 be MPNNs, or tuples of MPLang expressions, of the same type $d \to r$.

- We say that T_1 and T_2 are *equivalent* if they express the same GFMT.
- For a class $\mathbf{G}$ of graphs and a subset X of $\mathbb{R}^d$, we say that T_1 and T_2 are *equivalent over* $\mathbf{G}$ *and* X if the GFMTs expressed by T_1 and T_2 are equal on every graph G in $\mathbf{G}$ and every $\chi \in \mathit{Feat}(G, d, X)$ (see Sect. 3.1).

Example 4 illustrates the following general observation:

Proposition 1. *For every MPNN T there is an equivalent MPLang-expression that applies, in function applications, only activation functions used in T.*

Proof. (Sketch.) Since we can always substitute subexpressions of the form P_i by more complex expressions, MPLang is certainly closed under sequential composition. It thus suffices to verify that single MPNN layers L are expressible in MPLang. For each output component of L we devise a separate MPLang expression. Inspecting Eq. 2, we must argue for linear transformation; summation over neighbors; addition of a constant (component from the bias vector); and application of an activation function. Linear transformation, and addition of a constant, are expressible using the addition and scalar multiplication operators of MPLang. Summation over neighbors is provided by the $\Diamond$ operator. Application of an activation function is provided by function application in MPLang.

4 From MPLang to MPNN Under ReLU

In Proposition 1 we observed that MPLang readily provides all the operators that are implicitly present in MPNNs. MPLang, however, allows these operators to be combined arbitrarily in expressions, whereas MPNNs have a more rigid architecture. Nevertheless, at least under the ReLU activation function, we have the following strong result:

Theorem 1. *Every GFMT expressible in ReLU-MPLang is also expressible as a ReLU-MPNN.*

Crucial to proving results of this kind will be that the MPNN architecture allows the construction of concatenations of MPNNs. We begin by noting:

Lemma 1. *Let σ be an activation function. The class of GFMTs expressible as a single σ-MPNN layer is closed under concatenation and under parallel composition.*

Proof. (Sketch.) For parallel composition, we construct block-diagonal matrices from the matrices provided by the two layers. For concatenation, we can simply stack the matrices vertically. □

For $\sigma = \text{ReLU}$, we can extend the above Lemma to multi-layer MPNNs:

Lemma 2. *ReLU-MPNNs are closed under concatenation.*

Proof. Let L and K be two ReLU-MPNNs. Since ReLU is idempotent, every n-layer ReLU-MPNN is equivalent to an $n+1$-layer ReLU-MPNN. Hence we may assume that $L = L_1; \dots, L_n$ and $K = K_1; \dots; K_n$ have the same number of layers. Now $L \mid K = (L_1 \mid K_1); (L_2 \parallel K_2); \dots; (L_n \parallel K_n)$ if $n \geq 2$; if $n = 1$, clearly $L \mid K = L_1 \mid K_1$. Hence, the claim follows from Lemma 1. □

Note that a ReLU-MPNN layer can only output positive numeric values, since the result of ReLU is always positive. This explains why we must allow the identity function (id) in the last layer of a ReLU-MPNN (see the end of Sect. 3.4). Moreover, we can simulate intermediate id-layers in a ReLU-MPNN, thanks to the identity $x = \text{ReLU}(x) - \text{ReLU}(-x)$. Specifically, we have:

Lemma 3. *Let L be an id-layer and let K be a σ-layer. Then there exists a ReLU-layer L' and a σ-layer K' such that $L; K$ is equivalent to $L'; K'$.*

Proof. Let $L = (W_1, W_2, b, \text{id})$. We put

$$L' = (W_1, W_2, b, \text{ReLU}) \mid (-W_1, -W_2, -b, \text{ReLU})$$

which corresponds to a ReLU-layer by Lemma 1. Let $K = (A, B, c, \sigma)$. Consider the block matrices $A' = (A|-A)$ and $B' = (B|-B)$ (single-row block matrices, with two matrices stacked horizontally, not vertically). Now for K' we use (A', B', c, σ). □

We are now ready to prove Theorem 1. By Lemma 2, it suffices to focus on MPLang expressions, i.e., GFMTs of output arity one. So, our task is to construct, for every expression e in ReLU-MPLang, an equivalent ReLU-MPNN E. However, by Lemma 3, we are free to use intermediate id-layers in the construction of E. We proceed by induction on the structure of e. We skip the base cases and consider the inductive cases where e is of one of the forms ae_1, $e_1 + e_2$, $f(e_1)$ (with $f = \text{ReLU}$), or $\Diamond\, e_1$. By induction, we have MPNNs E_1 and E_2 for e_1 and e_2.

- If e is of the form ae_1, we set $E = E_1; (a, 0, 0, \text{id})$.
- If e is of the form $e_1 + e_2$, we set $E = (E_1 \mid E_2); ((1, 1), (0, 0), 0, \text{id})$. Here, $E_1 \mid E_2$ corresponds to a ReLU-MPNN by Lemma 2.
- If e is of the form $f(e_1)$, we set $E = E_1; (1, 0, 0, f)$.
- If e is of the form $\Diamond\, e_1$, we set $E = E_1; (0, 1, 0, \text{id})$.

5 Arbitrary Activation Functions

Theorem 1 only supports the ReLU function in MPLang expressions. On the other hand, the equivalent MPNN then only uses ReLU as well. If we allow arbitrary activation functions in MPNNs, can they then simulate also MPLang expressions that apply arbitrary functions? We can answer this question affirmatively, under the assumption that graphs have bounded degree and feature vectors come from a bounded domain. The proof of our Lemma 4 explains how we rely on the bounded-domain assumption. Moreover, also the degree has to be bounded, for otherwise we can still create unbounded values using $\Diamond(P_i)$.

Theorem 2. *Let p and d be natural numbers, let $\mathbb{G}_p$ be the class of graphs of degree at most p, and let $X \subseteq \mathbb{R}^d$ be bounded. For every GFMT T expressible in MPLang there exists an MPNN that is equivalent to T over $\mathbb{G}_p$ and X.*

The above theorem can be proven exactly as Theorem 1, once we can deal with the concatenation of two MPNN layers with possibly different activation functions. The following result addresses this task:

Lemma 4. *Let L and K be MPNN layers of type $d_L \to r_L$ and $d_K \to r_k$, respectively. Let $X_L \subseteq \mathbb{R}^{d_L}$ and $X_K \subseteq \mathbb{R}^{d_K}$ be bounded, and let p be a natural number. There exist two MPNN layers L' and K' such that*

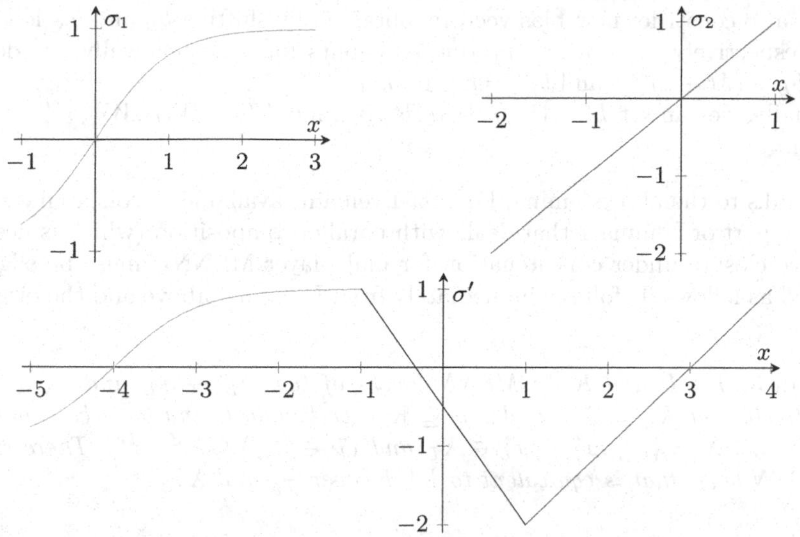

Fig. 1. Illustration of the proof of Lemma 4.

1. *L' and K' use the same activation function;*
2. *L' is equivalent to L over $\mathbb{G}_p$ and X_L;*
3. *K' is equivalent to K over $\mathbb{G}_p$ and X_K.*

Proof. Let $L = (W_{1L}, W_{2L}, b_L, \sigma_L)$ and $K = (W_{1K}, W_{2K}, b_K, \sigma_K)$. Let $w_{1,i}$, $w_{2,i}$ and b_i be the i-th row of W_{1L}, W_{2L} and b_L respectively. For each $i \in \{1, \ldots, r_L\}$ and for any $k \in \{1, \ldots, p\}$ consider the function

$$\lambda_i^k : \mathbb{R}^{(k+1)d_L} \to \mathbb{R} : (\vec{x_0}, \vec{x_1}, \ldots, \vec{x_k}) \mapsto w_{1,i} \cdot \vec{x_0} + w_{2,i} \cdot \vec{x_1} + \cdots + w_{2,i} \cdot \vec{x_k} + b_i.$$

Then for any $G \in \mathbb{G}_p$, any $\chi \in \mathit{Feat}(G, d, X_L)$, and $v \in V(G)$, each component of $L(G)(\chi)(v)$ will belong to the image of some function λ_i^k on X_L^{k+1}, with k the degree of v. Since X_L^{k+1} is bounded and λ_i^k is continuous, these images are also bounded and their finite union over $i \in \{1, \ldots, r\}$ and $k \in \{1, \ldots, p\}$ is also bounded. Let Y_1 be this union and let $M = \max Y_1$.

For K we can similarly define the functions κ_i^k and arrive at a bounded set $Y_K \subseteq \mathbb{R}$. We then define $m = \min Y_2$.

We will now construct a new activation function σ'. First define the functions $\sigma_L'(x) := \sigma_L(x + M_i + 1)$ for $x \in]-\infty, -1]$ and $\sigma_K'(x) := \sigma_K(x - m_i - 1)$ for $x \in [1, \infty[$. Notice how σ_L' is simply σ_L shifted to the left so that its highest possible input value, which is M, aligns with -1. Similarly, σ_K' is simply σ_K shifted to the right so that its lowest possible input value, which is m, aligns with 1. We then define σ' to be any continuous function that extends both σ_L' and σ_K'. An example of this construction can be seen in Fig. 1 with $\sigma_1 = \tanh$, $M = 3$, σ_2 the identity, and $m = -2$.

We also construct new bias vectors, obtained by shifting b_L and b_K left and right respectively to provide appropriate inputs for σ'. Specifically, we define $b'_L := b_L - (M+1)^{r\times 1}$ and $b'_K := b_K + (m+1)^{r\times 1}$.

Finally, we can set $L' = (W_{1L}, W_{2L}, b'_L, \sigma')$ and $K' = (W_{1K}, W_{2K}, b'_K, \sigma')$ as desired. □

Thanks to the above lemma, Lemma 1 remains available to concatenate layers. The part of Lemma 1 that deals with parallel composition (which is needed to prove closure under concatenation for multi-layer MPNNs) must be slightly adapted as follows. It follows immediately from Lemma 4 above and the original Lemma 1.

Lemma 5. *Let L and K be MPNN layers of type $d_L \to r_L$ and $d_K \to r_k$, respectively. Let $X_L \subseteq \mathbb{R}^{d_L}$ and $X_K \subseteq \mathbb{R}^{d_K}$ be bounded, and let p be a natural number. Let $X = \{(\vec{x_L}, \vec{x_K}) \mid \vec{x_L} \in X_L \text{ and } \vec{x_K} \in X_K\} \subseteq \mathbb{R}^{d_L+d_K}$. There exists an MPNN layer that is equivalent to $L \parallel K$ over $\mathbb{G}_p$ and X.*

6 Approximation by ReLU-MPNNs

Theorem 2 allows the use of arbitrary activation functions in the MPNN simulating an MPLang expression; these activation functions may even be different from the ones applied in the expression (see the proof of Lemma 4). What if we insist on MPNNs using a fixed activation function? In this case we can still recover our result, if we allow approximation. Moreover, we must slightly strengthen our assumption of feature vectors coming from a bounded domain, to coming from a compact domain.[1]

We will rely on a classical result in the approximation theory of neural networks [20,22], to the effect that continuous functions can be approximated arbitrarily well by piecewise linear functions, which can be modeled using ReLU.[2] In order to recall this result, we recall that the uniform distance between two continuous functions g and h from $\mathbb{R}$ to $\mathbb{R}$ on a compact domain Y equals $\rho_Y(g,h) = \sup_{x\in Y} |g(x) - h(x)|$.

Density Property. *Let Y be a compact subset of $\mathbb{R}$, let $f : \mathbb{R} \to \mathbb{R}$ be continuous on Y, and let $\epsilon > 0$ be a real number. There exists a positive integer n and real coefficients a_i, b_i, c_i, for $i = 1, \ldots, n$, such that $\rho_Y(f, f') \leq \epsilon$, where $f'(x) = \sum_{i=1}^{n} c_i \mathrm{ReLU}(a_i x - b_i)$.*

We want to extend the notion of uniform distance to GFMTs expressed in MPLang. For any MPLang expression e appropriate for input arity d, any class

---

[1] A subset of $\mathbb{R}$ or $\mathbb{R}^d$ is called compact if it is bounded and closed in the ordinary topology.

[2] The stated Density Property actually holds not just for ReLU, but for any nonpolynomial continuous function.

$\mathbf{G}$ of graphs, and any subset $X \subseteq \mathbb{R}^d$, the *image* of e over $\mathbf{G}$ and X is defined as the set

$$\{e(G)(\chi)(v) : G \in \mathbf{G} \;\&\; \chi \in \mathit{Feat}(G,d,X) \;\&\; v \in V(G)\}.$$

It is a subset of $\mathbb{R}$. We observe: (proof omitted)

Lemma 6. *For any natural number p and compact X, the image of e over $\mathbb{G}_p$ and X is a compact set.*

With p and X as in the lemma, and any two MPLang expression e_1 and e_2 appropriate for input arity d, the set

$$\{|e_1(G)(\chi)(v) - e_2(G)(\chi)(v)| : G \in \mathbb{G}_p \;\&\; \chi \in \mathit{Feat}(G,d,X) \;\&\; v \in V(G)\}$$

has a supremum. We define $\rho_{\mathbb{G}_p,X}(e_1,e_2)$, the uniform distance between e_1 and e_2 over $\mathbb{G}_p$ and X, to be that supremum.

The main result of this section can now be stated as follows. Note that we approximate MPLang expressions by ReLU-MPLang expressions. These can then be further converted to ReLU-MPNNs by Theorem 1.

Theorem 3. *Let p and d be natural numbers, and let $X \subseteq \mathbb{R}^d$ be compact. Let e be an MPLang expression appropriate for d, and let $\epsilon > 0$ be a real number. There exists a ReLU-MPLang expression e' such that $\rho_{\mathbb{G}_p,X}(e,e') \leq \epsilon$.*

Proof. By induction on the structure of e. We skip the base cases. In the inductive cases where e is of the form ae_1, $e_1 + e_2$, or $f(e_1)$, we consider any $G \in \mathbb{G}_p$, any $\chi \in \mathit{Feat}(G,d,X)$, and any $v \in V(G)$, but abbreviate $e(G)(\chi)(v)$ simply as e.

Let e be of the form ae_1. If $a = 0$ we set $e' = 0$. Otherwise, let e_1' be the expression obtained by induction applied to e_1 and ϵ/a. We then set $e' = ae_1'$. The inequality $|e - e'| \leq \epsilon$ is readily verified.

Let e be of the form $e_1 + e_2$. For $j = 1, 2$, let e_j' be the expression obtained by induction applied to e_j and $\epsilon/2$. We then set $e' = e_1' + e_2'$. The inequality $|e - e'| \leq \epsilon$ now follows from the triangle inequality.

Let e is of the form $f(e_1)$. By Lemma 6, the image of e_1 is a compact set $Y_1 \subseteq \mathbb{R}$. We define the closed interval $Y = [\min(Y_1) - \epsilon/2, \max(Y_1) + \epsilon/2]$. By the Density Property, there exists f' such that $\rho_Y(f,f') \leq \epsilon/2$. Since Y is compact, f' is uniformly continuous on Y. Thus there exists $\delta > 0$ such that $|f'(x) - f'(x')| < \epsilon/2$ whenever $|x - x'| < \delta$.

We now take e_1' to be the expression obtained by induction applied to e_1 and $\min(\delta, \epsilon/2)$. We see that the image of e_1' is contained in Y. Setting $e' = f(e_1')$, we verify that $|e - e'| = |f(e_1) - f'(e_1')| + |f'(e_1) - f'(e_1')| \leq \epsilon$ as desired.

Our final inductive case is when e is of the form $\Diamond\, e_1$. We again consider any $G \in \mathbb{G}_p$, any $\chi \in \mathit{Feat}(G,d,X)$, and any $v \in V(G)$, but this time abbreviate $e(G)(\chi)(v)$ as $e(v)$. Let e_1' be the expression obtained by induction applied to e_1 and ϵ/p. Setting $e' = \Diamond\, e_1'$, we verify, as desired:

$$\begin{aligned}|e(v) - e'(v)| &= |\sum_{u \in N(G)(v)} e_1(u) - \sum_{u \in N(G)(v)} e'_1(u)| \\ &\leq \sum_{u \in N(G)(v)} |e_1(u) - e'_1(u)| \\ &\leq p(\epsilon/p) \\ &= \epsilon.\end{aligned}$$

The penultimate step clearly uses that G has degree bound p. (This degree bound is also used in Lemma 6.)

7 Concluding Remarks

We believe that our approach has the advantage of modularity. For example, Theorem 1 is stated for ReLU, but holds for any activation function for which Lemmas 1 and 3 can be shown. We already noted that the Density Property holds not just for ReLU but for any nonpolynomial continuous activation function. It follows that for any activation function σ for which Lemmas 1 and 3 can be shown, every MPLang expression can be approximated by a σ-MPNN.

It would be interesting to see counterexamples that show that Theorems 2 and 3 do not hold without the restriction to bounded-degree graphs, or to features from a bounded or compact domain. Such counterexamples can probably be derived from known counterexamples in analysis or approximation theory.

Finally, in this work we have focused on the question whether MPLang can be simulated by MPNNs. However, it is also interesting to investigate the expressive power of MPLang by itself. For example, is the GFMT T_3 from Example 1 expressible in MPLang?

Acknowledgments. We thank the anonymous reviewers for suggesting improvements to the text. Jasper Steegmans is supported by the Special Research Fund (BOF) of Hasselt University.

References

1. Abboud, R., Ceylan, I., Grohe, M., Lukasiewicz, T.: The surprising power of graph neural networks with random node initialization. In: Zhou, Z.H. (ed.) Proceedings 30th International Joint Conference on Artificial Intelligence, pp. 2112–2118. ijcai.org (2021)
2. Abiteboul, S., Hull, R., Vianu, V.: Foundations of Databases. Addison-Wesley, Boston (1995)
3. Arora, R., Basu, A., Mianjy, P., Mukherjee, A.: Understanding deep neural networks with rectified linear units. In: Proceedings 6th International Conference on Learning Representations. OpenReview.net (2018)

4. Azizian, W., Lelarge, M.: Expressive power of invariant and equivariant graph neural networks. In: Proceedings 9th International Conference on Learning Representations. OpenReview.net (2021)
5. Balcilar, M., Héroux, P., et al.: Breaking the limits of message passing graph neural networks. In: Meila, M., Zhang, T. (eds.) Proceedings 38th International Conference on Machine Learning. Proceedings of Machine Learning Research, vol. 139, pp. 599–608 (2021)
6. Barceló, P., Kostylev, E., Monet, M., Pérez, J., Reutter, J., Silva, J.: The expressive power of graph neural networks as a query language. SIGMOD Rec. **49**(2), 6–17 (2020)
7. Barceló, P., Kostylev, E., Monet, M., Pérez, J., Reutter, J., Silva, J.: The logical expressiveness of graph neural networks. In: Proceedings 8th International Conference on Learning Representations. OpenReview.net (2020)
8. Brijder, R., Geerts, F., Van den Bussche, J., Weerwag, T.: On the expressive power of query languages for matrices. ACM Trans. Database Syst. **44**(4), 15:1–15:31 (2019)
9. Dannert, K.M., Grädel, E.: Semiring provenance for guarded logics. In: Madarász, J., Székely, G. (eds.) Hajnal Andréka and István Németi on Unity of Science. OCL, vol. 19, pp. 53–79. Springer, Cham (2021). https://doi.org/10.1007/978-3-030-64187-0_3
10. Geerts, F.: On the expressive power of linear algebra on graphs. Theory Comput. Syst. **65**(1), 179–239 (2021)
11. Geerts, F., Reutter, J.: Expressiveness and approximation properties of graph neural networks. In: ICLR. OpenReview.net (2022). (to appear)
12. Gilmer, J., et al.: Neural message passing for quantum chemistry. In: Precup, D., Teh, Y. (eds.) Proceedings 34th International Conference on Machine Learning. Proceedings of Machine Learning Research, vol. 70, pp. 1263–1272 (2017)
13. Goodfellow, I., Bengio, Y., Courville, A.: Deep Learning. MIT Press, Cambridge (2016)
14. Green, T., Karvounarakis, G., Tannen, V.: Provenance semirings. In: Proceedings 26th ACM Symposium on Principles of Database Systems, pp. 31–40 (2007)
15. Grohe, M.: The logic of graph neural networks. In: Proceedings 36th Annual ACM/IEEE Symposium on Logic in Computer Science, pp. 1–17. IEEE (2021)
16. Gurevich, Y.: Algebras of feasible functions. In: Proceedings 24th Symposium on Foundations of Computer Science, pp. 210–214. IEEE Computer Society (1983)
17. Gurevich, Y.: Logic and the challenge of computer science. In: Börger, E. (ed.) Current Trends in Theoretical Computer Science, pp. 1–57. Computer Science Press (1988)
18. Hamilton, W.: Graph Representation Learning. Synthesis Lectures on Artificial Intelligence and Machine Learning, Morgan & Claypool, San Rafael (2020)
19. Hella, L., et al.: Weak models of distributed computing, with connections to modal logic. Distrib. Comput. **28**(1), 31–53 (2013). https://doi.org/10.1007/s00446-013-0202-3
20. Leshno, M., Lin, V., Pinkus, A., Schocken, S.: Multilayer feedforward networks with a nonpolynomial activation function can approximate any function. Neural Netw. **6**(6), 861–867 (1993)
21. Morris, C., et al.: Weisfeiler and Leman go machine learning: the story so far. arXiv:2122.09992 (2021)
22. Pinkus, A.: Approximation theory of the MLP model in neural networks. Acta Numerica **8**, 143–195 (1999)

23. Rudin, W.: Principles of Mathematical Analysis, 3rd edn. McGraw Hill, New York (1976)
24. Russell, S., Norvig, P.: Artificial Intelligence: A Modern Approach, 4th edn. Pearson, London (2022)
25. Shalev-Shwartz, S., Ben-David, S.: Understanding Machine Learning: From Theory to Algorithms. Cambridge University Press, Cambridge (2014)

Assumption-Based Argumentation for Extended Disjunctive Logic Programming

Toshiko Wakaki(✉)

Shibaura Institute of Technology, 307 Fukasaku Minuma-ku,
Saitama-city, Saitama 337–8570, Japan
twakaki@shibaura-it.ac.jp

Abstract. We present the semantic correspondence between generalized assumption-based argumentation and extended disjunctive logic programming. In this paper, we propose an assumption-based framework (ABF) translated from an extended disjunctive logic program (EDLP), which incorporates explicit negation into Heyninck and Arieli's ABF induced by a disjunctive logic program to resolve problems remained in their approach. In our proposed ABF, we show how an argument is constructed from disjunctive rules. Then we show a 1-1 correspondence between answer sets of an EDLP P and stable argument extensions of the ABF translated from P with trivialization rules. Hereby thanks to introducing explicit negation, the semantic relationship between disjunctive default theories and assumption-based frameworks is obtained in our approach. Finally, after defining rationality postulates and consistency in our ABF, we show answer sets of a consistent EDLP can be captured by consistent stable extensions of the translated ABF with no trivialization rules, that is useful for ABA applications containing explicit negation.

Keywords: assumption-based argumentation · extended disjunctive logic programs · disjunctive default theories · arguments · argument extensions

1 Introduction

In our daily life, disjunctive information is often required in reasoning and argumentation to solve problems. In nonmonotonic reasoning, Gelfond et al. [16] proposed *disjunctive default logic* as a generalization of Reiter's default logic [19] to overcome problems of default logic in handling disjunctive information. They also showed that extended disjunctive logic programs (EDLPs) [15] which allow disjunctions with the connective "|" in rule heads as well as classical negation "¬" can be embedded into disjunctive default theories.

In contrast, in the context of formal argumentation, Beirlaen et al. [2] presented the extended ASPIC$^+$ framework where disjunctive reasoning is integrated in structured argumentation with defeasible rules [17] by incorporating

I. Varzinczak (Ed.): FoIKS 2022, LNCS 13388, pp. 35–54, 2022.
https://doi.org/10.1007/978-3-031-11321-5_3

reasoning by cases inference scheme [1]. In their framework, an argument is allowed to have a disjunctive conclusion, while disjunction is expressed by the classical connective "$\vee$". However, they did not show the relationship between their framework and other approaches in nonmonotonic reasoning such as a disjunctive default theory. Consider the following example shown by them.

Example 1. **(Kyoto protocol).** [2] *There are two candidates for an upcoming presidential election. The candidates had a debate in the capital. They were asked what measures are to be taken in order for the country to reach the Kyoto protocol objectives for reducing greenhouse gas emissions. The first candidate, a member of the purple party, argued that if she wins the election, she will reach the objectives by supporting investments in renewable energy. The second candidate, a member of the yellow party, argued that if she wins the election, she will reach the objectives by supporting sustainable farming methods. We have reasons to believe that, no matter which candidate wins the election, the Kyoto protocol objectives will be reached. If the purple candidate wins, she will support investments in renewable energy ($p \Rightarrow r$), which would in turn result in meeting the Kyoto objectives ($r \Rightarrow k$). Similarly, if the yellow candidate wins, she will support sustainable farming methods ($y \Rightarrow f$), which would result in meeting the Kyoto objectives ($f \Rightarrow k$). Since one of the two candidates is going to win ($p \vee y$), we can reason by cases to conclude that the Kyoto objectives will be reached (k).*

Beirlaen et al. [2] represented the information shown above in terms of the knowledge base $\mathcal{K}_1 = (\{p \Rightarrow r, r \Rightarrow k, y \Rightarrow f, f \Rightarrow k\}, \{p \vee y\})$ which consists of *defeasible* rules in ASPIC$^+$ [17] and the classical formula $p \vee y$ as facts. Their formulation derives the intended result k, however, a problem happens when K_1 is translated into a default theory. In fact, $\psi \Rightarrow \phi$ encodes the normal default $\frac{\psi:\phi}{\phi}$, then $\mathcal{K}_1$ is expressed by the default theory:

$$D_1 = \{\frac{p:r}{r}, \frac{r:k}{k}, \frac{y:f}{f}, \frac{f:k}{k}, \; p \vee y\}.$$

D_1 has a single extension, consisting of the disjunction $p \vee y$ and its logical consequences, where the four defaults "don't work". Thus the result is contrary to what we would expect. Instead of D_1, let us represent the information by the disjunctive default theory as follows:

$$D_2 = \{\frac{p:r}{r}, \frac{r:k}{k}, \frac{y:f}{f}, \frac{f:k}{k}, \; p \mid y\},$$

or the associated EDLP $P_1 = \{r \leftarrow p, not\ \neg r;\ k \leftarrow r, not\ \neg k; f \leftarrow y, not\ \neg f;\ k \leftarrow f, not\ \neg k;\ p \mid y \leftarrow\}$. Then D_2 has two extensions E_1 and E_2 s.t. $k \in E_i$ $(i=1,2)$, while P_1 has two answer sets $S_1 = \{p, r, k\}$ and $S_2 = \{y, f, k\}$, where $S_i \subseteq E_i (i=1,2)$. These results agree with our expectation that the Kyoto protocol objectives will be reached no matter which candidate wins the election. □

Recently, Heyninck and Arieli [13] proposed a generalized assumption-based framework (ABF, for short) induced by a disjunctive logic program (DLP), where disjunctions with the connective "$\vee$" are allowed to appear in rule heads. Though their ABF has a *contrariness* operator $\overline{}$ such that $\overline{not\ p} = p$ for every atom p, its language does not contain explicit negation. Then they showed a 1-1 correspondence between the *stable models* of a DLP and the stable *assumption* extensions of the ABF induced by a DLP. However, several issues are remained to be explored. First, they did not show how to construct an argument from disjunctive rules in a DLP nor took account of *argument* extensions of their ABF. Second, the semantic relationship between their ABF induced by a DLP and a disjunctive default theory is not considered in [13]. In logic programming, EDLPs are introduced to extend DLPs for knowledge representation, and in formal argumentation, generally assumption-based frameworks (ABFs) are capable of expressing explicit negation [12]. To the best of our knowledge, however, there is no study to show the semantic relationship between ABFs and EDLPs as well as the relationship between ABFs and disjunctive default theories.

The purpose of this paper is first to show how to construct an argument from disjunctive rules in (E)DLPs, and second to investigate the semantic relationship between ABFs and EDLPs (resp. disjunctive default theories). So far explicit negation has been used in various applications of argumentation. In structured argumentation, for example, as the ways in which arguments can be in conflict, ASPIC$^+$ allows the *rebutting* attack between two arguments having the mutually contradictory conclusions w.r.t. explicit negation along with undercutting and undermining attacks. In contrast, assumption-based argumentation (ABA) [9] whose language may contain explicit negation allows only attacks against the support of arguments as defined in terms of a notion of *contrary*. Since both systems incorporate explicit negation, conditions under which each system satisfies *rationality postulates* [5] were proposed to avoid anomalous results [10, 17].

As for recent ABA applications containing explicit negation, Schulz and Toni [21] proposed the approach of justifying answer sets of an extended logic program (which contains explicit negation) using argumentation. However counterexamples exist in their theorems [21, Theorems 1, 2] due to not taking account of rationality postulates [5, 10] or consistency in ABA as shown in [23].

To resolve the above-mentioned problems, this paper proposes an assumption-based framework (ABF) translated from an EDLP, which incorporates explicit negation into Heyninck et al.'s ABF induced by a DLP. Contribution of this study is as follows: First, we define an argument in the ABF translated from a given EDLP which is constructed from rules allowing disjunctions of the EDLP based on three inference rules provided in our ABF. Second, we show there is a one-to-one correspondence between *answer sets* of an EDLP P and stable *argument* extensions (resp. stable *assumption* extensions) of the ABF translated from P with trivialization rules. Third, hereby thanks to introducing explicit negation, the semantic relationship between disjunctive default theories and assumption-based frameworks is obtained on the basis of our ABF. To the best of our knowledge, this paper is the first one to show this relationship. Finally, since our ABF incorporates explicit negation $\neg$ along with disjunction $|$, we define rationality postulates and consistency in our ABFs to avoid anomalous outcomes as arise in [21] (see [23]). Then we show answer sets of a consistent EDLP can be captured by consistent stable extensions of the translated ABF with no trivialization rules. This is useful for ABA applications containing explicit negation (e.g. [21]).

The rest of this paper is as follows. Section 2 shows preliminaries. Section 3 presents an ABF translated from an EDLP and shows the definition of its argument, the semantic relationship between an EDLP (resp. a disjunctive default theory) and its translated ABF, and the relationship between a consistent EDLP and its translated ABF. Sections 4 discusses related work and concludes.

2 Preliminaries

2.1 Assumption-Based Argumentation

An ABA framework [4,9] is a tuple $\langle \mathcal{L}, \mathcal{R}, \mathcal{A}, \overline{} \rangle$, where $(\mathcal{L}, \mathcal{R})$ is a deductive system, consisting of a language $\mathcal{L}$ and a set $\mathcal{R}$ of inference rules of the form: $b_0 \leftarrow b_1, \ldots, b_m$ ($b_i \in \mathcal{L}$ for $0 \leq i \leq m$), $\mathcal{A} \subseteq \mathcal{L}$ is a (non-empty) set of *assumptions*, and $\overline{}$ is a total mapping from $\mathcal{A}$ into $\mathcal{L}$, which we call a contrariness operator. $\overline{\alpha}$ is referred to as the *contrary* of $\alpha \in \mathcal{A}$. We enforce that ABA frameworks are *flat*, namely assumptions in $\mathcal{A}$ do not appear in the heads of rules in $\mathcal{R}$. In ABA, an *argument* for a claim (or conclusion) $c \in \mathcal{L}$ supported by $K \subseteq \mathcal{A}$ ($K \vdash c$ in short) is defined as a (finite) tree with nodes labelled by sentences in $\mathcal{L}$ or by $\tau \notin \mathcal{L}$ denoting "true", where the root is labelled by c, and leaves are labelled either by τ or by assumptions in K. *attacks* is defined as follows.

- *An argument* $K_1 \vdash c_1$ *attacks an argument* $K_2 \vdash c_2$ iff $c_1 = \overline{\alpha}$ for $\exists \alpha \in K_2$.
- For $\Delta, \Delta' \subseteq \mathcal{A}$, and $\alpha \in \mathcal{A}$, Δ *attacks* α iff Δ enables the construction of an argument for conclusion $\overline{\alpha}$. Thus Δ *attacks* Δ' iff Δ *attacks* some $\alpha \in \Delta'$.

A set of arguments *Args* is *conflict-free* iff $\nexists\, A, B \in \mathit{Args}$ such that A *attacks* B. A set of assumptions Δ is *conflict-free* iff Δ does not attack itself.
Let $AF_{\mathcal{F}} = (AR, \mathit{attacks})$ be the abstract argumentation (AA) framework [6] generated from an ABA framework $\mathcal{F}$, where AR is the set of arguments s.t. $a \in AR$ iff an argument $a : K \vdash c$ is in $\mathcal{F}$, and $(a, b) \in \mathit{attacks}$ in $AF_{\mathcal{F}}$ iff a *attacks* b in $\mathcal{F}$ [8]. Let $\sigma \in$ {complete, preferred, grounded, stable, ideal} be the name of the argumentation semantics. Then the ABA semantics is given by σ argument extensions as well as by σ assumption extensions under σ semantics. In case σ=stable, $\mathit{Args} \subseteq AR$ is a stable argument extension iff it is conflict-free and *attacks* every $A \in AR \setminus \mathit{Args}$, while $\Delta \subseteq \mathcal{A}$ is a stable assumption extension iff it is conflict-free and *attacks* every $\alpha \in \mathcal{A} \setminus \Delta$. There is a 1-1 correspondence between σ assumption extensions and σ argument extensions. Let $\mathit{claim}(Ag)$ be the conclusion (or claim) of an argument Ag. Then the *conclusion* of a set of arguments $\mathcal{E}$ is defined as $\texttt{Concs}(\mathcal{E}) = \{c \in \mathcal{L} \mid c = \mathit{claim}(Ag) \text{ for } Ag \in \mathcal{E}\}$.

Rationality postulates [5] are stated in ABA as follows.

Definition 1. (Rationality postulates). [5,10] Let $\langle \mathcal{L}, \mathcal{R}, \mathcal{A}, \overline{} \rangle$ be a flat ABA framework, where $\overline{}$ has the property such that contraries of assumptions are not assumptions. A set $X \subseteq \mathcal{L}$ is said to be *contradictory* iff X is contradictory w.r.t. $\overline{}$, i.e. there exists an assumption $\alpha \in \mathcal{A}$ such that $\{\alpha, \overline{\alpha}\} \subseteq X$; or X is

contradictory w.r.t. $\neg$, i.e. there exists $s \in \mathcal{L}$ such that $\{s, \neg s\} \subseteq X$ if $\mathcal{L}$ contains an explicit negation operator $\neg$. Let $CN_{\mathcal{R}} : 2^{\mathcal{L}} \rightarrow 2^{\mathcal{L}}$ be a consequence operator. For a set $X \subseteq \mathcal{L}$, $CN_{\mathcal{R}}(X)$ is the smallest set such that $X \subseteq CN_{\mathcal{R}}(X)$, and for each rule $r \in \mathcal{R}$, if $body(r) \subseteq CN_{\mathcal{R}}(X)$ then $head(r) \in CN_{\mathcal{R}}(X)$. X is closed iff $X = CN_{\mathcal{R}}(X)$. A set $X \subseteq \mathcal{L}$ is said to be *inconsistent* iff its closure $CN_{\mathcal{R}}(X)$ is contradictory. X is said to be *consistent* iff it is not inconsistent. A flat ABA framework $\mathcal{F} = \langle \mathcal{L}, \mathcal{R}, \mathcal{A}, \overline{} \rangle$ is said to satisfy the *consistency-property* (resp. the *closure-property*) if for each complete extension $\mathcal{E}$ of $AF_{\mathcal{F}}$ generated from $\mathcal{F}$, $\texttt{Concs}(\mathcal{E})$ is consistent (resp. $\texttt{Concs}(\mathcal{E})$ is closed).
We say that a set of arguments $\mathcal{E}$ is consistent if $\texttt{Concs}(\mathcal{E})$ is consistent [23].

Heyninck and Arieli [12] proposed (generalized) assumption-based frameworks.

Definition 2. (Assumption-based frameworks). [12,13] Let $\mathcal{L}$ be a propositional language, $\psi, \phi \in \mathcal{L}$ be formulas and $\Gamma, \Gamma', \Lambda \subseteq \mathcal{L}$ be sets of formulas.

A (propositional) *logic* for a language $\mathcal{L}$ is a pair $\mathfrak{L}=(\mathcal{L}, \vdash)$, where $\vdash$ is a (Tarskian) consequence relation for $\mathcal{L}$, that is, a binary relation between sets of formulas and formulas in $\mathcal{L}$, which is reflexive (if $\psi \in \Gamma$ then $\Gamma \vdash \psi$), monotonic (if $\Gamma \vdash \psi$ and $\Gamma \subseteq \Gamma'$, then $\Gamma' \vdash \psi$), and transitive (if $\Gamma \vdash \psi$ and $\Gamma' \cup \{\psi\} \vdash \phi$, then $\Gamma \cup \Gamma' \vdash \phi$). The $\mathfrak{L}$-based transitive closure of a set Γ of $\mathcal{L}$-formulas is $\texttt{Cn}_{\mathfrak{L}}(\Gamma)=\{\psi \in \mathcal{L} \mid \Gamma \vdash \psi\}$ [12].

Let $\Gamma, \Lambda \subseteq \mathcal{L}$, $\Lambda \neq \emptyset$, and $\wp(\cdot)$ is the powerset operator. An *assumption-based framework* (or ABF, for short) is a tuple $\mathbf{ABF}= \langle \mathfrak{L}, \Gamma, \Lambda, - \rangle$, where $\mathfrak{L}=(\mathcal{L}, \vdash)$ is a propositional Tarskian logic, Γ (the strict assumptions) and Λ (the candidate or defeasible assumptions) are distinct countable sets of $\mathcal{L}$-formulas, and $- : \Lambda \rightarrow \wp(\mathcal{L})$ is a contrariness operator, assigning a finite set of $\mathcal{L}$-formulas to every defeasible assumption in Λ. In $\mathbf{ABF}$, *attacks* is defined as follows.

For $\Delta, \Theta \subseteq \Lambda$, and $\psi \in \Lambda$, Δ *attacks* ψ iff $\Gamma \cup \Delta \vdash \phi$ for some $\phi \in -\psi$.
Accordingly, Δ *attacks* Θ if Δ *attacks* some $\psi \in \Theta$.

The usual semantics in AA frameworks [6] is adapted to their ABFs as follows.

Definition 3. (ABF semantics).[4] Let $\mathbf{ABF}=\langle \mathfrak{L}, \Gamma, \Lambda, - \rangle$ be an assumption-based framework. A set of assumptions $\Delta \subseteq \Lambda$ is *closed* iff $\Delta=\{\alpha \in \Lambda \mid \Gamma \cup \Delta \vdash \alpha\}$. $\mathbf{ABF}$ is said to be *flat* iff every set of assumptions $\Delta \subseteq \Lambda$ is closed.
Let $\Delta \subseteq \Lambda$. Then Δ is *conflict-free* iff there is no $\Delta' \subseteq \Delta$ that attacks some $\psi \in \Delta$. Δ is *stable* iff it is closed, conflict-free and attacks every $\psi \in \Lambda \setminus \Delta$.

2.2 Disjunctive Logic Programs

A disjunctive logic program (DLP) [18] is a finite set of rules of the form[1]

$$p_1 \vee \ldots \vee p_k \leftarrow p_{k+1}, \ldots, p_m, not\ p_{m+1}, \ldots, not\ p_n, \tag{1}$$

[1] A disjunctive logic program (DLP) defined in this paper is different from a normal disjunctive program (NDP) defined in Sect. 2.3.

where $n \geq m \geq 0, k \geq 1$. Each p_i $(1 \leq i \leq n)$ is a ground atom and *not* means *negation as failure* (NAF). An atom preceded by *not* is called a NAF-atom. Let HB_P be the set of all ground atoms appearing in a DLP P called the Herbrand base. A set $M \subseteq HB_P$ satisfies a ground rule of the form (1) if $\{p_{k+1}, \ldots, p_m\} \subseteq M$ and $\{p_{m+1}, \ldots, p_n\} \cap M = \emptyset$ imply $\{p_1, \ldots, p_k\} \cap M \neq \emptyset$. M is a *model* of P if it satisfies every ground rule in P.
The *reduct* of P w.r.t. M is the DLP $P^M = \{p_1 \vee \ldots \vee p_k \leftarrow p_{k+1}, \ldots, p_m |$there is $(p_1 \vee \ldots \vee p_k \leftarrow p_{k+1}, \ldots, p_m, not\ p_{m+1}, \ldots, not\ p_n) \in P$ s.t. $\{p_{m+1}, \ldots, p_n\} \cap M = \emptyset\}$. Then M is a *stable model* if it is a $\subseteq$-minimal model of P^M [14,18].

In [13], Heyninck and Arieli defined $\mathcal{A}(P)$ as HB_P, and denoted by $\mathfrak{L} = \langle \mathcal{L}_{\mathtt{DLP}}, \vdash \rangle$ the logic based on the language $\mathcal{L}_{\mathtt{DLP}}$ consisting of disjunctions of atoms ($p_1 \vee \ldots \vee p_n$, for $n \geq 1$), NAF-atoms (*not* p for $p \in \mathcal{A}(P)$) and rules of the form (1), where $\vdash$ is constructed for $\mathcal{L}_{\mathtt{DLP}}$ by three inference rules: Modus Ponens (`MP`), Resolution (`Res`) and Reasoning by Cases (`RBC`)[2]. The ABF induced by a DLP P is defined by: $\mathbf{ABF}(P) = \langle \mathfrak{L}, P, \sim\mathcal{A}(P), - \rangle$, where $\mathfrak{L} = \langle \mathcal{L}_{\mathtt{DLP}}, \vdash \rangle$, $\sim\mathcal{A}(P) = \{not\ p |\ p \in \mathcal{A}(P)\}$, $-not\ p = \{p\}$ for every $p \in \mathcal{A}(P)$. Then they showed a one-to-one correspondence between stable models of a DLP P and stable extensions of $\mathbf{ABF}(P)$ as follows.

Proposition 1. *[13, Proposition 2 and Proposition 3]. Let P be a finite DLP. If M is a stable model of P, then $\overline{M} = \{not\ p \mid p \in (\mathcal{A}(P) \setminus M)\}$ is a stable extension* [3] *of* $\mathbf{ABF}(P)$. *Conversely if $\mathcal{E}$ is a stable extension of* $\mathbf{ABF}(P)$, *then $\underline{\mathcal{E}} = \{p \in \mathcal{A}(P) |\ not\ p \notin \mathcal{E}\}$ is a stable model of P.*

2.3 Extended Disjunctive Logic Programs

We consider a finite propositional extended disjunctive logic program (EDLP) [15] in this paper. An EDLP is a finite set of rules of the form:

$$L_1 | \ldots | L_k \leftarrow L_{k+1}, \ldots, L_m, notL_{m+1}, \ldots, notL_n, \tag{2}$$

where $n \geq m \geq 0, k \geq 1$. Each L_i $(1 \leq i \leq n)$ is a literal, that is, either a ground atom A (i.e. a propositional variable) or $\neg A$ preceded by classical negation $\neg$. *not* means NAF as before and *not* L is called a *NAF-literal*. The left (resp. right) part of $\leftarrow$ is the *head* (resp. the *body*). The symbol "|" is used to distinguish disjunction in the head of a rule from disjunction "$\vee$" used in classical logic. An EDLP is called a *normal disjunctive logic program* (NDP) if "$\neg$" does not occur in it, while an EDLP is called an *extended logic program* (ELP) if it contains no disjunction ($k = 1$). An ELP is called a *normal logic program* (NLP) if "$\neg$" does not occur in it. Let Lit_P be the set of all ground literals in the language of an EDLP P. When an EDLP P is an NDP (or NLP), Lit_P reduces to the

[2] The precise form of these inference rules will be described later in Remark 1.
[3] Stable extensions in [13] denote stable *assumption* extensions.

Herbrand base HB_P. The semantics of an EDLP is given by *answer sets* [15] (resp. *paraconsistent stable models* or *p-stable models* [4] [20]) defined as follows.

Definition 4. [15,20]. First, let P be a *not*-free EDLP (i.e., for each rule $m = n$). Then, $S \subseteq Lit_P$ is an *answer set* of P if S is a minimal set (w.r.t. $\subseteq$) satisfying the following two conditions:

(i) For each rule $L_1|\ldots|L_k \leftarrow L_{k+1},\ldots,L_m$ in P, if $\{L_{k+1},\ldots,L_m\} \subseteq S$, then $L_i \in S$ for some i $(1 \leq i \leq k)$.
(ii) If S contains a pair of literals L and $\neg L$, then $S = Lit_P$.

Second, let P be any EDLP and $S \subseteq Lit_P$. The *reduct* P^S of P w.r.t. S is a *not*-free EDLP $P^S = \{L_1|\ldots|L_k \leftarrow L_{k+1},\ldots,L_m \mid$ there is a rule of the form $L_1|\ldots|L_k \leftarrow L_{k+1},\ldots L_m, not\ L_{m+1},\ldots not\ L_n$ in P s.t. $\{L_{m+1},\ldots,L_n\} \cap S = \emptyset\}$. Then S is an answer set of P if S is the answer set of P^S.

On the other hand, p-stable models are regarded as answer sets defined without the condition (ii). An answer set is *consistent* if it is not Lit_P; otherwise it is *inconsistent*. An EDLP P is *consistent* if it has a consistent answer set; otherwise it is *inconsistent* under answer set semantics. In contrast, a p-stable model is *inconsistent* if it contains a pair of complementary literals; otherwise it is *consistent*. An EDLP P is *consistent* if it has a consistent p-stable model; otherwise it is *inconsistent* under paraconsistent stable model semantics.

Gelfond et al. [16] showed in the following theorem that a propositional EDLP P can be translated into a disjunctive default theory $emb_D(P)$ by replacing every rule in P of the form (2) with the disjunctive default

$$\frac{L_{k+1} \wedge \ldots \wedge L_m : \neg L_{m+1}, \ldots, \neg L_n}{L_1|\ldots|L_k}.$$

Theorem 1. *[16, Theorem 7.2]. Let P be a propositional EDLP. Then S is an answer set of P iff S is the set of all literals from an extension of $emb_D(P)$.*

Let P be an ELP, $P_{tr} \stackrel{\text{def}}{=} P \cup \{L \leftarrow p, \neg p \mid p \in Lit_P,\ L \in Lit_P\}$ be the ELP obtained from P by incorporating the *trivialization rules* [20], and $\mathcal{F}(P) = \langle \mathcal{L}_P, P, \mathcal{A}_P, \overline{} \rangle$ be the ABA framework translated from P, where $\mathtt{NAF}_P = \{not\ L \mid L \in Lit_P\}$, $\mathcal{L}_P = Lit_P \cup \mathtt{NAF}_P$, $\mathcal{A}_P = \mathtt{NAF}_P$, $\overline{not\ L} = L$ for $not\ L \in \mathcal{A}_P$. Let $\Delta_M = \{not\ L \mid L \in Lit_P \setminus M\}$ [5] for $M \subseteq Lit_P$. It was shown that answer sets (resp. paraconsistent stable models) of an ELP P can be captured by stable *argument* extensions of the ABA translated from the ELP P_{tr} (resp. P) as follows.

[4] In this paper, the term "p-stable models" is used as not abbreviation of *partial* stable model semantics by Przymusinski [18] but as that of *paraconsistent* stable model semantics by Sakama and Inoue [20].

[5] In [22], $\neg.CM$ is used rather than Δ_M to refer to the set $\{not\ L \mid L \in Lit_P \setminus M\}$. However for notational convenience, Δ_M is used instead of $\neg.CM$ to refer to this set in this paper.

Theorem 2. *[22, Theorem 3]. Let P be an ELP. Then M is a p-stable model of P iff there is a stable extension $\mathcal{E}$ of the ABA framework $\mathcal{F}(P)$ such that $M \cup \Delta_M = \mathtt{Concs}(\mathcal{E})$ (in other words, $M = \mathtt{Concs}(\mathcal{E}) \cap Lit_P$).*

Theorem 3. *[22, Theorem 4]. Let P be an ELP. Then S is an answer set of P iff there is a stable extension $\mathcal{E}_{tr}$ of the ABA framework $\mathcal{F}(P_{tr})$ such that $S \cup \Delta_S = \mathtt{Concs}(\mathcal{E}_{tr})$ (in other words, $S = \mathtt{Concs}(\mathcal{E}_{tr}) \cap Lit_P$).*

For a consistent ELP, the following Theorem holds.

Theorem 4. *[23, Theorem 5]. Let P be a consistent ELP. Then S is an answer set of P iff there is a consistent stable argument extension $\mathcal{E}$ of the ABA framework $\mathcal{F}(P)$ such that $S \cup \Delta_S = \mathtt{Concs}(\mathcal{E})$.*

3 ABA for Extended Disjunctive Logic Programming

3.1 Assumption-Based Frameworks Translated from EDLPs

We propose a generalized assumption-based framework (ABF) translated from an EDLP, which incorporates explicit negation in Heyninck et al.'s ABF induced by a DLP to resolve problems addressed in the introduction. An ABF translated from an EDLP is based on the logic constructed by three inference rules: Modus Ponens (MP), Resolution (Res) and Reasoning by Cases (RBC):

$$[\mathtt{MP}]\ \frac{\psi \leftarrow \phi_1, \ldots, \phi_n \qquad \phi_1 \quad \phi_2 \ \cdots \ \phi_n}{\psi}$$

$$[\mathtt{Res}]\ \frac{\psi'_1|\ldots|\psi'_m|\ell_1|\ldots|\ell_n|\psi''_1|\ldots|\psi''_k \qquad not\,\ell_1 \ \cdots \ not\,\ell_n}{\psi'_1|\ldots|\psi'_m|\psi''_1|\ldots|\psi''_k}$$

$$[\mathtt{RBC}]\ \frac{\begin{array}{cccc} \ell_1 & \ell_2 & & \ell_n \\ \vdots & \vdots & & \vdots \\ \psi & \psi & \cdots & \psi \end{array} \qquad \ell_1|\ldots|\ell_n}{\psi}$$

where $|$ is the connective of a disjunction, ℓ_i is a propositional literal, each $\phi_i \in \{\ell_i, not\ \ell_i\}$ is a propositional literal or its NAF-literal, and ψ, ψ_i are disjunctions of propositional literals using $|$. Since $\psi \leftarrow$ is identified with $\psi \leftarrow \tau$, [MP] implies Reflexivity: [Ref] $\frac{\psi\leftarrow}{\psi}$. In what follows, $\vdash_{\mathcal{R}}$ denotes derivability using three inference rules: [MP] (including [Ref]), [Res] and [RBC].

Definition 5. Given an EDLP P, we denote by $\mathfrak{L}=\langle \mathcal{L}_{\mathtt{EDLP}}, \vdash_{\mathcal{R}} \rangle$ the logic based on the language $\mathcal{L}_{\mathtt{EDLP}}$[6] which consists of disjunctions of propositional literals ($\ell_1|\ldots|\ell_n$, for $n \geq 1$, $\ell_i \in Lit_P$), NAF-literals ($not\ \ell$ for $\ell \in Lit_P$) and rules from P of the form (2).

[6] $\mathcal{L}_{\mathtt{EDLP}}$ (resp. $\mathcal{L}_{\mathtt{DLP}}$) is defined under a given EDLP (resp. DLP).

Remark 1: Heyninck and Arieli's ABF [13] is based on the logic $\mathfrak{L}=\langle \mathcal{L}_{\text{DLP}}, \vdash \rangle$, where $\vdash$ is constructed by three inference rules: [MP], [Res] and [RBC] having the restricted forms such that $|$ (resp. a literal ℓ_i) is replaced with $\vee$ (resp. an atom p_i), each $\phi_i \in \{p_i, not\ p_i\}$ is an atom or a NAF-atom, and ψ, ψ_i are disjunctions of atoms ($p_1 \vee \ldots \vee p_n$, for $n \geq 1$).

$\Sigma \vdash_{\mathcal{R}} \varphi$ holds iff φ is either in Σ or is derived from Σ using three inference rules above. According to Definition 2, $\Sigma \vdash_{\mathcal{R}} \varphi$ iff $\varphi \in \texttt{Cn}_{\mathfrak{L}}(\Sigma)$, where $\texttt{Cn}_{\mathfrak{L}}(\Sigma)$ is the $\mathfrak{L}$-based transitive closure of Σ (namely, the $\subseteq$-smallest set that contains $\Sigma \subseteq \mathcal{L}_{\text{EDLP}}$ and is closed under [MP], [Res] and [RBC]). Notice that for any $\varphi \in \texttt{Cn}_{\mathfrak{L}}(\Sigma)$, if φ is not of the form $\ell_1|\ldots|\ell_n$ (where $\ell_i \in Lit_P$), then $\varphi \in \Sigma$.

For a special ABF $\langle \mathfrak{L}, \Gamma, \Lambda, - \rangle$ such that each assumption from Λ has a unique contrary (i.e. $|-\alpha|=1$ [7] for $\forall \alpha \in \Lambda$), we may denote such an ABF by a tuple $\langle \mathfrak{L}, \Gamma, \Lambda, \bar{\ } \rangle$, where $\bar{\ }$ is the total mapping from Λ into $\mathcal{L}$ and $\overline{\alpha} \in \mathcal{L}$ is the *contrary* of $\alpha \in \Lambda$. In what follows, let $\texttt{NAF}_P=\{not\ \ell \mid \ell \in Lit_P\}$ and $\mathbb{L}_P=Lit_P \cup \texttt{NAF}_P$ for an EDLP P. We are now ready to define an ABF translated from an EDLP.

Definition 6. Let P be an EDLP. The assumption-based framework (ABF) translated from P is defined by: $\mathbf{ABF}(P)=\langle \mathfrak{L}, P, \mathcal{A}_P, \bar{\ } \rangle$, where $\mathfrak{L}=\langle \mathcal{L}_{\text{EDLP}}, \vdash_{\mathcal{R}} \rangle$, $\mathcal{A}_P = \texttt{NAF}_P = \{not\ \ell \mid \ell \in Lit_P\}$ and $\overline{not\ \ell}=\ell$ for every $not\ \ell \in \mathcal{A}_P$.

In the following, we show the ABF translated from an EDLP P is always *flat*.

Proposition 2. $\mathbf{ABF}$*(P)*$=\langle \mathfrak{L}, P, \mathcal{A}_P, \bar{\ } \rangle$ *translated from an EDLP P is flat.*

Proof. Let $\Delta \subseteq \mathcal{A}_P$. Since $\vdash_{\mathcal{R}}$ is reflexive, it holds that (i) $P \cup \Delta \vdash_{\mathcal{R}} \alpha$ for $\forall \alpha \in \Delta$. On the other hand, it holds that (ii) $P \cup \Delta \not\vdash_{\mathcal{R}} \beta$ for $\forall \beta \in \mathcal{A}_P \setminus \Delta$, since assumptions do not occur in the head of a rule from P. Then due to (i),(ii), it holds that $\Delta=\{\alpha \in \mathcal{A}_P \mid P \cup \Delta \vdash_{\mathcal{R}} \alpha\}$ for $\forall \Delta \subseteq \mathcal{A}_P$. Thus $\mathbf{ABF}(P)$ is flat. □

In $\mathbf{ABF}(P)$, *consistency* of a set of literals $X \subseteq \mathbb{L}_P$ is defined using a consequence operator $\texttt{CN}_P$ though in a standard ABA, $CN_{\mathcal{R}}$ is used as shown in Definition 1.

Definition 7. Let $\mathbf{ABF}(P)=\langle \mathfrak{L}, P, \mathcal{A}_P, \bar{\ } \rangle$ be the ABF translated from an EDLP P. For a set $X \subseteq \mathbb{L}_P = Lit_P \cup \texttt{NAF}_P$, X is said to be *contradictory* iff X is contradictory w.r.t. $\bar{\ }$, i.e. there exists an assumption $\alpha \in \mathcal{A}_P$ such that $\{\alpha, \overline{\alpha}\} \subseteq X$; or X is contradictory w.r.t. $\neg$, i.e. there exists $s \in \mathbb{L}_P$ such that $\{s, \neg s\} \subseteq X$.

Let $\texttt{CN}_P : \wp(\mathbb{L}_P) \rightarrow \wp(\mathbb{L}_P)$ be a consequence operator such that for $X \subseteq \mathbb{L}_P$,

$$\texttt{CN}_P(X) \stackrel{\text{def}}{=} \{\phi \in \mathbb{L}_P \mid P \cup X \vdash_{\mathcal{R}} \phi\} = \texttt{Cn}_{\mathfrak{L}}(P \cup X) \cap \mathbb{L}_P.$$

Then $\texttt{CN}_P(X)$ is said to be the closure of X. X is said to be *closed* w.r.t. $\texttt{CN}_P$ iff $X=\texttt{CN}_P(X)$. A set $X \subseteq \mathbb{L}_P$ is said to be *inconsistent* iff the closure $\texttt{CN}_P(X)$ is contradictory. X is said to be *consistent* iff it is not inconsistent.

[7] For a set S, $|S|$ denotes the cardinality of S.

3.2 Correspondence Between Answer Sets of an EDLP and Stable Assumption Extensions

Proposition 1 [13] for a DLP is generalized to Proposition 3 and Proposition 4 for an EDLP shown below. To this end, we prepare the following lemma.

Lemma 1. *Let P be a propositional NDP and P_D be the DLP translated from P which is obtained by replacing a rule: $p_1|\ldots|p_k \leftarrow p_{k+1},\ldots,p_m, not\ p_{m+1},\ldots, not\ p_n$ from P with the rule $p_1 \vee \ldots \vee p_k \leftarrow p_{k+1},\ldots,p_m, not\ p_{m+1},\ldots, not\ p_n$, where $HB_{P_D} = HB_P$. Then M is an answer set of P iff M is a stable model of P_D.*

Based on Lemma 1, Proposition 1 for a DLP under the logic $\mathfrak{L}=\langle \mathcal{L}_{\text{DLP}}, \vdash\rangle$ is mapped to Corollary 1 for an NDP under the logic $\mathfrak{L}=\langle \mathcal{L}_{\text{EDLP}}, \vdash_{\mathcal{R}}\rangle$ as follows.

Corollary 1. *Let P be an NDP and $\mathbf{ABF}(P)=\langle \mathfrak{L},P,\mathcal{A}_P,\bar{\ }\rangle$ where $\mathfrak{L}=\langle \mathcal{L}_{\text{EDLP}},\vdash_{\mathcal{R}}\rangle$. Then (i) If M is an answer set of P, then $\Delta=\{not\, p \mid p \in (HB_P \setminus M)\}$ is a stable assumption extension of $\mathbf{ABF}(P)$. (ii) Conversely if Δ is a stable assumption extension of $\mathbf{ABF}(P)$, then $M=\{p \in HB_P \mid not\ p \notin \Delta\}$ is an answer set of P.*

We show answer sets (resp. p-stable models) of an EDLP P are captured by stable assumption extensions of the ABF translated from P_{tr} (resp. P) as follows.

Proposition 3. *Let $\mathbf{ABF}(P) = \langle \mathfrak{L}, P, \mathcal{A}_P, \bar{\ }\rangle$ be the ABF translated from an EDLP P. If M is a p-stable model of P, $\Delta=\{not\ \ell \mid \ell \in (Lit_P \setminus M)\}$ is a stable assumption extension of $\mathbf{ABF}(P)$. Conversely if Δ is a stable assumption extension of $\mathbf{ABF}(P)$, $M=\{\ell \in Lit_P \mid not\ \ell \notin \Delta\}$ is a p-stable model of P.*

Proof. Let P^+ be the NDP which is obtained from an EDLP P by replacing each negative literal $\neg L$ in P with a newly introduced atom L'. For a set $S \subseteq Lit_P$, let S^+ be the set obtained by replacing each negative literal $\neg L$ in S with a newly introduced atom L'. Then the Herbrand base HB_{P^+} of P^+ is $(Lit_P)^+$.
(i) Suppose that M is a p-stable model of an EDLP P. Then
M is a p-stable model of an EDLP P iff M^+ is an answer set of an NDP P^+
iff $\Delta^+=\{not\, p \mid p \in (HB_{P^+} \setminus M^+)\}$ is a stable assumption extension of $\mathbf{ABF}(P^+)$
iff $\Delta=\{not\ \ell \mid \ell \in (Lit_P \setminus M)\}$ is a stable assumption extension of $\mathbf{ABF}(P)$.
(ii) The converse is also proved in a similar way to **(i)**. □

Proposition 4. *Let P be an EDLP and $\mathbf{ABF}(P_{tr}) = \langle \mathfrak{L}, P_{tr}, \mathcal{A}_P, \bar{\ }\rangle$ be the ABF translated from the EDLP $P_{tr}=P \cup \{L \leftarrow p, \neg p \mid p \in Lit_P,\ L \in Lit_P\}$, where $\mathbb{L}_{P_{tr}} = \mathbb{L}_P$. If S is an answer set of P, then $\Delta=\{not\ \ell \mid \ell \in (Lit_P \setminus S)\}$ is a stable assumption extension of $\mathbf{ABF}(P_{tr})$. Conversely if Δ is a stable assumption extension of $\mathbf{ABF}(P_{tr})$, then $S=\{\ell \in Lit_P \mid not\ \ell \notin \Delta\}$ is an answer set of P.*

Proof. Due to [20, Theorem 3.5], S is an answer set of an EDLP P iff S is p-stable model of P_{tr}. Then based on Proposition 3 and [20, Theorem 3.5], this proposition is easily proved. □

$\mathtt{CN}_P(\Delta)$ gives us the *conclusion* of an assumption extension Δ. In Proposition 5 below, we show the relationship between answer sets (or p-stable models) of an EDLP and the conclusions of assumption extensions of the translated ABF.

Lemma 2. *Let M be an answer set of an NDP P and let $\Delta_M = \{not\ L \mid L \in HB_P \setminus M\}$. Then $p \in M$ iff $p \in \mathtt{Cn}_{\mathfrak{L}}(P \cup \Delta_M)$ for $p \in HB_P$.*

Proof. Let P_D be the DLP translated from an NDP P as shown in Lemma 1. Based on [13, Corollary 1] and Lemma 1,

$p \in M$ iff $p \in \mathtt{Cn}_{\mathfrak{L}}(P_D \cup \Delta_M)$ for $p \in HB_P$, where $\mathfrak{L} = \langle \mathcal{L}_{\mathtt{DLP}}, \vdash \rangle$,
iff $p \in \mathtt{Cn}_{\mathfrak{L}}(P \cup \Delta_M)$, for $p \in HB_P$, where $\mathfrak{L} = \langle \mathcal{L}_{\mathtt{EDLP}}, \vdash_{\mathcal{R}} \rangle$. □

Lemma 3. *Let M be a p-stable model of an EDLP P and $\Delta_M = \{not\ \ell \mid \ell \in (Lit_P \setminus M)\}$. Then $\ell \in M$ iff $\ell \in \mathtt{Cn}_{\mathfrak{L}}(P \cup \Delta_M)$ for $\ell \in Lit_P$.*

Proposition 5. *Let P be an EDLP, M (resp. S) be a p-stable model (resp. an answer set) of P, and $\Delta_M = \{not\ \ell \mid \ell \in (Lit_P \setminus M)\}$ (resp. $\Delta_S = \{not\ \ell \mid \ell \in (Lit_P \setminus S)\}$) be the stable assumption extension of* **ABF***(P) (resp.* **ABF***(P_{tr})). Then it holds that, $\mathtt{CN}_P(\Delta_M) = M \cup \Delta_M$, and $\mathtt{CN}_{P_{tr}}(\Delta_S) = S \cup \Delta_S$.*

Proof.

1. Let M be a p-stable model of an EDLP P. Then for $\ell \in Lit_P$, it holds that $\ell \in M$ iff $\ell \in \mathtt{Cn}_{\mathfrak{L}}(P \cup \Delta_M)$ iff $\ell \in \mathtt{CN}_P(\Delta_M)$ according to Lemma 3. Besides $not\ \ell \in \Delta_M$ iff $not\ \ell \in \mathtt{CN}_P(\Delta_M)$. Hence $\mathtt{CN}_P(\Delta_M) = M \cup \Delta_M$ holds.
2. Let S be an answer set of P. Then S is a p-stable model of P_{tr} due to [20, Theorem 3.5]. Besides $\Delta_S = \{not\ \ell \mid \ell \in (Lit_P \setminus S)\}$ is the stable assumption extension of **ABF**(P_{tr}) due to Proposition 4. Hence by applying the result of the item 1 to a p-stable model S of P_{tr}, we obtain $\mathtt{CN}_{P_{tr}}(\Delta_S) = S \cup \Delta_S$. □

3.3 Arguments and Argument Extensions in ABFs Translated from EDLPs

We define arguments and attacks in the ABF translated from an EDLP P.

Definition 8. Let **ABF**$(P) = \langle \mathfrak{L}, P, \mathcal{A}_P, \overline{} \rangle$ be the ABF translated from an EDLP P, where $\mathfrak{L} = \langle \mathcal{L}_{\mathtt{EDLP}}, \vdash_{\mathcal{R}} \rangle$. $\Psi \in \mathcal{L}_{\mathtt{EDLP}}$ is said to be a defeasible consequence of P and $K \subseteq \mathcal{A}_P$ if $P \cup K \vdash_{\mathcal{R}} \Psi$ in which any assumption belonging to K is used to derive Ψ. K is said to be a support for Ψ w.r.t. P.

$P \cup K \vdash_{\mathcal{R}} \Psi$ addressed above is represented by a tree structure $\mathcal{T}_\Psi(K)$ as follows.

Definition 9. Let $\mathfrak{L} = \langle \mathcal{L}_{\mathtt{EDLP}}, \vdash_{\mathcal{R}} \rangle$ and **ABF**$(P) = \langle \mathfrak{L}, P, \mathcal{A}_P, \overline{} \rangle$ be the ABF translated from an EDLP P. Let $\mathcal{T}_\Psi(K)$ denote $P \cup K \vdash_{\mathcal{R}} \Psi$ where K is a support for Ψ w.r.t. P. In other words, $\mathcal{T}_\Psi(K)$ is a tree with the root node labelled by $\Psi \in \mathcal{L}_{\mathtt{EDLP}}$ defined as follows.

1. The cases using no inference rules:
   (1) For $not\ \ell \in \mathcal{A}_P$, there is a one-node tree $\mathcal{T}_\Psi(K)$ whose root node (i.e. the child) is labelled by $\Psi = not\ \ell$ and $K = \{not\ \ell\}$.

(2) For a rule $r \in P$, there is a one-node tree $\mathcal{T}_\Psi(K)$ whose root node (i.e. the child) is labelled by $\Psi = r$ and $K = \emptyset$.

2. The cases using inference rules:
(1) **i.** For a rule $\psi \leftarrow\ \in P$, by [Ref], there is a tree $\mathcal{T}_\psi(K)$ whose root node N is labelled by ψ and N has a unique child node, namely a one-node tree $\mathcal{T}_r(\emptyset)$ where $r = \psi \leftarrow$. Then $K = \emptyset$.
ii. For a rule $\psi \leftarrow \phi_1, \cdots, \phi_n$ in P, if for each ϕ_i $(1 \leq i \leq n)$, there exists a tree $\mathcal{T}_{\phi_i}(K_i)$ with the root node N_i labelled by ϕ_i, then by [MP], there is a tree $\mathcal{T}_\psi(K)$ with the root node N labelled by ψ and N has a child N_0 labelled by $r = \psi \leftarrow \phi_1, \cdots, \phi_n$ which is a one-node tree $\mathcal{T}_r(\emptyset)$ as well as n children N_i $(1 \leq i \leq n)$ where N_i is the root of a tree $\mathcal{T}_{\phi_i}(K_i)$. Then $K = \bigcup_i K_i$.
(2) Let $\Phi = \psi'_1 | \ldots | \psi'_m | \ell_1 | \ldots | \ell_n | \psi''_1 | \ldots | \psi''_k$ and $\Psi = \psi'_1 | \ldots | \psi'_m | \psi''_1 | \cdots | \psi''_k$, where $\ell_i \in Lit_P$ $(1 \leq i \leq n)$. If there is a tree $\mathcal{T}_\Phi(K')$ with the root node N_0 labelled by Φ, then by [Res], there is a tree $\mathcal{T}_\Psi(K)$ with the root node N labelled by Ψ and N has a child N_0 as well as n children $N_1, \ldots N_n$ each of which is a one-node tree $\mathcal{T}_{\phi_i}(\{\phi_i\})$ where $\phi_i = not\ \ell_i$ $(1 \leq i \leq n)$. Then $K = K' \cup \bigcup_{i=1}^{n} \{not\ \ell_i\}$.
(3) Let $(\ell_i \ldots \psi)$ [8] denote the reasoning for the case ℓ_i and $\mathcal{T}_{\ell_i}(\emptyset)$ be a one-node tree whose root is labelled by ℓ_i.
If the following conditions (**1**), (**2**) are satisfied: (**1**) there is a tree $\mathcal{T}_\Phi(K')$ whose root node N_0 is labelled by $\Phi = \ell_1 | \ldots | \ell_n$; (**2**) for each ℓ_i $(1 \leq i \leq n)$, there exists reasoning for a case ℓ_i such that $(\ell_i \ldots \psi)$, namely $P \cup \{\ell_i\} \cup K_i \vdash_\mathcal{R} \psi$ for $\exists K_i \subseteq \mathcal{A}_P$, which is represented by a tree $\mathcal{T}_\psi(K_i)$ constructed by newly introducing a tree $\mathcal{T}_{\ell_i}(\emptyset)$ in this definition;
then by [RBC], there is a tree $\mathcal{T}_\psi(K)$ with the root node N labelled by ψ and N has the child N_0 as well as n children $N_1, \ldots N_n$ where each N_i $(1 \leq i \leq n)$ is the root of a tree $\mathcal{T}_\psi(K_i)$ for the case ℓ_i. Then $K = K' \cup \bigcup_{i=1}^{n} \{K_i\}$.

In $\mathbf{ABF}(P)$ translated from an EDLP P, an argument is defined as a special tree $\mathcal{T}_\phi(K)$ whose root node is labelled by a literal or a NAF-literal $\phi \in \mathbb{L}_P$, and the attack relation *attacks* is defined as usual.

Definition 10. Let $\mathbf{ABF}(P) = \langle \mathfrak{L}, P, \mathcal{A}_P, \overline{} \rangle$ be the ABF translated from an EDLP P and $\phi \in \mathbb{L}_P = Lit_P \cup \mathtt{NAF}_P$. Then in $\mathbf{ABF}(P)$,

- an argument for a conclusion (or claim) ϕ supported by $K \subseteq \mathcal{A}_P$ ($K \vdash \phi$, for short) is a (finite) tree $\mathcal{T}_\phi(K)$ whose root node is labelled by $\phi \in \mathbb{L}_P$.
- $K_1 \vdash \phi_1$ *attacks* $K_2 \vdash \phi_2$ iff $\phi_1 = \overline{\alpha}$ for $\exists \alpha \in K_2$.

In $\mathbf{ABF}(P)$, the semantics is also given by argument extensions as follows.

[8] This is depicted vertically in the inference rule of [RBC].

Definition 11. Let $\mathbf{ABF}(P)$=$\langle \mathfrak{L}, P, \mathcal{A}_P, \overline{} \rangle$ be the ABF translated from an EDLP P, AR be the set of arguments generated from $\mathbf{ABF}(P)$, and $Args \subseteq AR$. Then $Args$ is *conflict-free* iff $\nexists A, B \in Args$ such that A *attacks* B. $Args$ is a *stable argument extension* iff it is conflict-free and attacks every argument in $AR \setminus Args$.

We show there is a 1-1 correspondence between stable argument extensions and stable assumption extensions of an ABF translated from an EDLP as follows.

Definition 12. Let $\mathbf{ABF}(P)$=$\langle \mathfrak{L}, P, \mathcal{A}_P, \overline{} \rangle$ be the ABF translated from an EDLP P, and AR be the set of all arguments that can be generated from $\mathbf{ABF}(P)$. $\mathsf{Asms2Args} : \wp(\mathcal{A}_P) \rightarrow \wp(AR)$ and $\mathsf{Args2Asms} : \wp(AR) \rightarrow \wp(\mathcal{A}_P)$ are functions s.t.

$\mathsf{Asms2Args}(\mathcal{A}sms)$=$\{K \vdash \phi \in AR \mid K \subseteq \mathcal{A}sms\}$,

$\mathsf{Args2Asms}(\mathcal{A}rgs)$=$\{\alpha \in \mathcal{A}_P \mid \alpha \in K$ for an argument $K \vdash \phi \in \mathcal{A}rgs\}$.

Theorem 5. *Let* $\mathbf{ABF}(P)$=$\langle \mathfrak{L}, P, \mathcal{A}_P, \overline{} \rangle$ *be the ABF translated from an EDLP P, and AR be the set of all arguments generated from* $\mathbf{ABF}(P)$. *If* $\mathcal{A}sms \subseteq \mathcal{A}_P$ *is a stable assumption extension, then* $\mathsf{Asms2Args}(\mathcal{A}sms)$ *is a stable argument extension, and if* $\mathcal{A}rgs \subseteq AR$ *is a stable argument extension, then* $\mathsf{Args2Asms}(\mathcal{A}rgs)$ *is a stable assumption extension.*

Proof. (Sketch) This is proved in a similar way to the proof in [22, Theorem 2].

3.4 Correspondence between Answer Sets of an EDLP and Stable Argument Extensions

First of all, we show there is a 1-1 correspondence between answer sets of an NDP P and stable argument extensions of the ABF translated from P as follows.

Theorem 6. *Let* $\mathbf{ABF}(P)$=$\langle \mathfrak{L}, P, \mathcal{A}_P, \overline{} \rangle$ *be the ABF translated from an NDP P, AR be the set of all arguments generated from* $\mathbf{ABF}(P)$ *and* $\mathcal{E} \subseteq AR$. *Then M is an answer set of an NDP P iff there is a stable argument extension $\mathcal{E}$ of* $\mathbf{ABF}(P)$ *such that* $M \cup \Delta_M = \mathtt{Concs}(\mathcal{E})$=$\mathtt{CN}_P(\Delta_M)$, *where* Δ_M=$\{not\ p \mid p \in (HB_P \setminus M)\}$ *is a stable assumption extension of* $\mathbf{ABF}(P)$.

Proof. Let $\mathtt{NAF}_P$=$\{not\ p \mid p \in HB_P\}$ and $\Delta_M \subseteq \mathtt{NAF}_P$. Firstly we show that due to the form of inference rules, $P \cup \Delta_M \vdash_{\mathcal{R}} not\ p$ iff $not\ p \in \Delta_M$. In other words,

$$not\ p \in \mathtt{Cn}_{\mathfrak{L}}(P \cup \Delta_M) \quad \text{iff} \quad not\ p \in \Delta_M. \tag{3}$$

$\Rightarrow$: Let M be an answer set of an NDP P. Then there exists the stable assumption extension Δ_M of $\mathbf{ABF}(P)$ such that Δ_M=$\{not\ p \mid p \in (HB_P \setminus M)\}$ due to Corollary 1(i). Hence due to Theorem 5, there exists the stable argument extension $\mathcal{E} = \mathsf{Asms2Args}(\Delta_M) = \{K \vdash \phi \mid K \subseteq \Delta_M,\ \phi \in \mathbb{L}_P = HB_P \cup \mathtt{NAF}_P\}$. Thus for M, Δ_M and $\mathcal{E}$=$\mathsf{Asms2Args}(\Delta_M)$, it holds that

$\mathtt{Concs}(\mathcal{E})$= $\{\phi \in \mathbb{L}_P \mid K \vdash \phi, K \subseteq \Delta_M,\ \mathbb{L}_P = HB_P \cup \mathtt{NAF}_P\}$

$=\{\phi \in \mathbb{L}_P \mid P \cup \Delta_M \vdash_{\mathcal{R}} \phi\} = \{\phi \in \mathbb{L}_P \mid \phi \in \mathtt{Cn}_{\mathfrak{L}}(P \cup \Delta_M)\} = M \cup \Delta_M.$

(due to Lemma 2 and (3))

⇐: Let $\mathcal{E}$ be a stable argument extension of **ABF**(P) translated from an NDP P. Then due to Theorem 5, there is the stable assumption extension Δ of **ABF**(P) s.t. $\Delta = \mathsf{Args2Asms}(\mathcal{E}) = \{\alpha \mid \alpha \in K \text{ for } K \vdash \phi \text{ in } \mathcal{E}\} = \bigcup_i K_i$ for $K_i \vdash \phi_i \in \mathcal{E}$. Moreover due to Corollary 1(ii), for this stable assumption extension Δ, there is the answer set M of the NDP P such that M=$\{p \in HB_P |\ not\ p \notin \Delta)\}$, while for this answer set M, Δ_M=$\{not\ p \in \mathtt{NAF}_P \mid p \notin M\}$ is the stable assumption extension of **ABF**(P) due to Corollary 1(i). Thus obviously $\Delta_M = \Delta$. Then for a stable argument extension $\mathcal{E}$, Δ=$\bigcup_i K_i$ s.t. $K_i \vdash \phi_i \in \mathcal{E}$, M=$\{p \in HB_P |\ not\ p \notin \Delta\}$ and $\Delta_M = \Delta$, it holds that

$$
\begin{aligned}
&\mathtt{Concs}(\mathcal{E}) = \{\phi \mid K \vdash \phi \in \mathcal{E} \textit{ for } \phi \in \mathbb{L}_P = HB_P \cup \mathtt{NAF}_P\}\\
&=\{\phi \in \mathbb{L}_P \mid P \cup \Delta \vdash_{\mathcal{R}} \phi \textit{ for } \Delta = \textstyle\bigcup_i K_i \textit{ where } K_i \vdash \phi_i \in \mathcal{E}\}\\
&=\{\phi \in \mathbb{L}_P \mid P \cup \Delta_M \vdash_{\mathcal{R}} \phi \textit{ for } M = \{p \in HB_P |\ not\ p \notin \Delta_M = \Delta\}\}\\
&=\{\phi \in \mathbb{L}_P \mid \phi \in \mathtt{Cn}_{\mathfrak{L}}(P \cup \Delta_M)\} = M \cup \Delta_M. \text{ (due to Lemma 2 and (3))} \quad \square
\end{aligned}
$$

Based on Theorem 6, we show that there is a 1-1 correspondence between answer sets (resp. p-stable models) of an EDLP P and stable *argument* extensions of **ABF**(P_{tr}) (resp. **ABF**(P)) as follows, though Propositions 4 and 3 show the similar correspondences for stable *assumption* extensions of the respective ABFs.

Theorem 7. *Let* **ABF**(P)=$\langle \mathfrak{L}, P, \mathcal{A}_P, \overline{\ \ }\rangle$ *be the ABF translated from an EDLP P, AR be the set of all arguments generated from* **ABF**(P) *and* $\mathcal{E} \subseteq AR$. *Then M is a p-stable model of an EDLP P iff there is a stable argument extension $\mathcal{E}$ of* **ABF**(P) *such that* $M \cup \Delta_M = \mathtt{Concs}(\mathcal{E})$=$\mathtt{CN}_P(\Delta_M)$, *where Δ_M is a stable assumption extension of* **ABF**(P).

Proof. (Sketch) Let P^+ be the NDP which is obtained from an EDLP P by replacing each negative literal $\neg L$ in P with a newly introduced atom L'. Then since Theorem 6 holds for the NDP P^+, this theorem for an EDLP P is proved in a similar way to Proposition 3. □

Theorem 8. *Let P be an EDLP,* **ABF**(P_{tr})=$\langle \mathfrak{L}, P_{tr}, \mathcal{A}_P, \overline{\ \ }\rangle$ *be the ABF translated from the EDLP* P_{tr}=$P \cup \{L \leftarrow p, \neg p \mid p \in Lit_P,\ L \in Lit_P\}$, *$AR$ be the set of all arguments generated from* **ABF**(P_{tr}) *and* $\mathcal{E}_{tr} \subseteq AR$, *where* $Lit_{P_{tr}} = Lit_P$, $\mathtt{NAF}_P$=$\{not\ \ell |\ \ell \in Lit_P\}$, $\mathcal{A}_P$=$\mathtt{NAF}_P$ *and* $\overline{not\ \ell}$=ℓ *for every* $not\ \ell \in \mathcal{A}_P$. *Then S is an answer set of an EDLP P iff there is a stable argument extension $\mathcal{E}_{tr}$ of* **ABF**(P_{tr}) *such that* $S \cup \Delta_S = \mathtt{Concs}(\mathcal{E}_{tr})$=$\mathtt{CN}_{P_{tr}}(\Delta_S)$, *where Δ_S is a stable assumption extension of* **ABF**(P_{tr}).

Proof. (Sketch) Based on [20, Theorem 3.5] and Theorem 7, this theorem for an EDLP P is proved. □

Theorems 7, 8 for an EDLP are the generalization of Theorems 2, 3 for an ELP. In what follows, we show the relationship to a disjunctive default theory [16].

Theorem 9. *Let P be an EDLP. Then S is the set of all literals from an extension for the disjunctive default theory $emb(P)$ if and only if there is a stable argument extension $\mathcal{E}_{tr}$ of* **ABF**(P_{tr}) *such that* $S = \mathtt{Concs}(\mathcal{E}_{tr}) \cap Lit_P$.

Proof. This theorem is easily proved based on Theorem 8 and Theorem 1. □

Example 2. **(Cont. Ex.1).** To solve the *Kyoto protocol problem* in argumentation, we construct $\mathbf{ABF}(P_1)$ from P_1. Its arguments and *attacks* are as follows:
$A_1\!:\!\{not\ y\}\vdash p$, $A_2\!:\!\{not\ p\}\vdash y$, $A_3\!:\!\{not\ y, not\neg r\}\vdash r$, $A_4\!:\!\{not\ p, not\neg f\}\vdash f$,
$A_5\!:\!\{not\ y, not\neg r, not\neg k\}\vdash k$, $A_6\!:\!\{not\ p, not\neg f, not\neg k\}\vdash k$,
$A_7\!:\!\{not\neg r, not\neg f, not\neg k\}\vdash k$, $A_8\!:\!\{not\ p\}\vdash not\ p$, $A_9\!:\!\{not\ y\}\vdash not\ y$,
$A_{10}\!:\!\{not\ r\}\vdash not\ r$, $A_{11}\!:\!\{not\ f\}\vdash not\ f$, $A_{12}\!:\!\{not\ k\}\vdash not\ k$,
$A_{13}: \{not\ \neg p\} \vdash not\ \neg p$, $A_{14}: \{not\ \neg y\} \vdash not\ \neg y$, $A_{15}: \{not\ \neg r\} \vdash not\ \neg r$,
$A_{16}: \{not\ \neg f\} \vdash not\ \neg f$, $A_{17}: \{not\ \neg k\} \vdash not\ \neg k$;
$attacks$=$\{(A_1, A_2), (A_1, A_4), (A_1, A_6), (A_1, A_8), (A_2, A_1), (A_2, A_3), (A_2, A_5),$
$(A_2, A_9), (A_3, A_{10}), (A_4, A_{11}), (A_5, A_{12}), (A_6, A_{12}), (A_7, A_{12})\}$.
Figure 1 shows A_3 which is constructed based on [Ref], [MP] and [Res]. Each A_i $(1 \leq i \leq 6)$ uses [Res]. Instead A_7 uses [RBC]. Arguments in $\mathbf{ABF}(P_1)$ coincide with those in $\mathbf{ABF}((P_1)_{tr})$. Then $\mathbf{ABF}((P_1)_{tr})$ (resp. $\mathbf{ABF}(P_1)$) has two stable argument extensions: $\mathcal{E}_1 = \{A_1, A_3, A_5, A_7, A_9, A_{11}\} \cup \{A_i | 13 \leq i \leq 17\}$,
$\mathcal{E}_2 = \{A_2, A_4, A_6, A_7, A_8, A_{10}\} \cup \{A_i | 13 \leq i \leq 17\}$,
where $\mathtt{Concs}(\mathcal{E}_1)$=$\{p,\ r,\ k,\ not\ y,\ not\ f\} \cup U$ and $\mathtt{Concs}(\mathcal{E}_2)$=$\{y,\ f,\ k,\ not\ p,\ not\ r\} \cup U$ for U=$\{not\ \neg p, not\ \neg y, not\ \neg r, not\ \neg f, not\ \neg k\}$. Hence the expected result is successfully obtained since $k \in \mathtt{Concs}(\mathcal{E}_i)$ for $\forall \mathcal{E}_i$ $(i$=$1,2)$.

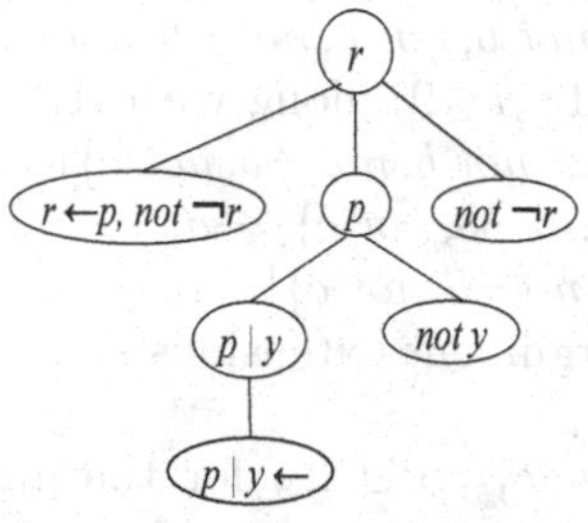

Fig. 1. A_3:$\{not\ y, not\neg r\}\vdash r$ in Ex.2

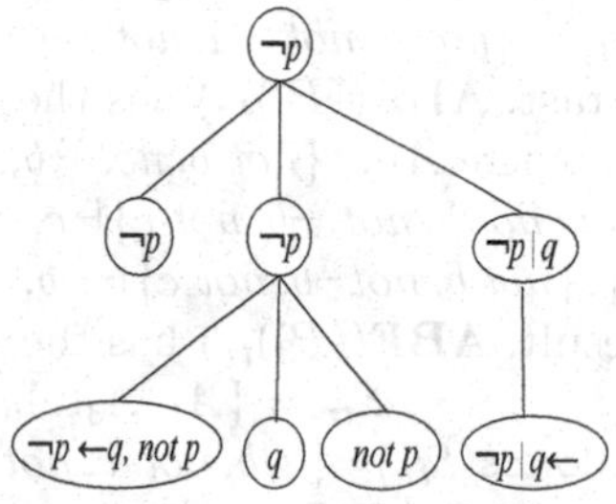

Fig. 2. A_1:$\{not\ p\} \vdash \neg p$ in Ex.3

Example 3. Consider P_2=$\{\neg p|q \leftarrow,\quad q \leftarrow \neg p, not\ \neg q,\quad \neg p \leftarrow q, not\ p\}$, where $Lit_{P_2} = \{p, q, \neg p, \neg q\}$. P_2 is not a *head-cycle-free* EDLP (HEDLP)[9] [3] but a general EDLP [11]. P_2 has the unique answer set $S = \{\neg p, q\}$, which is its unique p-stable model. In contrast, $\mathbf{ABF}(P_2)$ translated from P_2 has the unique stable assumption extension $\Delta_S = \{not\ p, not\ \neg q\}$. Besides $\mathbf{ABF}(P_2)$ has arguments:
$A_1: \{not\ p\} \vdash \neg p$, $A_2: \{not\ \neg q\} \vdash q$, $A_3: \{not\ q\} \vdash \neg p$, $A_4: \{not\ \neg p\} \vdash q$,
$A_5: \{not\ \neg q, not\ p\} \vdash \neg p$, $A_6: \{not\ p, not\ \neg q\} \vdash q$, $A_7: \{not\ \neg p, not\ p\} \vdash \neg p$,
$A_8: \{not\ q,\ not\ \neg q\} \vdash q$, $A_9: \{not\ \neg p\} \vdash not\ \neg p$, $A_{10}: \{not\ q\} \vdash not\ q$,
$A_{11}: \{not\ p\} \vdash not\ p$, $A_{12}: \{not\ \neg q\} \vdash not\ \neg q$,
and $attacks = \{(A_1, A_4), (A_1, A_7), (A_1, A_9), (A_2, A_3), (A_2, A_8), (A_2, A_{10}), (A_3, A_4),$
$(A_3,A_7), (A_3,A_9), (A_4,A_3), (A_4,A_8), (A_4,A_{10}), (A_5,A_4), (A_5,A_7), (A_5,A_9), (A_6,A_3),$

[9] This means that the knowledge expressed by the EDLP P_2 cannot be expressed by ELPs due to complexity results shown in [3,11].

$(A_6, A_8), (A_6, A_{10}), (A_7, A_4), (A_7, A_7), (A_7, A_9), (A_8, A_3), (A_8, A_8), (A_8, A_{10})\}$.

Figure 2 shows the tree structure of the argument $A_1 : \{not\ p\} \vdash \neg p$ which is constructed based on the inference rules: [Ref], [MP] and [RBC].
In this case, Arguments in $\mathbf{ABF}(P_2)$ coincide with those in $\mathbf{ABF}((P_2)_{tr})$. Thus $\mathbf{ABF}(P_2)$ (resp. $\mathbf{ABF}((P_2)_{tr})$) has the unique stable argument extension: $\mathcal{E}=\{A_1, A_2, A_5, A_6, A_{11}, A_{12}\}$, where $\texttt{Concs}(\mathcal{E})=\{\neg p, q, not\ p, not\ \neg q\}=S \cup \Delta_S$ for the answer set S and the stable assumption extensionΔ_S shown above.

Example 4. Consider the EDLP $P_3 = \{\neg a|b \leftarrow not\ \neg b, \quad a \leftarrow not\ c\}$, It has the unique answer set $S=\{a, b\}$, while it has two p-stable models $M_1=\{a, b\}=S$, $M_2=\{a, \neg a\}$, where M_1 is consistent but M_2 is inconsistent. Let $\mathbf{ABF}(P_3)$ be the ABF translated from P_3, which has arguments and *attacks* as follows:

$A_1 : \{not\ c\} \vdash a, \quad A_2 : \{not\ b, not\ \neg b\} \vdash \neg a, \quad A_3 : \{not\ \neg a, not\ \neg b\} \vdash b,$
$A_4 : \{not\ a\} \vdash not\ a, \quad A_5 : \{not\ b\} \vdash not\ b, \quad A_6 : \{not\ c\} \vdash not\ c,$
$A_7 : \{not\ \neg a\} \vdash not\ \neg a, \quad A_8 : \{not\ \neg b\} \vdash not\ \neg b, \quad A_9 : \{not\ \neg c\} \vdash not\ \neg c,$

attacks$=\{(A_1, A_4), (A_2, A_3), (A_2, A_7), (A_3, A_2), (A_3, A_5)\}$.
Then $\mathbf{ABF}(P_3)$ has two stable argument extensions $\mathcal{E}_1, \mathcal{E}_2$ as follows.

$$\mathcal{E}_1 = \{A_1, A_3, A_6, A_7, A_8, A_9\}, \qquad \mathcal{E}_2 = \{A_1, A_2, A_5, A_6, A_8, A_9\},$$

where $\texttt{Concs}(\mathcal{E}_1)= \{a,\ b,\ not\ c,\ not\ \neg a,\ not\ \neg b,\ not \neg c\} = M_1 \cup \Delta_{M_1}$,
$\texttt{Concs}(\mathcal{E}_2)= \{a,\ \neg a,\ not\ b,\ not\ c,\ not\ \neg b,\ not \neg c\} = M_2 \cup \Delta_{M_2}$,
for p-stable models M_1, M_2 of P_3 and stable assumption extensions Δ_{M_1}, Δ_{M_2} s.t. $\Delta_{M_1}= \{not\ c, not\ \neg a, not\ \neg b, not \neg c\}$, $\Delta_{M_2}=\{not\ b, not\ c, not\ \neg b, not \neg c\}$.
In contrast, $\mathbf{ABF}((P_3)_{tr})$ has the arguments A_i $(1 \leq i \leq 9)$ along with A_j $(10 \leq j \leq 15)$, where $A_{10} : \{not\ b, not\ \neg b, not\ c\} \vdash a$, $A_{11} : \{not\ b, not\ \neg b, not\ c\} \vdash b$,

$A_{12} : \{not\ b, not\ \neg b, not\ c\} \vdash c, \quad A_{13} : \{not\ b, not \neg b, not\ c\} \vdash \neg a,$
$A_{14} : \{not\ b, not \neg b, not\ c\} \vdash \neg b, \quad A_{15} : \{not\ b, not \neg b, not\ c\} \vdash \neg c.$

As a result, $\mathbf{ABF}((P_3)_{tr})$ has the unique stable argument extensions $\mathcal{E}_{tr}$:

$$\mathcal{E}_{tr} = \{A_1, A_3, A_6, A_7, A_8, A_9\}=\mathcal{E}_1,$$

where $\texttt{Concs}(\mathcal{E}_{tr})=\{a, b, not\ c, not\ \neg a, not\ \neg b, not \neg c\}=S \cup \Delta_S$ for the unique answer set S of P_3 and the unique stable assumption extension Δ_S of $\mathbf{ABF}((P_3)_{tr})$.

Remark 2: In Example 2, [Ref] is used to construct each A_i $(1 \leq i \leq 7)$. Thus we need [Ref]. Without [MP], we cannot build A_i $(3 \leq i \leq 7)$, which means we cannot infer k. Thus we need [MP]. In Example 3, [RBC] is used to construct A_1, A_2, A_5, A_6. Hence without [RBC], we cannot construct these arguments, which means $\mathbf{ABF}(P_2)$ has no stable extension. Thus we need [RBC]. In Example 4, without [Res], we cannot construct A_3 along with A_2. Thus we need [Res].

3.5 Correspondence Between Answer Sets of a Consistent EDLP and Consistent Stable Argument Extensions

Rationality postulates are defined in $\mathbf{ABF}(P)$ translated from an EDLP P like Definition 1. In what follows, we show that such $\mathbf{ABF}(P)$ always satisfies the *closure-property* (or direct consistency postulate [5]) under the stable semantics.

Definition 13. (Rationality postulates). Given an EDLP P, $\mathbf{ABF}(P)$= $\langle \mathfrak{L}, P, \mathcal{A}_P, \overline{} \rangle$ is said to satisfy the *consistency-property* (resp. the *closure-property*) under the σ semantics if for each σ argument extension $\mathcal{E}$ of the AA framework $AF_{\mathcal{F}}$ generated from $\mathcal{F}$=$\mathbf{ABF}(P)$, $\mathtt{Concs}(\mathcal{E})$ is consistent (resp. $\mathtt{Concs}(\mathcal{E})$ is closed w.r.t. $\mathtt{CN}_P$).

Theorem 10. *Let $\mathcal{F}$ be $\mathbf{ABF}(P)$ =$\langle \mathfrak{L}, P, \mathcal{A}_P, \overline{} \rangle$ translated from an EDLP P and $\mathcal{E}$ be a stable argument extension of $AF_{\mathcal{F}}$ generated from $\mathcal{F}$=$\mathbf{ABF}(P)$.*

1. *$\mathcal{F}$ satisfies* the closure-property *under the stable semantics.*
2. *$\mathcal{F}$ satisfies* the consistency-property *under the stable semantics*
   iff for every $\mathcal{E}$, $\mathtt{Concs}(\mathcal{E})$ is consistent
   iff for every $\mathcal{E}$, $\mathtt{Concs}(\mathcal{E})$ is not contradictory w.r.t. explicit negation $\neg$.

Proof.

1. Let M be a p-stable model of P and $\mathcal{E}$ be a stable argument extension of $\mathcal{F}$ =$\mathbf{ABF}(P)$ satisfying $M \cup \Delta_M = \mathtt{Concs}(\mathcal{E})$. Then $\mathtt{CN}_P(M \cup \Delta_M) = \{\phi \in \mathbb{L}_P | P \cup M \cup \Delta_M \vdash_{\mathcal{R}} \phi\} = \Delta_M \cup \{\ell \in Lit_P | P \cup M \cup \Delta_M \vdash_{\mathcal{R}} \ell\}$. (4)
   Due to Lemma 3 and the transitive closure property of $\mathtt{Cn}_{\mathfrak{L}}$, for $\ell \in Lit_P$, $\ell \in M$ iff $\ell \in \mathtt{Cn}_{\mathfrak{L}}(P \cup \Delta_M)$ iff $\ell \in \mathtt{Cn}_{\mathfrak{L}}(P \cup \Delta_M \cup \bigcup_{\ell \in M}\{\ell\}) = \mathtt{Cn}_{\mathfrak{L}}(P \cup \Delta_M \cup M)$.
   Hence $\ell \in M$ iff $P \cup M \cup \Delta_M \vdash_{\mathcal{R}} \ell$ for $\ell \in Lit_P$. (5)
   Then (5) means that M=$\{\ell \in Lit_P | P \cup M \cup \Delta_M \vdash_{\mathcal{R}} \ell\}$. As a result, (4) leads to $\mathtt{CN}_P(M \cup \Delta_M)$=$M \cup \Delta_M$, namely, $\mathtt{CN}_P(\mathtt{Concs}(\mathcal{E}))$=$\mathtt{Concs}(\mathcal{E})$. Thus $\mathcal{F}$ satisfies the closure-property under the stable semantics.
2. Based on the result of the item 1, the item 2 is easily proved. □

Given an EDLP P, the notions of consistent argument extensions and consistency in $\mathbf{ABF}(P)$ are defined like [23, Definitions 6, 7] as follows.

Definition 14. (Consistent argument extensions). Given an EDLP P, let $\mathcal{E}$ be a σ argument extension of $\mathbf{ABF}(P)$=$\langle \mathfrak{L}, P, \mathcal{A}_P, \overline{} \rangle$. Then $\mathcal{E}$ is said to be *consistent* if $\mathtt{Concs}(\mathcal{E})$ is not contradictory w.r.t. $\neg$; otherwise it is *inconsistent.*

Definition 15. (Consistency in ABFs translated from EDLPs). Given an EDLP P, $\mathbf{ABF}(P)$=$\langle \mathfrak{L}, P, \mathcal{A}_P, \overline{} \rangle$ is said to be *consistent* under σ semantics if $\mathbf{ABF}(P)$ has a consistent σ argument extension (or a consistent σ assumption extension); otherwise it is *inconsistent.*

We show that there is a 1-1 correspondence between answer sets of a consistent EDLP P and the consistent stable argument extensions of $\mathbf{ABF}(P)$ translated from P, which is a generalization of Theorem 4 for a consistent ELP.

Theorem 11. *Let P be a* consistent *EDLP and* $\mathbf{ABF}(P)$=$\langle \mathfrak{L}, P, \mathcal{A}_P, \overline{} \rangle$ *be the ABF translated from P. Then S is an answer set of P iff there is a* consistent *stable argument extension $\mathcal{E}$ of $\mathbf{ABF}(P)$ such that $S \cup \Delta_S = \mathtt{Concs}(\mathcal{E})$=$\mathtt{CN}_P(\Delta_S)$, where Δ_S is the consistent stable assumption extension of $\mathbf{ABF}(P)$.*

Proof.(Sketch) We can prove that S is a consistent answer set of an EDLP P iff there is a consistent p-stable model S of P. (6)

Now let P a consistent EDLP. Then based on both Theorem 7 for $\mathbf{ABF}(P)$ and (6), this theorem is proved in a similar way to the proof of [23, Theorem 5]. □

Corollary 2. *Let P be a consistent EDLP. The following holds.* **(i)** *$\mathcal{E}$ is a consistent stable extension of $\mathbf{ABF}(P)$ iff $\mathcal{E}$ is a stable extension of $\mathbf{ABF}(P_{tr})$.* **(ii)** *$\mathbf{ABF}(P_{tr})$ satisfies the rationality postulates under the stable semantics.*

Example 5. **(Innocent unless proved guilty)**. Consider the EDLP P_4 [18], which states that everyone is pronounced not guilty unless proven otherwise:

$$P_4 = \{innocent \mid guilty \leftarrow charged, \quad \neg guilty \leftarrow not\ proven, \quad charged \leftarrow\}.$$

Let i, g, c, p stand for *innocent, guilty, charged, proven.* Then P_4 has the unique answer set $S = \{c, i, \neg g\}$, while it has two p-stable models $M_1 = \{c, i, \neg g\}$ and $M_2 = \{c, g, \neg g\}$, where $M_1 = S$ is consistent but M_2 is inconsistent.

To solve this problem in argumentation, we construct $\mathbf{ABF}(P_4)$ from P_4, which has arguments:

$A_1 : \{\} \vdash c, \quad A_2 : \{not\ g\} \vdash i, \quad A_3 : \{not\ i\} \vdash g, \quad A_4 : \{not\ p\} \vdash \neg g,$
$A_5 : \{not\ i\} \vdash not\ i, \quad A_6 : \{not\ g\} \vdash not\ g, \quad A_7 : \{not\ c\} \vdash not\ c,$
$A_8 : \{not\ p\} \vdash not\ p, \quad A_9 : \{not\ \neg i\} \vdash not\ \neg i, \quad A_{10} : \{not\ \neg g\} \vdash not\ \neg g,$
$A_{11} : \{not\ \neg c\} \vdash not\ \neg c, \quad A_{12} : \{not\ \neg p\} \vdash not\ \neg p,$

and $attacks = \{(A_1, A_7), (A_2, A_3), (A_2, A_5), (A_3, A_2), (A_3, A_6), (A_4, A_{10})\}$ where $|attacks| = 6$. Then $\mathbf{ABF}(P_4)$ has two stable argument extensions, $\mathcal{E}_1$ and $\mathcal{E}_2$:

$$\mathcal{E}_1 = \{A_1, A_2, A_4, A_6, A_8, A_9, A_{11}, A_{12}\},$$
$$\mathcal{E}_2 = \{A_1, A_3, A_4, A_5, A_8, A_9, A_{11}, A_{12}\},$$

where $\mathtt{Concs}(\mathcal{E}_1) = \{c, i, \neg g, not\ g, not\ p, not\ \neg i, not\ \neg c, not\ \neg p\}$,
$\mathtt{Concs}(\mathcal{E}_2) = \{c, g, \neg g, not\ i, not\ p, not\ \neg i, not\ \neg c, not\ \neg p\}$. Thus $\mathcal{E}_1$ is consistent but $\mathcal{E}_2$ is inconsistent, where $\mathtt{Concs}(\mathcal{E}_1) \cap Lit_\Pi = \{c, i, \neg g\} = S$. Hence $\mathbf{ABF}(P_4)$ is consistent under the stable semantics. In contrast, $\mathbf{ABF}((P_4)_{tr})$ has the unique stable argument extension $\mathcal{E}_1$ due to six additionally introduced arguments to $A_i (1 \leq i \leq 12)$ and $|attacks| = 26$.

Using $\mathbf{ABF}(P_4)$, we can decide that the attorney-at-law having the argument A_4 for the claim $\neg g$ wins and the prosecutor having A_3 for g loses since $A_4 \in \mathcal{E}_1$ and $A_3 \notin \mathcal{E}_1$ for its unique consistent extension $\mathcal{E}_1$. Therefore $\neg g$ is decided.

4 Related Work and Conclusion

Beirlaen et al.'s extended ASPIC$^+$ framework [2] as well as Heyninck and Arieli's ABF induced by a DLP [13] can handle disjunctive information in argumentation. However the semantic correspondence to a disjunctive default theory is not shown in both [2] and [13]. In contrast, thanks to introducing explicit negation, our approach can capture not only the answer set semantics of an EDLP but also the semantics of a disjunctive default theory as shown in Theorem 8 and Theorem 9. Besides Beirlaen et al. [2] allowed arguments to have disjunctive conclusions, while Heyninck et al. [13] did not define arguments. In contrast, in our ABFs, arguments are defined but they are not allowed to have disjunctive conclusions (i.e. claims). As one of practical advantages of our approach, even if disjunctive information exists, we can directly use dialectic proof procedures [7,8] since the AA framework [6] can be generated from our ABF treating disjunctive information.

To sum up, as for argument extensions, Theorem 2 and Theorem 3 for an ELP (resp. Theorem 4 for a consistent ELP) in standard ABA frameworks are broaden to Theorem 7 and Theorem 8 for an EDLP (resp. Theorem 11 for a consistent EDLP) in generalized ABA frameworks, i.e. ABFs translated from EDLPs. Similarly as for assumption extensions, Proposition 1 for ABF induced by a DLP is generalized to Proposition 3 and Proposition 4 for the respective ABFs translated from EDLPs.

In (extended) disjunctive logic programming, the existence of disjunction generally increases the expressive power of logic programs while brings computational penalty [11]. By analogy, argumentation in ABFs translated from (E)DLPs increases the expressive power of ABF while it would introduce additional complexity. The analysis of complexity is left for future work.

Acknowledgments. The author would like to thank Chiaki Sakama and the anonymous reviewers of the paper for their valuable comments and suggestions.

References

1. Beirlaen, M., Heyninck, J., Straßer, C: Reasoning by cases in structured argumentation. In: Proceedings of the 2017 ACM Symposium on Applied Computing, pp. 989–994, ACM (2017)
2. Beirlaen, M., Heyninck, J., Straßer, C: A critical assessment of Pollock's work on logic-based argumentation with suppositions. In: Proceedings of the 17th International Workshop on Non-Monotonic Reasoning (NMR-2018), pp. 63–72 (2018)
3. Ben-Eliyahu, R., Dechter, R.: Propositional semantics for disjunctive logic programs. Ann. Math. Artif. Intell. **12**(1–2), 53–87 (1994)
4. Bondarenko, A., Dung, P.M., Kowalski, R.A., Toni, F.: An abstract, argumentation-theoretic approach to default reasoning. Artif. Intell. **93**, 63–101 (1997)
5. Caminada, M., Amgoud, L.: On the evaluation of argumentation formalisms. Artif. Intell. **171**(5–6), 286–310 (2007)
6. Dung, P.M.: On the acceptability of arguments and its fundamental role in nonmonotonic reasoning, logic programming and n-person games. Artif. Intell. **77**, 321–357 (1995)
7. Dung, P.M., Kowalski, R.A., Toni, F.: Dialectic proof procedures for assumption-based, admissible argumentation. Artif. Intell. **170**(2), 114–159 (2006)
8. Dung, P.M., Mancarella, P., Toni, F.: Computing ideal sceptical argumentation. Artif. Intell. **171**, 642–674 (2007)
9. Dung, P.M., Kowalski, R.A., Toni, F.: Assumption-based argumentation. In: Simari, G., Rahwan, I. (eds) Argumentation in Artificial Intelligence, pp. 199–218, Springer, Boston (2009). https://doi.org/10.1007/978-0-387-98197-0_10
10. Dung, P.M., Thang, P.M.: Closure and consistency in logic-associated argumentation. J. Artif. Intell. Res. **49**, 79–109 (2014)
11. Eiter, T., Gottlob, G.: Complexity results for disjunctive logic programming and application to nonmonotonic logics. In: Proceedings of the International Logic Programming Symposium (ILPS 1993), pp. 266–278. MIT Press (1993)

12. Heyninck, J., Arieli, O.: On the semantics of simple contrapositive assumption-based argumentation frameworks. In: Proceedings of Computational Models of Argument (COMMA-2018), pp. 9–20, IOS Press (2018)
13. Heyninck, J., Arieli, O.: An argumentative characterization of disjunctive logic programming. In: Moura Oliveira, P., Novais, P., Reis, L.P. (eds.) EPIA 2019. LNCS (LNAI), vol. 11805, pp. 526–538. Springer, Cham (2019). https://doi.org/10.1007/978-3-030-30244-3_44
14. Gelfond, M., Lifschitz, V.: The stable model semantics for logic programming. In: Proceedings of ICLP/SLP-1988, pp. 1070–1080. MIT Press, Cambridge (1988)
15. Gelfond, M., Lifschitz, V.: Classical negation in logic programs and disjunctive databases. New Gener. Comput. **9**, 365–385 (1991)
16. Gelfond, M., Lifschitz, V., Przymusinska, H., Truszczynski, M.: Disjunctive Defaults. In: Proceedings of International Conference on Principles of Knowledge Representation and Reasoning (KR-1991), pp. 230–237 (1991)
17. Prakken, H.: An abstract framework for argumentation with structured arguments. Argum. Comput. **1**(2), 93–124 (2010)
18. Przymusinski, T.C.: Stable semantics for disjunctive programs. New Gener. Comput. **9**, 401–424 (1991)
19. Reiter, R.: A Logic for default reasoning. Artif Intell. **13**, 81–132 (1980)
20. Sakama, C., Inoue, K.: Paraconsistent stable semantics for extended disjunctive programs. J. Logic Comput. **5**(3), 265–285 (1995)
21. Schulz, C., Toni, F.: Justifying answer sets using argumentation. Theory Pract. Logic Program. **16**(1), 59–110 (2016)
22. Wakaki, T.: Assumption-based argumentation equipped with preferences and its application to decision-making, practical reasoning, and epistemic reasoning. J. Comput. Intell. **33**(4), 706–736 (2017)
23. Wakaki, T.: Consistency in assumption-based argumentation. In: Proceedings of Computational Models of Argument (COMMA-2020), pp. 371–382, IOS Press (2020)

A Graph Based Semantics for Logical Functional Diagrams in Power Plant Controllers

Aziz Sfar[1,2](✉), Dina Irofti[2], and Madalina Croitoru[1]

[1] GraphIK, INRIA, LIRMM, CNRS and University of Montpellier, Montpellier, France
medaziz.sfar@gmail.com

[2] PRISME Department, EDF R&D, Paris, France

Abstract. In this paper we place ourselves in the setting of formal representation of functional specifications given in logical diagrams (*LD*) for verification and test purposes. Our contribution consists in defining a formal structure that explicitly encodes the semantics and behavior of a *LD*. We put in a complete transformation procedure of the non-formal *LD* specifications into a directed state graph such that properties like oscillatory behavior become formally verifiable on *LDs*. We motivate and illustrate our approach with a scenario inspired from a real world power plant specification.

Keywords: System Validation · Functional Specifications · Logic Functional Diagram · Graph based Knowledge Representation and Reasoning

1 Introduction

A power plant is a complex system and its functional behavior is described, for each of its subsystems, using logical diagrams. The logical diagrams are coded and uploaded into the controllers. During the power plant life-cycle (around 60 years and even more), the controllers' code needs to be updated and verified. Engineers generate scenarios in order to verify the new code. However, the scenarios generation is far from being a simple procedure because of the system's complexity. Indeed, the power plant contains a few hundred subsystems, and the behavior of each subsystem is described in a few hundred pages of logical diagrams. Knowing that a logical diagram page contains on average 10 logic blocks, a quick calculation shows that a power plant behavior can be described by a few hundred of thousands of logic blocks. Another nontrivial problem for scenario generation for such systems is caused by the loops existing between the logic blocks, i.e. the input of some logic blocks depends on their outputs, which can cause cyclic behaviors. These are indefinite variations of signals in the controller without a change occurring on its input parameters.

Logical diagram specifications lack the formal semantics that allow the use of formal methods for properties verification and test scenarios generation.

I. Varzinczak (Ed.): FoIKS 2022, LNCS 13388, pp. 55–74, 2022.
https://doi.org/10.1007/978-3-031-11321-5_4

Done through manual procedures, these tasks are tedious. In this paper, we tackle the problem of lack of semantics of logical diagram specifications. To solve this problem, we propose a formal graph model called the Sequential Graph of State/Transition ($SGST$) and we define a transformation method of logical diagrams into the proposed graph. On the $SGST$, we show how to formally verify that the functional behavior described by the logical diagram specification is deterministic. In fact, the specification model is supposed to provide a description of the expected behavior of the controller. If the expected behavior itself is non-deterministic, then test generation based on that behavior does not make sense. This problem can be generated by the presence of loop structures in the logical diagrams that may prevent the behavior (i.e. the expected outputs) from converging. The convergence property has to be verified before getting to test generation. Verifying this property directly on the logical diagram, which is a mix of logical blocks and connections presented in a non-formal diagram, is not easy to achieve. This task is possible in theory, as the logical diagrams can be reduced to combinatorial circuits. In literature, a combinatorial circuit [7] is a collection of logic gates for which the outputs can be written as Boolean functions of the inputs. In [11] it is shown that a cyclic circuit can be combinatorial, and a method based on binary decision diagrams is proposed to obtain the truth table of the circuit. The problem of where to cut the loops in the circuits and how to solve this loops has also been addressed in other studies [12], and applied in particular on the Esterel synchronous programming language [6]. Another algorithm for analysis cyclic circuits based on minimising the set of input assignments to cover all the combinatorial circuit has been proposed in [8]. Identifying oscillatory behavior due the combinatorial loops in the circuit has also been studied (see [2] and references therein). However, all studies cited here are mainly based on simulation rather than formal verification on models. The focus of these works is entirely dedicated to the verification of the cyclic behavior of the circuits and not to test purposes. Yet, several studies have already been published for the matter of both formal properties verification and test sequences generation. For instance, in their survey [5] Lee and Yannakakis address the techniques and challenges of black box tests derived from design specifications given in the form of finite state machines (Mealy machines). In [13], the author extends the test sequences generation to timed state machines inspired from the theory of timed automata [1]. These results and many others (such as formal verification of properties [9]) are applicable on state/transition graphs and can by no means be directly used on logical diagrams. In order to take advantage of the already established techniques, we focus our study on transforming logical diagrams into formal state graphs. Prvosot [10], has proposed transformation procedures of Grafcet specifications into Mealy machines, allowing the application of the previously mentioned formal methods of verification and test generation. However, Grafcets and logical diagrams are completely different representation models. A model transformation of logical diagrams into state graphs has been conducted by Electricté de France (EDF) [3] for cyclic behavior verification purposes; we note that [3] is not suitable for test generation and does not take into consideration the behavior of timer blocks. We inspired our work from both [10] and [3] to develop a formal state graph representation of the exhaustive behavior encoded in the logical diagram, the *SGST*. The proposed graph allows the verification of the cyclic behavior (called convergence in this paper) and potentially the formal verification of other properties. It also provides the ground to obtain the equivalent Mealy machine on which the existing formal test generation results can be applied.

This paper is organized as follows. The second section introduces the Logical Diagram specification with an example. A formal definition of the proposed *SGST* is given in Sect. 3. Section 4 details the model transformation procedure from logical diagrams to *SGST* graphs. In Sect. 5 we show how the behavior convergence property could be formally verified on the *SGST*. A discussion and a conclusion are given in the last section.

2 Motivating Example and Preliminary Notions

The main objective of a logical controller is to fulfill the set of requirements that it was built for. After their definition, the requirements are transformed into a functional description of the expected behavior called functional specification. A two level verification is needed to validate the controllers: first, the model is compared with respect to the specification, and second, the physical controller is tested to verify the conformity with respect to the specification (see functional validation and system validation in [4]). For both aspects, the specification model is the key point and the basis of the procedure, therefore it has to be well established and comprehended. In this section, we introduce Logical Diagram specifications used for power plant controllers, we define its composing elements and explain how it describes the functional behavior of the controller.

2.1 Logical Diagrams

Logical diagrams are specification models used to describe control functions in power plants. They contain a number of interconnected logic blocks that define how a system should behave under a set of input values.

Figure 1 illustrates a logical diagram extracted from a larger real world controller's logic specification in a power plant. It has five inputs (denoted by i_1 to i_5), one output (denoted o_1) and logic blocks: either blocks corresponding to logic gates or status blocks (corresponding to memory and on-delay timer blocks described below). The gates in Fig. 1 are: two NOT gates followed by two AND gates and two OR gates. They correspond to the conjunction ($\cdot$), disjunction ($+$), and negation ($^{-}$) Boolean operators, respectively (e.g. the output of an OR gate with two inputs is equal to 0 if and only if both inputs are equal to 0 etc.). The on-delay timer block gives the value 1 at its output if its input maintains the value 1 for 2 s; 2 s being the characteristic delay θ of the timer shown in the T block in Fig. 1. The memory block is a set (E) /reset (H) block: if the E input is equal to 1, then the output is equal to 1; if the H input is equal to 1, then the output of the block is 0. If both E and H inputs are equal to 1, the output is equal to 0 since the memory in this example gives priority to the reset H over the set E. This priority is indicated in the block symbol by the letter p. A 0 at both inputs keeps the output of the memory block at the same last given value.

The timer and the memory are blocks whose outputs not only depend on the values at their inputs, but also on their last memorized status. In this paper we call them ***status blocks***. Each of them possesses a finite set of status values and evolves between them. A status block output value {0, 1} is associated to each possible status. In the case of the example of Fig. 1, the memory block M_2 has two possible status values M_1 and M_0 where the status M_1 gives a logic value of 1 at the output of the block M_2 and M_0 status corresponds to the logic value 0. The on-delay timer block T_1 has 3 statuses denoted TD_0, TI_0 and TA_1, where the associated block output values are 0, 0 and 1, respectively. We also note on this example the presence of a loop structure (containing the block T_1, an OR block and the memory block M_2).

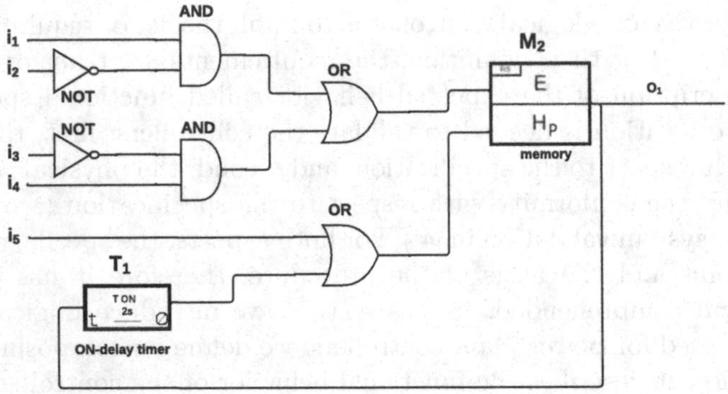

Fig. 1. Example of a logical diagram specification.

More formally, a logical diagram specification is composed of I, the set of inputs of the diagram, O, the set of outputs of the diagram and B, the set of the logic blocks of the diagram. The **logic blocks** B connect the outputs O to the inputs I and define the function that relates them. $B = B^S \cup B^{LG}$, namely:

- the logic gates B^{LG}: these are the AND, OR and NOT blocks in the diagram. Each of them is equivalent to a Boolean expression over its entries using the Boolean operators (+), (.) and ($^-$) for AND, OR and NOT, respectively.
- the status blocks B^S: these are blocks that have a status that evolves between a set of values. The evolution of a status of a block $b^s \in B^S$ depends on the values at its entries and the last value of its status. A logic value at the output of the status block is associated to each of these status values.

Definition 1 (P_{status} set). *We denote by P_{status} the set of all the possible status values of the blocks B^S of the logical diagram.*

We note that the status values that a block $b^s \in B^S$ can take are in a subset of P_{status}. Some insights are given in the following example.

Example 1 (Illustration of P_{status} set on the motivating example given in Fig. 1). For an on-delay timer status block (such as the block denoted T_1 in Fig. 1), $P^{TON}_{status} = \{TD\_0, TI\_0, TA\_1\}$; for a memory status block (such as M_2 block in Fig. 1), $P^{M}_{status} = \{M\_0, M\_1\}$. The associated block output logic value of a status value 'S_X' is indicated in its name by the numeric 'X'. The P_{status} set for the example illustrated in Fig. 1 is $P_{status} = P^{TON}_{status} \cup P^{M}_{status}$.

Logic Variables *Vars***.** Logic gates B^{LG} in the diagram can be developed into Boolean expressions over logic variables *Vars* by substituting them with their equivalent Boolean operator. Basically, we end up having outputs O and entries of B^S blocks that are equal to Boolean expressions on *Vars*.

Definition 2 (*Vars* and Exp_{Vars} sets). *We define $Vars = I \cup O_{B^S}$ by the set of logic variables, that includes I, the set of input variables of the logical diagram and O_{B^S}: the set of output variables of status blocks B^s in the diagram.*

We denote by Exp_{Vars} the infinite set of all possible Boolean expressions on logic variables in *Vars*. For example, $(o_{b^s_k} + i_k) \in Exp_{Vars}$. In the reminder of this paper, we will use the following mathematical notations on sets. Let A be a set of elements:

$A^k = \overbrace{A \times A \times ... \times A}^{k}$ is the set of all ordered k-tuples of elements of A. Given $e = (a_1, ..., a_k) \in A^k$, we denote $e(i)$ the ith element of e, i.e. $e(i) = a_i$. Given $e = (a_1, ..., a_k) \in A^k$, we denote $ord_e(a_k) = k$ the order k of a_k in e.

2.2 Test Generation for Logical Diagrams

Let us explain how these diagrams are supposed to be read and subsequently implemented in a physical system (i.e. the logic controller[1]). The diagrams are evaluated in evaluation cycles repeated periodically. Within each evaluation cycle the status blocks B^S are sequentially evaluated in accordance to a defined order ω while logic gates are evaluated from left to right.

The logic specification diagrams are implemented using a low level programming language into logical controllers. In order to check the conformity of the code with respect to the diagram, test beds are generated. The tests function in a black box manner: we check the conformance of the observed output values to the expected ones for different input values.

As one can see, even for a simple diagram like the one given in Fig. 1, finding an exhaustive testing strategy is not obvious. A simple solution for scenario test generation is through simulations of the diagram for each and every possible combination on the inputs $i_1 ... i_5$. This poses practical difficulties for two main reasons. On one hand, manual exhaustive test generation is a tedious, time-consuming task that has to be done to hundreds and hundreds of logical diagram specifications uploaded on logic controllers. On the other hand, a loop structure in the logical diagram could cause oscillation problems. This means that logic

[1] We refer to the implemented logical diagram specification as a logic controller.

values that circulate in a loop could keep changing indefinitely when passing through the blocks of the loop. This is a non desired phenomenon as it might prevent the controller's outputs from converging for a fixed set of input values. To overcome these difficulties, we propose (1) a graph state model called ***sequential graph of state/transition*** (*SGST*) and (2) a transformation procedure of the logical diagrams into the *SGST*. In this new graph, the nodes represent the states of the logical controller. The edges are labelled with the Boolean conditions over logic variables *Vars*. For instance, using the procedure we propose in this paper, we obtain for the logical diagram shown in Fig. 1 the corresponding sequential graph of state/transition given by Fig. 2. Throughout this paper, the logical diagram in Fig. 1 will be our case study.

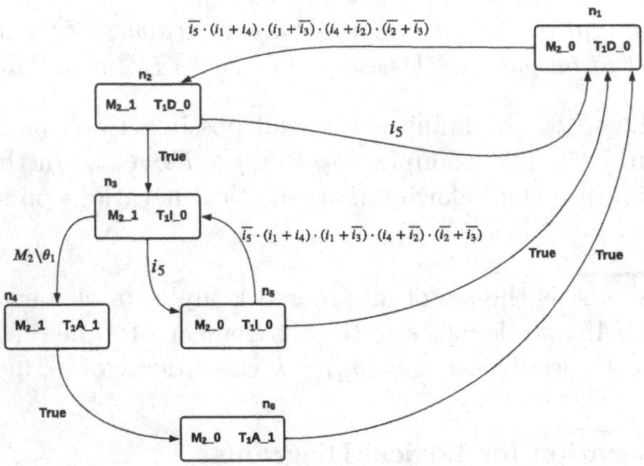

Fig. 2. The *SGST* corresponding to the non-formal logical diagram given by Fig. 1, obtained using our transformation procedure.

3 The Sequential Graph of State/Transition (*SGST*)

A *sequential graph of state transition* (*SGST*) is a combinatorial structure that explicitly represents all the possible evaluation steps within evaluation cycles of the logical diagram by the controller.

Formally, the **Sequential Graph of State/Transition** (SGST) is an directed graph defined by the tuple (N, E) where N is the set of nodes and $E \subseteq N \times N$ is the set of directed edges. Nodes and edges of the graph are both labeled using the labeling functions l_N and l_E, respectively.

Definition 3 (l_N function). *For a given set of status blocks B^S, the* ***labeling function of the nodes of the SGST graph*** *l_N is defined as l_N: $N \mapsto (P_{status})^L$, where $L = Card(B^S)$. This function assigns, for each status block $b^s \in B^S$ in the logical diagram, a status value to the node $n \in N$ in the SGST.*

Definition 4 (l_E function). *For a given set of logic variables Vars, the* ***labeling function of the edges of the SGST graph*** $l_E : E \mapsto Exp_{Vars}$*, assigns a logical expression including logic variables from Vars to* $e \in E$ *in the SGST.*

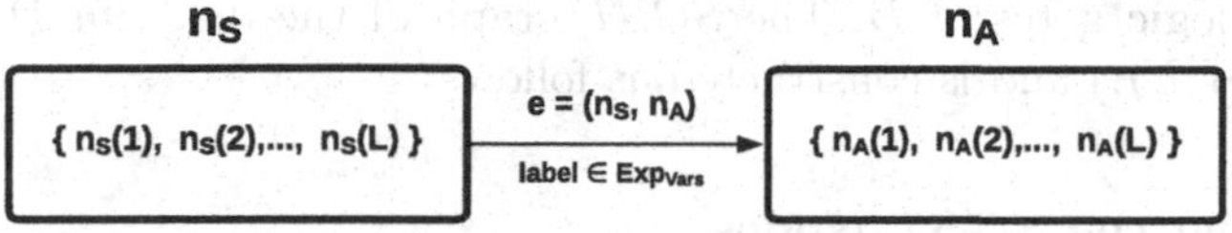

Fig. 3. SGST graph representation: example of two states n_S and n_A linked with a transition e. An edge e of the *SGST* graph links two states, from the starting node n_S to the arrival node n_A. The edge e is labelled with a Boolean expression *label* from the Exp_{Vars} set. The starting and arrival nodes are labelled with a set of L status values, where L is the total number of status blocks in the logical diagram, i.e. $Card(B^S) = L$. The set Exp_{Vars} contains Boolean expressions on logic variables in *Vars*. The set $Vars = I \cup O_{B^S}$ contains both the input variables I of the logical diagram, and the output variables O_{B^S} of status blocks B^S in the logical diagram.

Definition 5 ($Eval_{logic}$ function). *We define* $Eval_{logic} : P_{status} \mapsto \{0, 1\}$ *as the* ***logic evaluation function*** *that returns the equivalent logic value of a status value. In a node* $n \in N$*, the logic value at the output of a status block* b_i^s *is* $o_{b_i^s} = Eval_{logic}(n(i))$ *where* $n(i)$ *is the status value of the block* b_i^s *in the node* n*.*

Some notions from Definitions 2–5 are illustrated in Fig. 3. We remind that a logic value at the output o_{b^s} is associated to each status value of b^s assigned to a node $n \in N$. Therefore, a node n containing the status values of all blocks B^S encodes the logic values at each of their outputs.

Definition 6 (n^{Seq} logical sequence). *For a given set of status blocks* B^S*, we define* n^{Seq} *of a node* $n \in N$ *as the logical sequence on status blocks output variables* o_{B^S}*. It is a logical expression associated to the set of status values in* N*. This expression uses only the conjunction operator AND (.) and the complement operator (¯) and involves all the status output variables* $o_{b^s} \in O_{B^S}$ *of status blocks* $b^s \in B^S$*:*

$$n^{Seq} = \prod_{k=1}^{L} (Eval_{logic}(n(k)) \cdot o_{b_k^s} + \overline{Eval_{logic}(n(k))} \cdot \overline{o_{b_k^s}});\ where\, L = Card(B^S).$$

Example 2 (Illustration of n^{Seq} on the motivating example given by Fig. 1). In the *SGST* example of Fig. 2, the node n_1 encapsulates the status values $\{M_2\_0, T_1D\_0\}$. These status values correspond to the logic values 0 and 0 at the outputs of the blocks M_2 and T_1, respectively. The logical sequence n_1^{Seq} of the node n_1 is

$$n_1^{Seq} = (\overbrace{Eval_{logic}(M_2\_0)}^{0} \cdot o_{M_2} + \overbrace{\overline{Eval_{logic}(M_2\_0)}}^{1} \cdot \overline{o_{M_2}}) \cdot (\overbrace{Eval_{logic}(T_1D\_0)}^{0} \cdot o_{T_1} + \overbrace{\overline{Eval_{logic}(T_1D\_0)}}^{1} \cdot \overline{o_{T_1}}) = \overline{o_{M_2}} \cdot \overline{o_{T_1}}.$$

4 *SGST* Construction

Given a logical diagram D, we consider I_D (the set of input variables of D) and B_D (the set of logic blocks of D). As explained in the previous sections, $B_D = {B_D}^S \cup {B_D}^{LG}$, where ${B_D}^S$ is the set of status blocks of D and ${B_D}^{LG}$ is the set of logic gates of D. The $SGST$ graph of the diagram D is denoted $SGST_D : (N_D, E_D)$ and is constructed as follows.

4.1 Building the *SGST* Nodes

We will construct one node for each possible combination of status values between the status blocks. Let us start by defining the set of all the possible combinations of status values of blocks $b^s \in B_D^S$. Let n_T be the number of on-delay timer blocks and n_M the number of memory blocks in the logical diagram D. We define the set of all combinations of status values as $C_{status} = (P^M_{status})^{n_M} \times (P^{TON}_{status})^{n_T}$. The number of nodes of the $SGST_D$ is $Card(N_D) = Card(C_{status}) = 3^{n_T} \times 2^{n_M}$; $Card(P^{TON}_{status}) = 3$, $Card(P^M_{status}) = 2$. To each of these nodes we attribute a combination of status values using the l_N function: $\forall n_i \in N_D$, $l_N(n_i) = c_i$ where $c_i \in C_{status}, i \in \{1..Card(N_D)\}$. In the example of Fig. 1, $P^{TON}_{status} = \{TD\_0, TI\_0, TA\_1\}$, $P^M_{status} = \{M\_0, M\_1\}$, C_{status} =$\{(M\_0, TD\_0), (M\_1, TD\_0), (M\_1, TI\_0), (M\_1, TA\_1), (M\_0, TI\_0), (M\_0, TA\_1)\}$. $Card(C_{status}) = 6$, the $SGST_D$ has therefore six nodes n_1 to $n_6 \in N_D$ labeled $c_1 ... c_6 \in C_{status}$, as shown in Fig. 2.

4.2 Building the *SGST* Edges

Edges that link the nodes in the graph are labelled with a logical expressions over *Vars*. If for a set of logic values of *Vars*, a Boolean expression that labels an edge starting from a node n_S and arriving to a node n_A is True, a change of status values of a B^S block in the diagram takes place. The $SGST_D$ of a logical diagram D is developed to represent the evolution of states of the diagram in a formal model. The way this evolution works is defined by the evaluation process of the diagram by the controller. This evaluation is done in periodic cycles:

1. Reading and saving the values of all input variables i_k, where $k \in \mathbb{N}$.
2. Running a **sequential** evaluation algorithm on status blocks: at this point, each status block is evaluated, one after another, in accordance to the logic values at their entries and their last evaluated status value. The logic values at the entries of status blocks are obtained by the evaluation of logic gates connected to these entries in a left to right direction. We denote by ω the order of the evaluation sequence of status blocks.
3. Evaluating outputs. Outputs o_k are Boolean expressions of input variables I_D and status blocks output variables O_{B^S}.

We build the edges of the $SGST_D$ that link the nodes following the sequential evaluation of status blocks that we just established. This sequential evaluation dictates that only one status block is evaluated at a time. In other words, status blocks are not evaluated simultaneously. The result of evaluation of a status block is used in the evaluation of the next status block in the ordered sequence ω. This is translated in the graph by building edges that only connect nodes that have the same status values for all status blocks except for one. We call these nodes neighboring nodes.

Proposition 1. *Let n_1 and n_2 be two nodes of N_D and $l_N(n_1) = (\mu_1, \mu_2, ..., \mu_L)$ and $l_N(n_2) = (\lambda_1, \lambda_2, ..., \lambda_L)$ be their status values. n_1 and n_2 are two neighboring nodes and can possibly be linked by an edge in the $SGST_D$ if $\exists c \in \{1, ..., L\}$, with $L = Card(B_D^S)$, satisfying the following two conditions:*

- *$\forall k \in \{1, ..., L\} \backslash c$, $\mu_k = \lambda_k$; we note that $\mu_k = n_1(k)$ is the status value of b_k^s in n_1 and $\lambda_k = n_2(k)$ is the status value of b_k^s in n_2 where b_k^s are status blocks in B_D^S.*
- *$n_1(c) = \mu_c \neq \lambda_c = n_2(c)$ where $b_c^s \in B_D^S$ is the only status block that changes value from μ_c in node n_1 to λ_c in node n_2.*

Roughly speaking, Proposition 1 tells us that two nodes in the $SGST$ graph can be neighbours only if all their status values are identical except one. To conclude, an edge of the graph is equivalent to a change of the status value of a single status block between two neighboring nodes n_S and n_A linked by that edge. We refer to this change of value as an evolution *evol* and we define $EVOL_{b^s}$ as the set of all evolution possibilities of $b^s \in B_D^S$ between its status values $P_{status}^{b^s}$.

Definition 7 (*evol* **tuple**)**.** *An* ***evolution*** *evol $\in EVOL_{b^s}$ is defined by the tuple (s_i, s_f, C_{evol}), with:*

- *s_i: the initial status value of the evolution evol; $s_i \in P_{status}^{b^s}$*
- *s_f: the final status value of the evolution evol; $s_f \in P_{status}^{b^s}$*
- *C_{evol}: the evolution condition; this is a Boolean expression deducted from the logical diagram. The evolution from s_i to s_f can only occur if this expression is True. We note that $C_{evol} \in Exp_{Vars}$.*

In order to construct the edges of the $SGST_D$, first the Boolean expressions of the status block entries have to be calculated (A). Second, the evolution sets $EVOL_{b^s}$ of every status block b^s have to be determined using the calculated expressions (B). Finally, edges of the graph are constructed based on the determined evolution sets of status blocks (C).

A. Developing the logical expressions at the entries of status blocks: We remind that the status value of a block $b^s \in B_D^S$ is calculated based on its previous status value and the logic values at the entries of the block. The logic values at these entries are obtained by evaluating the elements connected to them. We develop these connections into Boolean expressions. The evaluation of a Boolean expression associated to an entry of a status block gives the logic value of that entry. These expressions are developed as follows:

Let $x_{b_k^s}$ be an entry of a status block $b_k^s \in B_D^S$. $x_{b_k^s}$ could be connected to one of the following elements:

Algorithm 1. Evolution construction algorithm for status blocks of memory type

Input: Boolean expressions (E,H) ▷ E and H are the two entries of the memory block
Output: evolution set $EVOL_{b^s}$; Reminder: $evol \in EVOL_{b^s}, evol = (s_i, s_f, C_{evol})$
for all $(s_i, s_f) \in P^M_{status} \times P^M_{status}$ **do**
  if $(s_i, s_f) = (M\_1, M\_0)$ **then**
    $C_{evol} \leftarrow H$
  else if $(s_i, s_f) = (M\_0, M\_1)$ **then**
    $C_{evol} \leftarrow E \cdot \overline{H}$
  end if
  $evol \leftarrow (s_i, s_f, C_{evol})$
  add *evol* to $EVOL_M$
end for

- An input $i_j \in I_D$: in this case the logic value of this entry is equal to the logic value of the input variable $x_{b^s_k} = i_j$;
- The output of a status block b^s_j, with $j \neq k$: here, the entry takes the logic value of the output of the block b^s_j denoted by $o_{b^s_j}$. Then, $x_{b^s_k} = o_{b^s_j}$;
- The output of a logic gate b_{LG}: we denote by $o_{b_{LG}}$ the output of the logic gate b_{LG}; then, $x_{b^s_k} = o_{b_{LG}}$. The output $o_{b_{LG}}$ of the logic gate b_{LG} can be developed into a Boolean expression that uses the logic operator of the block b_{LG} over its entries. Entries of b_{LG} that are connected to the output of another logic gate are further developed into Boolean expressions and so on. This recursive development continues through all the encountered logic gates and stops at logic inputs I_D and status block outputs O_{B^S}.

Example 3. The logical diagram of Fig. 1 has two status blocks T_1 and M_2 with outputs denoted o_{T_1} and o_{M_2}, respectively. The block T_1 has a single entry x_T directly connected to the output o_M: $x_T = o_{M_2}$. The block M_2 has two input terminals denoted E and H. The entry H is connected to the output of an 'OR' gate that we call or_1, $H = or_1$. The variable or_1 can be developed into the following expression: $or_1 = i_5 + o_{T_1}$. The expression of the entry H is therefore $H = i_5 + o_T$ Similarly, the input terminal E is connected to an 'OR' logic gate $E = or_2$. We denote by x_1 and x_2 the input terminals of this 'OR' gate. $or_2 = x_1 + x_2$. Both x_1 and x_2 are connected to logic gates. They are therefore developed into Boolean expression in their turn. Following this process, we obtain $E = i_1 \cdot \overline{i_2} + \overline{i_3} \cdot i_4$.

B. Building the evolution sets $EVOL_{b^s}$ of every status blocks $b^s \in B^S_D$: The evolution possibilities of each status block are determined by the nature of the status block (i.e. memory blocks or timer blocks). Knowing the Boolean expressions at the entries of a block $b^s \in B^S_D$ we define the algorithms that construct all the evolution possibilities $EVOL_{b^s}$ of the block: Algorithm 1 corresponds to the evolution set construction for memory blocks, and Algorithm 2

Algorithm 2. Evolution construction algorithm for status blocks of timer type

Input: Boolean expressions X ▷ X is the entry of the block
Output: evolution set $EVOL_{b^s}$; $evol \in EVOL_{b^s}, evol = (s_i, s_f, C_{evol})$
for all $(s_i, s_f) \in P^{TON}_{status} \times P^{TON}_{status}$ **do**
  if $(s_i, s_f) = (TD\_0, TI\_0)$ **then**
    $C_{evol} \leftarrow X$
  else if $(s_i, s_f) = (TI\_0, TD\_0)$ **then**
    $C_{evol} \leftarrow \overline{X}$
  else if $(s_i, s_f) = (TI\_0, TA\_1)$ **then**
    $C_{evol} \leftarrow X \cdot X/\theta$
  else if $(s_i, s_f) = (TA\_1, TD\_0)$ **then**
    $C_{evol} \leftarrow \overline{X}$
  end if
  $evol \leftarrow (s_i, s_f, C_{evol})$
  add $evol$ to $EVOL_M$
end for

constructs the evolution set for a timer block. We note that, in the case of timers, in addition to the logic value at the entry of the block, the status and output value of timer blocks also depend on time. After receiving a stimulus (i.e. a rinsing or falling edge), a timer changes its status automatically after a time period during which the stimulus action is maintained. In the case of an on-delay timer with a characteristic delay θ, if its input X is set to 1 for a period $\Delta_t > \theta$, the timer goes to the activated status TA giving the value 1 at its output instead of 0 in its deactivated status TD. We introduce another logic variable $X/\theta \in Vars$ such that: $\begin{cases} X/\theta = 1 & \text{if X holds the value 1 for a period } t > \theta \\ X/\theta = 0 & \text{otherwise} \end{cases}$

Example 4. Applying Algorithms 1 and 2 on the diagram example of Fig. 1, we obtain the evolution sets of the two status blocks in the diagram. For the timer T_1, we obtain $T_1 = \{evol_1, evol_2, evol_3, evol_4\}$, with:
$evol_1 = (T_1D\_0, T_1I\_0, O_{M_2})$; $evol_2 = (T_1I\_0, T_1D\_0, \overline{O_{M_2}})$;
$evol_3 = (T_1I\_0, T_1A\_1, O_{M_2} \cdot O_{M_2\backslash\theta})$; $evol_4 = (T_1A\_1, T_1D\_0, \overline{O_{M_2}})$. For the memory M_2, we obtain $EVOL_{M_2} = \{evol_1, evol_2\}$, with:
$evol_1 = (M_2\_0, M_2\_1, \overline{O_{T_1}} \cdot \overline{e_5} \cdot (i_1 + i_4) \cdot (i_1 + \overline{i_3}) \cdot (i_4 + \overline{i_2}) \cdot (\overline{i_2} + \overline{i_3}))$;
$evol_2 = (M_2\_1, M_2\_0, O_{T_1} + e_5)$.

C. Building the edges E_D of the $SGST_D$: Having built the nodes of the graph and determined all the evolution sets $EVOL_{b^s}$ of status blocks B^S_D, we build the edges that connect these nodes. We remind that the controller evaluates its status blocks in an ordered sequence ω. In other words, status blocks are not evaluated simultaneously; they are evaluated one at a time. A node in the $SGST_D$ encapsulates the status values of all the blocks B^S_D. Two nodes in N_D can have one or many different status values. The sequential evaluation of the controller is reproduced in the graph by building edges that only link neighboring nodes that have the exact same status values of all blocks B^S_D except for one (see Proposition 1). An edge linking two neighboring nodes corresponds to an evolution $evol \in EVOL_{b^s}$ of a single status block b^s. We propose Algorithm 5 for building the edges of the graph $SGST_D$ of a logical diagram D. For each node $n_k \in N_D$ of the $SGST$, the algorithm generates all the possible outgoing edges corresponding to all the evolution possibilities of all status blocks B_D^S from the node n_k. Algorithms 3 and 4 are used in Algorithm 5 for neighboring nodes recognition and nodes logical sequences generation.

Algorithm 3. Test whether n_S and n_A are neighboring nodes (see Proposition 1)

Input: $n_S, n_A \in N_D$
Output1: $AreNeighbors \in \{True, False\}$
Output2: c the index of the status block whose status value is changed from n_S to n_A
$AreNeighbors \leftarrow True$
$differences \leftarrow 0$ ▷ number of different status values between n_S and n_A
for all $k = 1$ to $Card(B_D^S)$ **do**
  if $n_S(k) \neq n_A(k)$ **then**
    $differences \leftarrow differences + 1$
    if $differences > 1$ **then**
      $AreNeighbors \leftarrow False$
      break loop
    end if
    $c \leftarrow k$ ▷ c is the index of the block that changes status from n_S to n_A
  end if
end for

5 Reasoning with the *SGST*

In this section, we show how the convergence of the expected behavior of the controller described by its logical diagram could be verified using the equivalent $SGST$ of that diagram. The $SGST$ is composed of a set of nodes and edges that reproduce the information encoded in the logical diagram in a formal and explicit description. A ***node*** in the $SGST$ corresponds to a possible state of the controller, i.e. a possible combination of status values. An ***edge*** corresponds to an evolution of a single status block. That is a change of the status value of a block $b^s \in B_D^S$. The outgoing edges E_{n_j} of a node $n_j \in N_D$ in the $SGST_D$ graph of a diagram D, are all the theoretical evolution possibilities of all the status blocks from the node n_j. In practice, only one of these outgoing edges $e \in E_{n_j}$, is **traversed** depending on the values of the input variables I_D of the diagram. A ***traversal of an edge*** (n_S, n_A) is the effective transition of the controller's state from the node n_S to the node n_A by running the correspondent status block evolution of the traversed edge.

We remind that a **full evaluation cycle** of the logical diagram is held periodically by the controller. In each evaluation cycle of the diagram, status blocks

Algorithm 4. Build n_S^{seq} the logical values sequence of outputs $O_{B_D^S}$ equivalent to status values in n_S (see Definition 6)

Input: $n_S \in N_D$
Output: n_S^{seq}
$n_S^{seq} \leftarrow True$
for all $k = 1$ to $Card(B_D^S)$ **do**
$n_S^{seq} \leftarrow n_S^{seq} \cdot (Eval_{logic}(n_S(b_k^s)) \cdot o_{b_k^s} + \overline{Eval_{logic}(n_S(k))} \cdot \overline{o_{b_k^s}})$
end for

are evaluated one after another according to an order ω until each and every block $b^s \in B^S$ is evaluated once and only once.

In the *SGST*, for a set of input values I_v a full evaluation cycle corresponds to a chain of successive edges traversed one after another in respect to the order of evaluation ω. In some cases, many successive evaluation sequences ω may have to be run to finally **converge to a node**. However in other cases, even after multiple evaluation sequences, this convergence may never be reached; Traversal of edges could be endlessly possible for a set of input values I_v.

The convergence of status values is the property that we are going to study in the rest of this paper. If we consider the real world case of power plants, the convergence property has to be verified on the logical diagrams before implementing them in the controllers. The non convergence of the evaluation cycles of the diagram for a set of input values I_v leads to the physical output signals of the controller alternating continuously between 0 and 1 which is a non-desired phenomenon. In the *SGST* graph, this corresponds to a circuit of nodes being visited over and over again indefinitely. We will define trails and circuits in the graph then propose a formal criteria of behavior convergence on the *SGST*.

5.1 Traversal of the *SGST*: Trails

In the *SGST*, nodes are visited by traversing the edges that link them. A sequence of visited nodes in the graph is called a trail τ and is defined as follows:

Definition 8 (Trail τ). *For a given $SGST_D = (N_D, E_D)$, a **trail** $\tau \in (N_D)^k$, with $k \in \mathbb{N}$, is an ordered set of nodes $(n_1, n_2, ..., n_k)$ where each pair of successive nodes n_i and n_{i+1}, with $i \in \{1..k-1\}$, are neighboring nodes.*

A trail is therefore a series of state changes along neighboring nodes in the *SGST* graph. From the *SGST* we can form an infinite number of trails. However, only a finite subset of these trails could be effectively traversed in practice. This is due to the order ω of the evaluation of status blocks. We call trails that are conform to the order ω **determined trails**. These trails correspond to the progressive traversal of **viable edges** in the *SGST* for a set of input values I_v.

Definition 9 (Viability of an edge). *Let $e = (n_S, n_A) \in E_D$ be an edge in the SGST linking the start node n_S to the arrival node n_A. The edge e is said*

Algorithm 5. Edge construction Algorithm

Input1: $EVOL_{b^s}$ $\forall b^s \in B_D^S$
Input2: the set of nodes N_D of the $SGST_D$
Output: the set of edges E_D of the $SGST_D$; $e \in E, e = (n_S, n_A, label)$
for all $(n_S, n_A) \in N_D \times N_D$ **do**
  AreNeighbors, c =Algorithm 3(n_S, n_A) ▷ c is the index of the status block whose status value is changed from n_S to n_A
  if $AreNeighbors = True$ **then**
    n_S^{seq}=Algorithm 4(n_S)
    for all $evol \in EVOL_{b_c^s}$ **do** ▷ b_c^s is the block that changes status from n_S ▷ to n_A; $EVOL_{b_c^s}$ is the evolution set of b_c^s
      if $s_i(evol) = n_S(c)$ and $s_f(evol) = n_A(c)$ **then** ▷ Find the evolution of b_c^s ▷ that corresponds to ▷ $n_S(c) \mapsto n_A(c)$
        $expression = C_{evol}(evol)$
        if $expression \wedge n_S^{Seq} \neq False$ **then** ▷ check if the expression of the evolution is contradictory ▷ to the logic values of status blocks outputs $O_{B_D^S}$ ▷ given by the sequence n_S^{Seq} of n_S
          $e = (n_S, n_A)$
          $l_E(e) = C_{evol}(evol)$
          add e to the set of edges E_D
          break loop
        end if
      end if
    end for
  end if
end for

to be viable for a set of input values I_v if the Boolean expression label(e) is True for the values I_v. The traversal of the edge e changes the state of the controller from n_S to n_A by changing the status value of a single block $b^s \in B_D^S$. We denote by $b^s_{n_S \mapsto n_A}$ the status block b^s whose value was altered by going from n_S to n_A.

A node n_k visited in the middle of a determined trail τ, can have multiple outgoing edges in the $SGST$ that are viable at the same time for a set of input values I_v. Only one of the viable edges is traversed in accordance to the edge traversal determination rule $(n_{k-1}, n_k) \mapsto n_{k+1}$: the next viable edge (n_k, n_{k+1}) to be traversed is the one that alters the status value of the block $b^s_{n_k \mapsto n_{k+1}}$ of the lowest order in ω after $b^s_{n_{k-1} \mapsto n_k}$ the block whose status value changed from n_{k-1} to n_k.

Proposition 2 (edge traversal determination rule). *Let $\omega \in (B^S)^L$, be an order of the evaluation sequence of status blocks, $L = Card(B^S)$. Let us suppose that for a set of input values I_v, the controller is placed in the state of node n_k, coming from the previous node n_{k-1}; between these two nodes, the status value*

of the block $b^s_{n_{k-1}\mapsto n_k}$ *has changed. Let* N_{next} *be the set of all the reachable nodes from* n_k *by the viable edges* $e = (n_k, n_{next})$*, with* $n_{next} \in N_{next}$*.*
Then, the next node $n_{k+1} \in N_{next}$ *to be effectively visited in the trail is satisfying:* $ord_\omega(b^s_{n_{k-1}\mapsto n_k}) < ord_\omega(b^s_{n_k\mapsto n_{k+1}}) < ord_\omega(b^s_{n_k\mapsto n_{next}})\ \forall n_{next} \in N_{next}\backslash\{n_{k+1}\}$*, where* $\forall n_j \in N_D;\ \forall(n_k, n_j) \in E_D$ *then* $ord_\omega(b^s_{n_k\mapsto n_j})$ *is the order of evaluation of the block* $b^s_{n_k\mapsto n_j}$ *in* ω*;* $b^s_{n_k\mapsto n_j}$ *is the status changing block from* n_k *to* n_j*.*

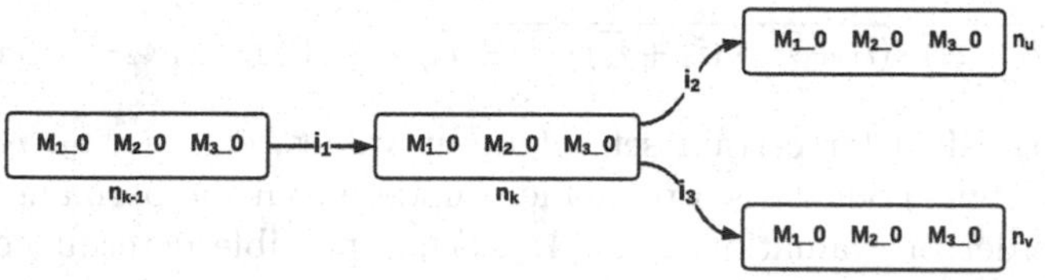

Fig. 4. Trail building in an SGST graph. Two possible trails are valid for the same input values I_v in this example, depending on the order of evaluation of the three status blocks, M_1, M_2, M_3.

Example 5. Let us consider a logical diagram with three status blocks of memory type $B^S = \{M_1, M_2, M_3\}$; the corresponding *SGST* graph is illustrated in Fig. 4, and has four nodes and three edges. We fix the set of inputs values $(i_1, i_2, i_3) = (1, 1, 1)$. We suppose that the last visited node is n_k coming from n_{k-1}. We note that, from the node n_k, both edges $e_2 = (n_k, n_u)$ and $e_3 = (n_k, n_v)$, labeled i_2 and i_3, respectively, are viable for the input values $(1, 1, 1)$. The following node of the trail $\tau = (n_{k-1}, n_k)$ is determined in accordance with Proposition 2. We first consider $\omega = (M_1, M_2, M_3)$ the order of evaluation of the three status blocks. The last traversed edge is $e_1 = (n_{k-1}, n_k)$ with a change on the status value of the block M_1 of order $ord_\omega(M_1) = 1$ in the evaluation sequence ω. Edge e_2 alters the status value of the block M_2 of order $ord_\omega(M_2) = 2$ while the edge e_3 alters the status value of the block M_3 of order $ord_\omega(M_3) = 3$. Since $ord_\omega(M_1) = 1 < ord_\omega(M_2) = 2 < ord_\omega(M_3) = 3$, the next status block to be evaluated after M_1 is M_2, so the next node in the trail τ is $n_{k+1} = n_u$. In this case, $\tau = (n_{k-1}, n_k, n_u)$. However, if we consider $\omega = (M_3, M_2, M_1)$, i.e. $ord_\omega(M_3) = 1$, $ord_\omega(M_2) = 2$, $ord_\omega(M_1) = 3$, then the last evaluated block M_1 in the trail (n_{k-1}, n_k) is of order 3 which is the last order in the evaluation sequence ω. For the same input values $(i_1, i_2, i_3) = (1, 1, 1)$, the next block to be evaluated from n_k is this time M_3 of the order 1, which corresponds to edge $e_3 = (n_k, n_v)$. In this case, $\tau = (n_{k-1}, n_k, n_v)$.

We make the assumption that the initial node of a determined trail is a permanent node. A **permanent node**, unlike a transitional node, is a node in which the controller's state can remain permanently for a certain set of input values.

Definition 10 (permanent node). *Let* $n_k \in N_D$ *be a node, and* $E_{n_k} \subset E_D$ *the set of outgoing edges from the node* n_k*. We say that the node* n_k *is a* ***permanent***

***node** if* $\exists\, I_v$*, a set of input values, satisfying the holding on condition of the node* n_k*:* $C_{Hold} = \prod_{e_i \in E_{n_k}} \overline{label(e_i)}$.

Example 6. For the *SGST* graph given in Fig. 2, node n_2 has two outgoing edges labeled i_5 and *True*. The holding on condition of node n_2 is $C_{Hold} = \overline{i_5} \cdot \overline{True} = False$. This condition is False for any set of input values I_v; thus, n_2 is not a permanent node. The node n_1 has only one outgoing edge labeled $\overline{i_5} \cdot (i_1 + i_4) \cdot (i_1 + \overline{i_3}) \cdot (i_4 + \overline{i_2}) \cdot (\overline{i_2} + \overline{i_3})$. The holding on condition of node n_1 is

$$C_{Hold} = \overline{\overline{i_5} \cdot (i_1 + i_4) \cdot (i_1 + \overline{i_3}) \cdot (i_4 + \overline{i_2}) \cdot (\overline{i_2} + \overline{i_3})} = i_5 + i_2 \cdot i_3 + i_2 \cdot \overline{i_4} + i_3 \cdot \overline{i_1} + \overline{i_1} \cdot \overline{i_4},$$

and can be satisfied for certain sets of input values, e.g. $(i_1,\, i_2,\, i_3,\, i_4,\, i_5) = (0,0,0,0,1)$. Thus, node n_1 is a permanent node. From the permanent node n_1, and for an order of evaluation $\omega = (M_2, T_1)$, a possible determined trail that could be effectively traversed would be $\tau_1 = (n_1, n_2, n_3, n_4, n_6, n_1)$ for the order $\omega = (M_2, T_1)$ and the set of input values $(i_1,\, i_2,\, i_3,\, i_4,\, i_5) = (1,0,0,1,0)$.

5.2 Formal Verification of the Convergence Property in the *SGST*

In practice, we say that a signal converges if its periodic evaluation by the logic controller gives a constant value over a long time range during which the input signals I remain constant. A non convergent Boolean signal is a signal that keeps oscillating between 0 and 1 over multiple evaluation cycles of the logic controller while input values are unchanged. In the *SGST*, oscillating Boolean signals correspond to an indefinite visiting of the same subset of nodes over and over again. This causes an indefinite change of status values, which results in its turn to an indefinite change of logic values at the output of status blocks.

Definition 11 (Circuits in the *SGST*). *We define a **circuit** in a SGST graph as a finite series of nodes* $(n_1,\, n_2,\, ...,\, n_m)$ *such that the consecutive nodes* n_k *and* n_{k+1} *are neighboring nodes and* $n_1 = n_m$*.*

However, a determined trail in the *SGST* graph could contain a circuit of nodes without necessarily traversing it indefinitely. Indeed, a trail could correspond to a one-time traversal of a circuit to leave it as soon as it visits the same node twice, as shown by Example 7.

Example 7. We consider the *SGST* graph given by Fig. 5. The status blocks of the *SGST* are $B^S = \{M_1,\, M_2\}$. It contains three possible circuits (n_1, n_2, n_1), (n_1, n_3, n_1) and $(n_1, n_2, n_1, n_3, n_1)$. We suppose the evaluation order $\omega = (M_2, M_1)$. We fix a set of input values $(i_1,\, i_2,\, i_3,\, i_4) = (1,1,1,0)$. Using Proposition 2, we obtain the trail $\tau = (n_1, n_3, n_1, n_2)$, starting from the permanent node n_1. We can observe that τ contains the circuit $(n_1,\, n_3,\, n_1)$, but this circuit is quit to the node n_2. However, if we fix the set of input values at $(i_1,\, i_2,\, i_3,\, i_4) = (1,1,0,0)$, and start from node n_1, we obtain the trail $\tau =(n_1,\, n_3,\, n_1,\, n_3)$ that is equivalent to an indefinite traversal of the circuit $(n_1,\, n_3,\, n_1)$.

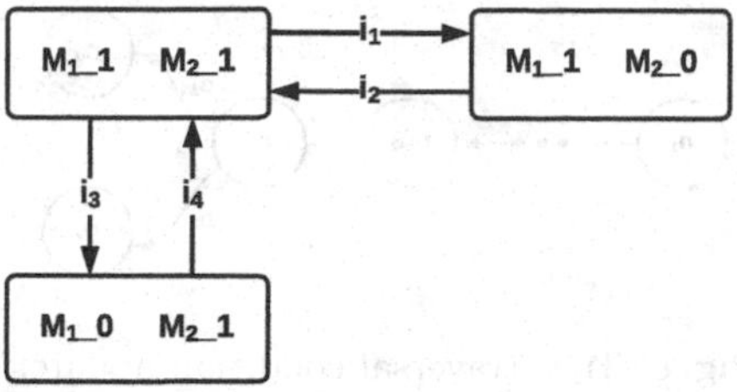

Fig. 5. Example of multiple circuits in an *SGST* graph.

Generally, if for an input I_v the progressive calculation of a the nodes of a trail τ results in visiting twice the same successive neighboring node couple (n_k, n_{k+1}), then τ corresponds to a circuit of nodes that can be indefinitely visited and the outputs of the blocks whose status values are changed in that trail are oscillating.

Definition 12 (Convergence property in a trail). *We denote by* T^{ω}_{SGST} *the set of all determined trails in the SGST that can be effectively traversed for an evaluation order* ω. *A trail* $\tau = (n_1, n_2, ..., n_m) \in T^{\omega}_{SGST}$ *is* ***convergent*** *if* $e_k \neq e_j \ \forall e_k = (n_k, n_{k+1})$, $e_j = (n_j, n_{j+1})$ *two tuples of neighboring nodes in* τ.

Definition 13 (Convergent logical diagram). *We say that a logical diagram* D *is convergent for all the sets of input values if all the determined trails of its* $SGST_D$ *are convergent.*

We propose a method for searching all the determined trails T^{ω}_{SGST} for an evaluation order ω. Trails are determined by giving their symbolic Boolean condition of traversal instead of the sets of input values. This means that a determined trail $\tau \in T^{\omega}_{SGST}$ is defined by the sequence of its nodes $\tau = (n_1, n_2, ..., n_{m-1}, n_m)$ and its traversal condition $C_\tau = \prod_{k \in \{1..m-1\}} label(e_k = (n_k, n_{k+1}))$.

Starting from each permanent node in the *SGST* we calculate all the possible trails based on the trail determination rule for an order ω (Proposition 2). From each node we explore all the possible outgoing edges by negating the condition labels of edges alternating the blocks of the least order. Each label of an explored edge is added to C_τ. For instance, let us suppose that a trail reaches a node n_k coming from n_{k-1} and that n_k that has two outgoing edges $e_1 = (n_k, n_u)$ and $e_2 = (n_k, n_v)$. $\tau = (n_1, n_2, ..., n_{k-1}, n_k)$, $C_\tau = \prod_{j \in \{1..k-1\}} label(e_k = (n_j, n_{j+1}))$ (Fig. 6).

We suppose that $ord_\omega(b^s_{n_{k-1} \mapsto n_k}) < ord_\omega(b^s_{n_k \mapsto n_u}) < ord_\omega(b^s_{n_k \mapsto n_v})$, for the order ω. Since the status block altered by e_1 is of a lower order than the one altered by e_2, if $label(e_1) = True$ then e_1 is the next movement in τ, but if $label(e_1) = False$ and $label(e_2) = True$ then the next movement in τ is e_2. Thus, two determined trails τ_1 and τ_2 can branch off from the determined trail τ at n_k such that $C_{\tau_1} = C_\tau \cdot label(e_1)$ and $C_{\tau_2} = C_\tau \cdot \overline{label(e_1)} \cdot label(c_2)$. Both new trails continue the course and branch off to more possible trails at each bifurcation. Path exploration of a trail can stop in one of the following scenarios:

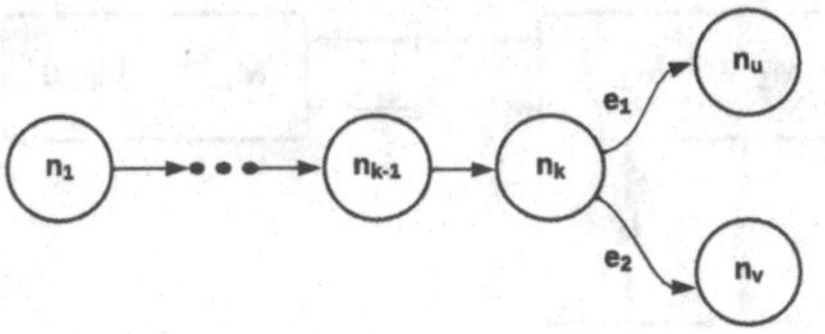

Fig. 6. Trail traversal condition update.

- If the last encountered node n_k is a permanent node. Here, τ is a determined trail that puts the controller in the state of the node n_k starting from the state of the initial node n_1 for all the input values I_v that satisfy C_τ.
- If for the last encountered node n_k, the update of the traversal condition $C_\tau \cdot label(e = (n_{k-1}, n_k))$ is False. This means that the trail is not possible due to a contradiction of the condition labels of the graph edges crossed by the trail.
- If the last two couple of nodes (n_{k-1}, n_k) have already been visited in τ. In this case τ corresponds to a circuit of nodes that can be effectively traversed an infinite number of times for the inputs values I_v satisfying condition C_τ.

Example 8. The *SGST* example of Fig. 2 has only one permanent node n_1. Starting from n_1, only two determined trails are possible in the case of the evaluation order $\omega = (T_1, M_2)$:
$\tau_1 = (n_1, n_1)$, $C_{\tau_1} = i_5 + i_2 \cdot i_3 + i_2 \cdot \overline{i_4} + i_3 \cdot \overline{i_1} + \overline{i_1} \cdot \overline{i_4}$, $\tau_2 = (n_1, n_2, n_3, n_4, n_6, n_1, n_2)$, $C_{\tau_2} = \overline{i_5} \cdot (i_1 + i_4) \cdot (i_1 + \overline{i_3}) \cdot (i_4 + \overline{i_2}) \cdot (\overline{i_2} + \overline{i_3})$. τ_2 does not converge meaning that for any set of input values I_v that satisfies C_{τ_2} the nodes of τ_2 are visited indefinitely which causes oscillating signals in the controller.

6 Discussion

In this paper, we proposed a formal model, called the *SGST* graph, representing the possible states of a controller programmed based on a logical diagram specification. We show how to transform the logical diagram into the corresponding *SGST* graph, and how to verify the convergence property, i.e. verify that the controller does not have undesired oscillatory behavior.

Making sure that the behavior described by the logical diagram converges is crucial for test generation and for the overall verification and validation procedure. However, this is not the sole goal of transforming logical diagrams into *SGST* graphs. We developed the *SGST* to take a step in the application of the existing formal testing methods on logical diagrams. For the time being, generating tests derived from logical diagram specifications of power plants logical controllers is still handled manually or simulation-based. So, we designed the *SGST* to provide an explicit formal and exhaustive representation of the evolution steps between possible states of the controller described by a logical diagram. A test scenario is a sequence of these steps which are modeled with edges in the

SGST. Therefore, the test sequences generation could be transposed into the application of existing graph traversal techniques such as the Chinese postman tour [14]. The existing test generation [5,13] and selection techniques are based on finite state machines specifications. We consider the *SGST* to be an important intermediate step to move from non-formal diagrams to state machines. A coded solution of the developed method has shown that the construction of the *SGST* is feasible for small to medium sized logical diagrams and provides very accurate graph representations. However, the *SGST* generation for diagrams with tens of memory and timer blocks is quite expensive in terms of complexity. This is due to the number of maximal states being in the worst case exponential to the number of blocks which can quickly lead to an explosion of the graph size.

We are currently working on methods to resolve the complexity problems such as the partition of the diagram into sub-diagrams and transforming them in *SGSTs* then synchronizing them. Further work on complexity as well as transforming *SGST* graphs into state machines to apply test generation results in the literature for controllers described with logical diagrams will be handled in the future.

References

1. Alur, R., Dill, D.L.: A theory of timed automata. Theor. Comput. Sci. **126**(2), 183–235 (1994)
2. Fayyazi, M., Kirsch, L.: Efficient simulation of oscillatory combinational loops. In: Proceedings of the 47th Design Automation Conference, pp. 777–780 (2010)
3. Jean-françois Hery, J.C.L.: Stabilité de la spécification logique du contrôle-commande - méthodologie et mise en œuvre. Technical report, EDF R&D (2019)
4. Power Plants IEC: Instrumentation and control important to safety-general requirements for systems. IEC 61513. International Electrotechnical Commission (2011)
5. Lee, D., Yannakakis, M.: Principles and methods of testing finite state machines-a survey. Proc. IEEE **84**(8), 1090–1123 (1996)
6. Lukoschus, J., Von Hanxleden, R.: Removing cycles in Esterel programs. EURASIP J. Embed. Syst. **2007**, 1–23 (2007)
7. Malik, S.: Analysis of cyclic combinational circuits. IEEE Trans. Comput. Aided Des. Integr. Circ. Syst. **13**(7), 950–956 (1994)
8. Neiroukh, O., Edwards, S., Song, X.: An efficient algorithm for the analysis of cyclic circuits, vol. 2006, p. 6–pp, April 2006. https://doi.org/10.1109/ISVLSI.2006.18
9. Peled, D., Vardi, M.Y., Yannakakis, M.: Black box checking. In: Wu, J., Chanson, S.T., Gao, Q. (eds.) Formal Methods for Protocol Engineering and Distributed Systems. IAICT, vol. 28, pp. 225–240. Springer, Boston (1999). https://doi.org/10.1007/978-0-387-35578-8_13
10. Provost, J., Roussel, J.M., Faure, J.M.: Translating Grafcet specifications into mealy machines for conformance test purposes. Control. Eng. Pract. **19**(9), 947–957 (2011)
11. Riedel, M.D.: Cyclic combinational circuits. California Institute of Technology (2004)

12. Shiple, T.R., Berry, G., Touati, H.: Constructive analysis of cyclic circuits. In: Proceedings ED&TC European Design and Test Conference, pp. 328–333 (1996)
13. Springintveld, J., Vaandrager, F., D'Argenio, P.R.: Testing timed automata. Theor. Comput. Sci. **254**(1–2), 225–257 (2001)
14. Thimbleby, H.: The directed Chinese postman problem. Softw. Pract. Exp. **33**(11), 1081–1096 (2003)

Database Repair via Event-Condition-Action Rules in Dynamic Logic

Guillaume Feuillade, Andreas Herzig(✉), and Christos Rantsoudis

IRIT, Univ. Paul Sabatier, CNRS, Toulouse, France
andreas.herzig@irit.fr

Abstract. Event-condition-action (ECA) rules equip a database with information about preferred ways to restore integrity. They face problems of non-terminating executions and only procedural semantics had been given to them up to now. Declarative semantics however exist for a particular class of ECA rules lacking the event argument, called active integrity constraints. We generalise one of these semantics to ECA rules and couch it in a simple dynamic logic with deterministic past.

Keywords: ECA rules · database repair · dynamic logic

1 Introduction

To restore the integrity of a database violating some constraints is an old and notoriously difficult problem. One of the main difficulties is that there are typically several possible repairs: the integrity constraints alone do not provide enough clues which the 'right' repair is. A natural idea is to add more information to the integrity constraints. The most influential proposal was to move from classical, static integrity constraints to so-called Event-Condition-Action (ECA) rules. Such rules indicate how integrity should be restored: when the *event* occurs and the *condition* is satisfied then the *action* is triggered which, intuitively, makes the violating condition false [4,13,15]. A relational database together with a set of ECA rules make up an *active database* [13,24]. Such databases were studied intensely in the database literature in the last four decades. The existing semantics are mainly procedural and chain rule applications: the actions of some ECA rule trigger other ECA rules, and so on. As argued in [7], *"their lack of declarative semantics makes it difficult to understand the behavior of multiple ECAs acting together and to evaluate rule-processing algorithms in a principled way"*. They are moreover plagued by termination problems.

Up to now only few declarative, logical semantics were given to ECA rules. Most of them adopted the logic programming paradigm [6,14,19,22,23]; Bertossi and Pinto considered the Situation Calculus where ECA rules are described in a first-order language plus one second-order axiom and where repairs are performed by means of auxiliary actions [5]. Our aim in this paper is to associate

I. Varzinczak (Ed.): FoIKS 2022, LNCS 13388, pp. 75–92, 2022.
https://doi.org/10.1007/978-3-031-11321-5_5

a semantics to ECA rules that is based in dynamic logic. A main reason for our choice is that the latter has built-in constructions allowing us to reason about terminating executions. Another reason is that this allows us to start from existing declarative semantics for a simplified version of ECA rules. The latter lack the event argument and are simply condition-action couples and were investigated in the database and AI literature since more than 15 years under the denomination *active integrity constraints*, abbreviated AIC [7,10–12,20]. AICs *"encode explicitly both an integrity constraint and preferred basic actions to repair it, if it is violated"* [12]. Their semantics notably avoids problems with non-terminating executions that plague ECA rules. Syntactically, an AIC is a couple of the form

$$r = \langle \mathbf{C}(r), \mathbf{A}(r) \rangle$$

where the condition $\mathbf{C}(r)$ is a conjunction of literals $L_1 \wedge \cdots \wedge L_n$ (that we can think of as the negation of a classical, static integrity constraint) and the action $\mathbf{A}(r)$ is a set of assignments of the form either $+p$ or $-p$. It is supposed that each of the elements of $\mathbf{A}(r)$ makes one of the literals false: $+p \in \mathbf{A}(r)$ implies $\neg p \in \mathbf{C}(r)$, and $-p \in \mathbf{A}(r)$ implies $p \in \mathbf{C}(r)$. For instance, the AIC $\langle$`Bachelor` $\wedge$ `Married`, `-Bachelor`$\rangle$ says that when `Bachelor` $\wedge$ `Married` is true then the action `-Bachelor` should be executed, that is, `Bachelor` should be deleted. Given a set of AICs, a *repair* of a database $\mathcal{D}$ is a set of AIC actions whose execution produces a database that is consistent with the integrity constraints. Their semantics is clear when there is only one AIC: if the set of AICs is the singleton $R = \{\langle \mathbf{C}(r), \mathbf{A}(r) \rangle\}$ and $\mathcal{D} \models \mathbf{C}(r)$ then each assignment $\alpha \in \mathbf{A}(r)$ is a possible repair of $\mathcal{D}$. When there are several AICs then their actions may interfere in complex ways, just as ECA rules do: the execution of $\mathbf{A}(r)$ makes $\mathbf{C}(r)$ false but may have the side effect that the constraint part $\mathbf{C}(r')$ of another AIC r' that was false before the execution becomes satisfied, and so on. The consistency restoring actions should therefore be arranged in a way such that overall consistency is obtained. Moreover, the database should be changed minimally only. Several semantics achieving this were designed, among which preferred, founded, and justified repairs [12] as well as more recent prioritised versions [8]. Implementations of such repairs were also studied, and the account was extended in order to cope with various applications [16–18].

In this paper we examine whether and how the existing logical semantics of AICs can be generalised to ECA rules. For that reason (and also because the term 'rule' has a procedural connotation) we will henceforth use the term *ECA constraints* instead of ECA rules. Syntactically, we consider ECA constraints of the form

$$\kappa = \langle \mathbf{E}(\kappa), \mathbf{C}(\kappa), \mathbf{A}(\kappa) \rangle$$

where $\langle \mathbf{C}(\kappa), \mathbf{A}(\kappa) \rangle$ is an AIC and $\mathbf{E}(\kappa)$ is a boolean formula built from assignments of the form either $+p$ or $-p$. The latter describes the assignments that must have occurred for the AIC to trigger. Our convention is to call *events* assignments that happened in the past and led to a violation state, and to call *actions* assignments to be performed in the future in order to repair the violation.

Example 1. Let $\mathtt{em}_{e,d}$ stand for "e is an employee of department d". The set of ECA constraints

$$\mathcal{K}_{\mathtt{em}} = \{ \langle +\mathtt{em}_{e,d_1}, \mathtt{em}_{e,d_1} \wedge \mathtt{em}_{e,d_2}, \{-\mathtt{em}_{e,d_2}\} \rangle, \\ \langle +\mathtt{em}_{e,d_2}, \mathtt{em}_{e,d_1} \wedge \mathtt{em}_{e,d_2}, \{-\mathtt{em}_{e,d_1}\} \rangle \}$$

implements a 'priority to the input' policy repairing the functionality constraint for the employee relation: when the integrity constraint $\neg\mathtt{em}_{e,d_1} \vee \neg\mathtt{em}_{e,d_2}$ is violated and $\mathtt{em}_{e,d_1}$ was made true in the last update action then the latter is retained and $\mathtt{em}_{e,d_2}$ is made false in order to enforce the constraint; symmetrically, $\mathtt{em}_{e,d_1}$ is made false when $+\mathtt{em}_{e,d_2}$ is part of the last update action.

Observe that if the last update action of our example made neither $\mathtt{em}_{e,d_1}$ nor $\mathtt{em}_{e,d_2}$ true, i.e., if $\neg+\mathtt{em}_{e,d_1} \wedge \neg+\mathtt{em}_{e,d_2}$ holds, then the database already violated the integrity constraints before the last update action. In such cases ECA constraints are of no help: the violation should have been repaired earlier. Moreover, if the last update action of our example made $\mathtt{em}_{e,d_1}$ and $\mathtt{em}_{e,d_2}$ simultaneously true, i.e., if $+\mathtt{em}_{e,d_1} \wedge +\mathtt{em}_{e,d_2}$ holds, then the ECA constraints of $\mathcal{K}_{\mathtt{em}}$ authorise both $-\mathtt{em}_{e,d_1}$ and $-\mathtt{em}_{e,d_2}$ as legal repairs.

A logical account of ECA constraints not only requires reasoning about integrity constraints and actions to repair them when violated, but also reasoning about the event bringing about the violation. Semantically, we have to go beyond the models of AIC-based repairs, which are simply classical valuations (alias databases): in order to account for the last update action we have to add information about the events that led to the present database state. Our models are therefore couples made up of a classical valuation $\mathcal{D}$ and a set of assignments $\mathcal{H}$. The intuition is that the assignments in $\mathcal{H}$ are among the set of assignments that led to $\mathcal{D}$; that is, $\mathcal{H}$ is part of the last update action that took place and brought the database into its present state $\mathcal{D}$. For these models, we show that the AIC definitions of founded repair and well-founded repair can be generalised in a natural way to ECA constraints. We then provide a dynamic logic analysis of ECA constraints. More precisely, we resort to a dialect of dynamic logic: Dynamic Logic of Propositional Assignments DL-PA [2,3]. In order to take events into account we have to extend the basic logic by connectives referring to a deterministic past. We are going to show that this can be done without harm, i.e., without modifying the formal properties of the logic; in particular, satisfiability, validity and model checking stay in PSPACE. Our extension is appropriate to reason about ECA-based repairs: for several definitions of ECA repairs we give DL-PA programs whose executability characterises the existence of a repair (Theorems 1–4). Furthermore, several other interesting reasoning problems can be expressed in DL-PA, such as uniqueness of a repair or equivalence of two different sets of constraints.

Just as most of the papers in the AIC literature we rely on grounding and restrict our presentation to the propositional case. This simplifies the presentation and allows us to abstract away from orthogonal first-order aspects. A full-fledged analysis would require a first-order version of DL-PA, which is something that has not been studied yet.

The paper is organised as follows. In Sect. 2 we recall AICs and their semantics. We then define ECA constraints (Sect. 3) and their semantics in terms of databases with histories (Sect. 4). In Sect. 5 we introduce a version of dynamic logic with past and in Sect. 6 we capture well-founded ECA repairs in DL-PA. In Sect. 7 we show that other interesting decision problems can be expressed. Section 8 concludes. Proofs are contained in the long version.[1]

2 Background: AICs and Their Semantics

We suppose given a set of propositional variables $\mathbb{P}$ with typical elements $p, q, \ldots$ Just as most of the papers in the AIC literature we suppose that $\mathbb{P}$ is finite. A *literal* is an element of $\mathbb{P}$ or a negation thereof. A *database*, alias a *valuation*, is a set of propositional variables $\mathcal{D} \subseteq \mathbb{P}$.

An *assignment* is of the form either $+p$ or $-p$, for $p \in \mathbb{P}$. The former sets p to true and the latter sets p to false. We use $\alpha, \alpha', \ldots$ for assignments. For every α, the assignment $\bar{\alpha}$ is the opposite assignment, formally defined by $\overline{+p} = -p$ and $\overline{-p} = +p$. The set of all assignments is

$$\mathbb{A} = \{+p \ : \ p \in \mathbb{P}\} \cup \{-p \ : \ p \in \mathbb{P}\}.$$

An *update action* is some subset of $\mathbb{A}$. For every update action $A \subseteq \mathbb{A}$ we define the set of variables it makes true and the set of variables it makes false:

$$A^+ = \{p \ : \ +p \in A\},$$
$$A^- = \{p \ : \ -p \in A\}.$$

An update action A is *consistent* if $A^+ \cap A^- = \emptyset$. It is *relevant w.r.t. a database* $\mathcal{D}$ if for every $p \in \mathbb{P}$, if $+p \in A$ then $p \notin \mathcal{D}$ and if $-p \in A$ then $p \in \mathcal{D}$. Hence relevance implies consistency. The *update* of a database $\mathcal{D}$ by an update action A is a partial function $\circ$ that is defined if and only if A is consistent, and if so then

$$\mathcal{D} \circ A = (\mathcal{D} \setminus A^-) \cup A^+.$$

Proposition 1. *Let $A = \{\alpha_1, \ldots, \alpha_n\}$ be a consistent update action. Then $\mathcal{D} \circ A = \mathcal{D} \circ \{\alpha_1\} \circ \cdots \circ \{\alpha_n\}$, for every ordering of the α_i.*

An *active integrity constraint* (AIC) is a couple $r = \langle \mathbf{C}(r), \mathbf{A}(r) \rangle$ where $\mathbf{C}(r)$ is a conjunction of literals and $\mathbf{A}(r) \subseteq \mathbb{A}$ is an update action such that if $+p \in \mathbf{A}(r)$ then $\neg p \in \mathbf{C}(r)$ and if $-p \in \mathbf{A}(r)$ then $p \in \mathbf{C}(r)$. The condition $\mathbf{C}(r)$ can be thought of as the negation of an integrity constraint. For example, in $r = \langle \texttt{Bachelor} \wedge \texttt{Married}, -\texttt{Bachelor} \rangle$ the first element is the negation of the integrity constraint $\neg\texttt{Bachelor} \vee \neg\texttt{Married}$ and the second element indicates the way its violation should be repaired, namely by deleting `Bachelor`.

[1] https://www.irit.fr/~Andreas.Herzig/P/Foiks22Long.pdf.

In the rest of the section we recall several definitions of repairs via a given set of AICs R. The formula

$$\mathsf{OK}(R) = \bigwedge_{r \in R} \neg \mathbf{C}(r)$$

expresses that none of the AICs in R is applicable: all the static constraints hold. When $\mathsf{OK}(R)$ is false then the database has to be repaired.

The first two semantics do not make any use of the active part of AICs. First, an update action A is a *weak repair* of a database $\mathcal{D}$ via a set of AICs R if it is relevant w.r.t. $\mathcal{D}$ and $\mathcal{D} \circ A \models \mathsf{OK}(R)$. The latter means that A makes the static constraints true. Second, A is a *minimal repair* of $\mathcal{D}$ via R if it is a weak repair that is set inclusion minimal: there is no weak repair A' of $\mathcal{D}$ via R such that $A' \subset A$.[2] The remaining semantics all require that each of the assignments of an update action is supported by an AIC in some way.

First, an update action A is a *founded weak repair* of a database $\mathcal{D}$ via a set of AICs R if it is a weak repair of $\mathcal{D}$ via R and for every $\alpha \in A$ there is an $r \in R$ such that

- $\alpha \in \mathbf{A}(r)$,
- $\mathcal{D} \circ (A \setminus \{\alpha\}) \models \mathbf{C}(r)$.

Second, A is a *founded repair* if it is both a founded weak repair and a minimal repair [10].[3]

A weak repair A of $\mathcal{D}$ via R is *well-founded* [7] if there is a sequence $\langle \alpha_1, \dots, \alpha_n \rangle$ such that $A = \{\alpha_1, \dots, \alpha_n\}$ and for every α_i, $1 \leq i \leq n$, there is an $r_i \in R$ such that

- $\alpha_i \in \mathbf{A}(r_i)$,
- $\mathcal{D} \circ \{\alpha_1, \dots, \alpha_{i-1}\} \models \mathbf{C}(r_i)$.

A *well-founded repair* is a well-founded weak repair that is also a minimal repair.[4]

For example, for $\mathcal{D} = \emptyset$ and $R = \{\langle \neg p \wedge \neg q, \{+q\} \rangle\}$ any update action $A \subseteq \{+p \ : \ p \in \mathbb{P}\}$ containing $+p$ or $+q$ is a weak repair of $\mathcal{D}$ via R; the minimal repairs of $\mathcal{D}$ via R are $\{+p\}$ and $\{+q\}$; and the only founded and well-founded repair of $\mathcal{D}$ via R is $\{+q\}$. An example where founded repairs and well-founded repairs behave differently is $\mathcal{D} = \emptyset$ and $R = \{r_1, r_2, r_3\}$ with

$$r_1 = \langle \neg p \wedge \neg q, \emptyset \rangle,$$
$$r_2 = \langle \neg p \wedge q, \{+p\} \rangle,$$
$$r_3 = \langle p \wedge \neg q, \{+q\} \rangle.$$

[2] Most papers in the AIC literature call these simply *repairs*, but we prefer our denomination because it avoids ambiguities.

[3] Note that a founded repair is different from a minimal founded weak repair: whereas the latter is guaranteed to exist in case a founded weak repair does, the former is not; in other words, it may happen that founded weak repairs exist but founded repairs do not (because none of the founded weak repairs happens to also be a minimal repair in the traditional sense).

[4] Again, the existence of a well-founded weak repair does not guarantee the existence of a well-founded repair.

Then $A = \{+p, +q\}$ is a founded repair: the third AIC supports $+q$ because $\mathcal{D} \circ \{+p\} \models \mathbf{C}(r_3)$; and the second AIC supports $+p$ because $\mathcal{D} \circ \{+q\} \models \mathbf{C}(r_2)$. However, there is no well-founded weak repair because the only applicable rule from $\mathcal{D}$ (viz. the first one) has an empty active part.

Two other definitions of repairs are prominent in the literature. *Grounded repairs* generalise founded repairs [7]. As each of them is both a founded and well-founded repair, we skip this generalisation. We also do not present *justified repairs* [12]: their definition is the most complex one and it is not obvious how to transfer it to ECA constraints. Moreover, as argued in [7] they do not provide the intuitive repairs at least in some cases.

3 ECA Constraints

We consider two kinds of *boolean formulas*. The first, static language $\mathcal{L}_{\mathbb{P}}$ is built from the variables in $\mathbb{P}$ and describes the condition part of ECAs. The second, dynamic language $\mathcal{L}_{\mathbb{A}}$ is built from the set of assignments $\mathbb{A}$ and describes the event part of ECAs, i.e., the last update action that took place. The grammars for the two languages are therefore

$$\mathcal{L}_{\mathbb{P}} : \varphi ::= p \mid \neg\varphi \mid \varphi \vee \varphi,$$
$$\mathcal{L}_{\mathbb{A}} : \varphi ::= \alpha \mid \neg\varphi \mid \varphi \vee \varphi,$$

where p ranges over $\mathbb{P}$ and α over the set of all assignments $\mathbb{A}$. We call the elements of $\mathbb{A}$ *history atoms.*

An *event-condition-action (ECA) constraint* combines an event description in $\mathcal{L}_{\mathbb{A}}$, a condition description in $\mathcal{L}_{\mathbb{P}}$, and an update action in $2^{\mathbb{A}}$: it is a triple

$$\kappa = \langle \mathbf{E}(\kappa), \mathbf{C}(\kappa), \mathbf{A}(\kappa) \rangle$$

where $\mathbf{E}(\kappa) \in \mathcal{L}_{\mathbb{A}}$, $\mathbf{C}(\kappa) \in \mathcal{L}_{\mathbb{P}}$, and $\mathbf{A}(\kappa) \subseteq \mathbb{A}$. The event part of an ECA constraint describes the last update action. It does so only partially: only events that are relevant for the triggering of the rule are described.[5]

Sets of ECA constraints are noted $\mathcal{K}$. We take over from AICs the formula $\mathsf{OK}(\mathcal{K})$ expressing that none of the ECA constraints is applicable:

$$\mathsf{OK}(\mathcal{K}) = \bigwedge_{\kappa \in \mathcal{K}} \neg\mathbf{C}(\kappa).$$

Example 2. [20, Example 4.6] Every manager of a project carried out by a department must be an employee of that department; if employee e just became the manager of project p or if the project was just assigned to department d_1 then the constraint should be repaired by making e a member of d_1. Moreover, if e has just been removed from d_1 then the project should either be removed from d_1, too, or should get a new manager. Together with the ECA version of

[5] We observe that in the literature, $\mathbf{C}(\kappa)$ is usually restricted to conjunctions of literals. Our account however does not require this.

the functionality constraint on the em relation of Example 1 we obtain the set of ECA constraints

$$\begin{aligned}\mathcal{K}_{\mathtt{mg}} = \{ &\langle +\mathtt{em}_{e,d_1}, \mathtt{em}_{e,d_1} \wedge \mathtt{em}_{e,d_2}, \{-\mathtt{em}_{e,d_2}\}\rangle,\\ &\langle +\mathtt{em}_{e,d_2}, \mathtt{em}_{e,d_1} \wedge \mathtt{em}_{e,d_2}, \{-\mathtt{em}_{e,d_1}\}\rangle,\\ &\langle +\mathtt{mg}_{e,p} \vee +\mathtt{pr}_{p,d_1}, \mathtt{mg}_{e,p} \wedge \mathtt{pr}_{p,d_1} \wedge \neg\mathtt{em}_{e,d_1}, \{+\mathtt{em}_{e,d_1}\}\rangle,\\ &\langle -\mathtt{em}_{e,d_1}, \mathtt{mg}_{e,p} \wedge \mathtt{pr}_{p,d_1} \wedge \neg\mathtt{em}_{e,d_1}, \{-\mathtt{mg}_{e,p}, -\mathtt{pr}_{p,d_1}\}\rangle \}.\end{aligned}$$

Example 3. Suppose some IoT device functions (fun) if and only if the battery is loaded (bat) and the wiring is in order (wire); this is captured by the equivalence $\mathtt{fun} \leftrightarrow (\mathtt{bat} \wedge \mathtt{wire})$, or, equivalently, by the three implications $\mathtt{fun} \to \mathtt{bat}$, $\mathtt{fun} \to \mathtt{wire}$, and $(\mathtt{bat} \wedge \mathtt{wire}) \to \mathtt{fun}$. If we assume a 'priority to the input' repair strategy then the set of ECA constraints has to be

$$\begin{aligned}\mathcal{K}_{\mathtt{iot}} = \{ &\langle +\mathtt{fun}, \mathtt{fun} \wedge \neg\mathtt{bat}, \{+\mathtt{bat}\}\rangle,\\ &\langle -\mathtt{bat}, \mathtt{fun} \wedge \neg\mathtt{bat}, \{-\mathtt{fun}\}\rangle,\\ &\langle +\mathtt{fun}, \mathtt{fun} \wedge \neg\mathtt{wire}, \{+\mathtt{wire}\}\rangle,\\ &\langle -\mathtt{wire}, \mathtt{fun} \wedge \neg\mathtt{wire}, \{-\mathtt{fun}\}\rangle,\\ &\langle +\mathtt{bat}, \mathtt{bat} \wedge \mathtt{wire} \wedge \neg\mathtt{fun}, \{+\mathtt{fun}\}\rangle,\\ &\langle +\mathtt{wire}, \mathtt{bat} \wedge \mathtt{wire} \wedge \neg\mathtt{fun}, \{+\mathtt{fun}\}\rangle,\\ &\langle -\mathtt{fun}, \mathtt{bat} \wedge \mathtt{wire} \wedge \neg\mathtt{fun}, \{-\mathtt{bat}, -\mathtt{wire}\}\rangle \}.\end{aligned}$$

Observe that when the last update action made bat and wire simultaneously true then $\mathcal{K}_{\mathtt{iot}}$ allows to go either way. We can exclude this and force the repair to fail in that case by refining the last three ECA constraints to

$$\begin{aligned}&\langle +\mathtt{bat} \wedge +\mathtt{wire} \wedge -\mathtt{fun}, \mathtt{bat} \wedge \mathtt{wire} \wedge \neg\mathtt{fun}, \emptyset\rangle,\\ &\langle +\mathtt{bat} \wedge \neg -\mathtt{fun}, \mathtt{bat} \wedge \mathtt{wire} \wedge \neg\mathtt{fun}, \{+\mathtt{fun}\}\rangle,\\ &\langle +\mathtt{wire} \wedge \neg -\mathtt{fun}, \mathtt{bat} \wedge \mathtt{wire} \wedge \neg\mathtt{fun}, \{+\mathtt{fun}\}\rangle,\\ &\langle +\mathtt{bat} \wedge \neg +\mathtt{wire} \wedge -\mathtt{fun}, \mathtt{bat} \wedge \mathtt{wire} \wedge \neg\mathtt{fun}, \{-\mathtt{wire}\}\rangle,\\ &\langle +\mathtt{wire} \wedge \neg +\mathtt{bat} \wedge -\mathtt{fun}, \mathtt{bat} \wedge \mathtt{wire} \wedge \neg\mathtt{fun}, \{-\mathtt{bat}\}\rangle.\end{aligned}$$

4 Semantics for ECA Constraints

As ECA constraints refer to the past, their semantics requires more than just valuations, alias databases: we have to add the immediately preceding update action that caused the present database state. Based on such models, we examine several definitions of AIC-based repairs in view of extending them to ECA constraints.

4.1 Databases with History

A *database with history*, or h-database for short, is a couple $\Delta = \langle \mathcal{D}, \mathcal{H} \rangle$ made up of a valuation $\mathcal{D} \subseteq \mathbb{P}$ and an update action $\mathcal{H} \subseteq \mathbb{A}$ such that $\mathcal{H}^+ \subseteq \mathcal{D}$ and

$\mathcal{H}^- \cap \mathcal{D} = \emptyset$. The intuition is that $\mathcal{H}$ is the most recent update action that brought the database into the state $\mathcal{D}$. This explains the two above conditions: they guarantee that if $+p$ is among the last assignments in $\mathcal{H}$ then $p \in \mathcal{D}$ and if $-p$ is among the last assignments in $\mathcal{H}$ then $p \notin \mathcal{D}$. Note that thanks to this constraint $\mathcal{H}$ is consistent: $\mathcal{H}^+$ and $\mathcal{H}^-$ are necessarily disjoint.

The *update* of a history $\mathcal{H}$ by an event A is defined as

$$\mathcal{H} \circ A = A \cup \{\alpha \in \mathcal{H} \ : \ A \cup \{\alpha\} \text{ is consistent}\}.$$

For example, $\{+p, +q\} \circ \{-p\} = \{-p, +q\}$. The history $\mathcal{H} \circ A$ is consistent if and only if A and $\mathcal{H}$ are both consistent.

Let $\mathbb{M}$ be the class of all couples $\Delta = \langle \mathcal{D}, \mathcal{H} \rangle$ with $\mathcal{D} \subseteq \mathbb{P}$ and $\mathcal{H} \subseteq \mathbb{A}$ such that $\mathcal{H}^+ \subseteq \mathcal{D}$ and $\mathcal{H}^- \cap \mathcal{D} = \emptyset$. We interpret the formulas of $\mathcal{L}_{\mathbb{P}}$ in $\mathcal{D}$ and those of $\mathcal{L}_{\mathbb{A}}$ in $\mathcal{H}$, in the standard way. For example, consider the h-database $\langle \mathcal{D}, \mathcal{H} \rangle$ with $\mathcal{D} = \{\mathtt{em}_{e,d_1}, \mathtt{em}_{e,d_2}\}$ and $\mathcal{H} = \{+\mathtt{em}_{e,d_1}\}$. Then we have $\mathcal{D} \models \mathtt{em}_{e,d_1} \wedge \neg \mathtt{em}_{e,d_3}$ and $\mathcal{H} \models +\mathtt{em}_{e,d_1} \wedge \neg{+\mathtt{em}_{e,d_2}}$.

Now that we can interpret the event part and the condition part of an ECA constraint, it remains to interpret the action part. This is more delicate and amounts to designing the semantics of repairs based on ECA constraints. In the rest of the section we examine several possible definitions. For a start, we take over the two most basic definitions of repairs from AICs: weak repairs and minimal repairs of an h-database $\langle \mathcal{D}, \mathcal{H} \rangle$ via a set of ECA constraints $\mathcal{K}$.

- An update action A is a *weak repair* of $\langle \mathcal{D}, \mathcal{H} \rangle$ via $\mathcal{K}$ if A is relevant w.r.t. $\mathcal{D}$ and $\mathcal{D} \circ A \models \mathsf{OK}(\mathcal{K})$;
- An update action A is a *minimal repair* of $\langle \mathcal{D}, \mathcal{H} \rangle$ via $\mathcal{K}$ if A is a weak repair that is set inclusion minimal: there is no weak repair A' of $\langle \mathcal{D}, \mathcal{H} \rangle$ via $\mathcal{K}$ such that $A' \subset A$.

4.2 Founded and Well-Founded ECA Repairs

A straightforward adaption of founded repairs to ECA constraints goes as follows. Suppose given a candidate repair $A \subseteq \mathbb{A}$. For an ECA constraint κ to support an assignment $\alpha \in A$ given an h-database $\langle \mathcal{D}, \mathcal{H} \rangle$, the constraint $\mathbf{E}(\kappa)$ about the immediately preceding event should be satisfied by $\mathcal{H}$ together with the rest of changes imposed by A, i.e., we should also have $\mathcal{H} \circ (A \setminus \{\alpha\}) \models \mathbf{E}(\kappa)$.[6] This leads to the following definition: a weak repair A is a *founded weak ECA repair* of $\langle \mathcal{D}, \mathcal{H} \rangle$ via a set of ECAs $\mathcal{K}$ if for every $\alpha \in A$ there is a $\kappa \in \mathcal{K}$ such that

- $\alpha \in \mathbf{A}(\kappa)$,
- $\mathcal{D} \circ (A \setminus \{\alpha\}) \models \mathbf{C}(\kappa)$,
- $\mathcal{H} \circ (A \setminus \{\alpha\}) \models \mathbf{E}(\kappa)$.

[6] A naive adaption would only require $\mathcal{H} \models \mathbf{E}(\kappa)$. However, the support for α would be much weaker; see also the example below.

Once again, a *founded ECA repair* is a founded weak ECA repair that is also a minimal repair.

Moving on to well-founded repairs, an appropriate definition of ECA-based repairs should not only check the condition part of constraints in a sequential way, but should also do so for their triggering event part. Thus we get the following definition: a weak repair A of $\langle \mathcal{D}, \mathcal{H} \rangle$ via the set of ECA constraints $\mathcal{K}$ is a *well-founded weak ECA repair* if there is a sequence of assignments $\langle \alpha_1, \ldots, \alpha_n \rangle$ such that $A = \{\alpha_1, \ldots, \alpha_n\}$ and such that for every α_i, $1 \leq i \leq n$, there is a $\kappa_i \in \mathcal{K}$ such that

- $\alpha_i \in \mathbf{A}(\kappa_i)$,
- $\mathcal{D} \circ \{\alpha_1, \ldots, \alpha_{i-1}\} \models \mathbf{C}(\kappa_i)$,[7]
- $\mathcal{H} \circ \{\alpha_1\} \circ \cdots \circ \{\alpha_{i-1}\} \models \mathbf{E}(\kappa_i)$.

As always, a *well-founded ECA repair* is defined as a well-founded weak ECA repair that is also a minimal repair.

Example 4 (Example 2, ctd.). Consider the ECA constraints $\mathcal{K}_{\mathtt{mg}}$ of Example 2 and the h-database $\langle \mathcal{D}, \mathcal{H} \rangle$ with $\mathcal{D} = \{\mathtt{mg}_{e,p}, \mathtt{pr}_{p,d_1}, \mathtt{em}_{e,d_2}\}$ and $\mathcal{H} = \{+\mathtt{mg}_{e,p}\}$, that is, e just became manager of project p. There is only one intended repair: $A = \{+\mathtt{em}_{e,d_1}, -\mathtt{em}_{e,d_2}\}$. Based on the above definitions it is easy to check that A is a founded ECA repair (both assignments in A are sufficiently supported by $\mathcal{D}$ and $\mathcal{H}$) as well as a well-founded ECA repair of $\langle \mathcal{D}, \mathcal{H} \rangle$ via $\mathcal{K}_{\mathtt{mg}}$.

Remark 1. A well-founded weak ECA repair of $\langle \mathcal{D}, \mathcal{H} \rangle$ via $\mathcal{K}$ is also a weak repair. It follows that $\mathcal{D} \circ A \models \mathsf{OK}(\mathcal{K})$, i.e., the repaired database satisfies the static integrity constraints.

In the rest of the paper we undertake a formal analysis of ECA repairs in a version of dynamic logic. The iteration operator of the latter allows us to get close to the procedural semantics while discarding infinite runs. We follow the arguments in [7] against founded and justified AICs and focus on well-founded ECA repairs.

5 Dynamic Logic of Propositional Assignments with Deterministic Past

In the models of Sect. 4 we can actually interpret a richer language that is made up of boolean formulas built from $\mathbb{P} \cup \mathbb{A}$. We further add to that hybrid language modal operators indexed by programs. The resulting logic DL-PA$^{\pm}$ extends Dynamic Logic of Propositional Assignments DL-PA [2,3].

[7] This is equivalent to $\mathcal{D} \circ \{\alpha_1\} \circ \cdots \circ \{\alpha_{i-1}\} \models \mathbf{C}(\kappa_i)$ because A is consistent (cf. Proposition 1).

5.1 Language of DL-PA$^{\pm}$

The hybrid language $\mathcal{L}_{\text{DL-PA}^{\pm}}$ of DL-PA with past has two kinds of atomic formulas: propositional variables of the form p and history atoms of the form $+p$ and $-p$. Moreover, it has two kinds of expressions: formulas, noted φ, and programs, noted π. It is defined by the following grammar

$$\varphi ::= p \mid +p \mid -p \mid \neg\varphi \mid \varphi \vee \varphi \mid \langle\pi\rangle\varphi,$$
$$\pi ::= A \mid \pi;\pi \mid \pi \cup \pi \mid \pi^* \mid \varphi?,$$

where p ranges over $\mathbb{P}$ and A over subsets of $\mathbb{A}$. The formula $\langle\pi\rangle\varphi$ combines a formula φ and a program π and reads "π can be executed and the resulting database satisfies φ". Programs are built from sets of assignments by means of the program operators of PDL: $\pi_1;\pi_2$ is sequential composition, $\pi_1 \cup \pi_2$ is nondeterministic composition, π^* is unbounded iteration, and $\varphi?$ is test. Note that the expression $+p$ is a formula while the expression $\{+p\}$ is a program. The program $\{+p\}$ makes p true, while the formula $+p$ expresses that p has just been made true. The languages $\mathcal{L}_{\mathbb{P}}$ and $\mathcal{L}_{\mathbb{A}}$ of Sect. 3 are both fragments of the language of DL-PA$^{\pm}$.

For a set of propositional variables $P = \{p_1, \ldots, p_n\}$, we abbreviate the atomic program $\{-p_1, \ldots, -p_n\}$ by $-P$. The program π^+ abbreviates $\pi;\pi^*$. For $n \geq 0$, we define n-ary sequential composition $;_{i=1,\ldots,n} \pi_i$ by induction on n

$$;_{i=1,\ldots,0} \pi_i = \top?,$$
$$;_{i=1,\ldots,n+1} \pi_i = (;_{i=1,\ldots,n} \pi_i); \pi_{n+1}.$$

A particular case of such arbitrary sequences is n-times iteration of π, expressed as $\pi^n = ;_{i=1,\ldots,n} \pi$. We also define $\pi^{\leq n}$ as $(\top?{\cup}\pi)^n$. (It could as well be defined as $\bigcup_{0 \leq i \leq n} \pi^i$.) Finally and as usual in dynamic logic, the formula $[\pi]\varphi$ abbreviates $\neg\langle\pi\rangle\neg\varphi$.

Given a program π, $\mathbb{P}_\pi$ is the set of propositional variables occurring in π. For example, $\mathbb{P}_{+p \wedge \langle -q \rangle \neg -r?} = \{p, q, r\}$.

5.2 Semantics of DL-PA$^{\pm}$

The semantics is the same as that of ECA constraints, namely in terms of databases with history. The interpretation of a formula is a subset of the set of all h-databases $\mathbb{M}$, and the interpretation of a program is a relation on $\mathbb{M}$. More precisely, the interpretation of formulas is

$$\Delta \models p \;\text{ if }\; \Delta = \langle \mathcal{D}, \mathcal{H} \rangle \text{ and } p \in \mathcal{D},$$
$$\Delta \models +p \;\text{ if }\; \Delta = \langle \mathcal{D}, \mathcal{H} \rangle \text{ and } +p \in \mathcal{H},$$
$$\Delta \models -p \;\text{ if }\; \Delta = \langle \mathcal{D}, \mathcal{H} \rangle \text{ and } -p \in \mathcal{H},$$
$$\Delta \models \langle\pi\rangle\varphi \;\text{ if }\; \Delta ||\pi|| \Delta' \text{ and } \Delta' \models \varphi \text{ for some } \Delta',$$

and as expected for the boolean connectives; the interpretation of programs is

$$\begin{aligned}
\Delta||A||\Delta' &\text{ if } A \text{ is consistent, } \Delta = \langle \mathcal{D}, \mathcal{H} \rangle, \text{ and } \Delta' = \langle \mathcal{D} \circ A, \mathcal{H} \circ A \rangle, \\
\Delta||\pi_1;\pi_2||\Delta' &\text{ if } \Delta||\pi_1||\Delta'' \text{ and } \Delta''||\pi_2||\Delta' \text{ for some } \Delta'', \\
\Delta||\pi_1 \cup \pi_2||\Delta' &\text{ if } \Delta||\pi_1||\Delta' \text{ or } \Delta||\pi_2||\Delta', \\
\Delta||\pi^*||\Delta' &\text{ if } \Delta||\pi^n||\Delta' \text{ for some } n \geq 0, \\
\Delta||\varphi?||\Delta' &\text{ if } \Delta \models \varphi \text{ and } \Delta' = \Delta.
\end{aligned}$$

Hence the interpretation of a set of assignments A updates both the database $\mathcal{D}$ and the history $\mathcal{H}$ by A.[8]

We say that a program π is *executable* at an h-database Δ if $\Delta \models \langle \pi \rangle \top$, i.e., if there is an h-database Δ' such that $\Delta||\pi||\Delta'$. Clearly, consistency of $A \subseteq \mathbb{A}$ is the same as executability of the program A at every h-database Δ.

The definitions of validity and satisfiability in the class of models $\mathbb{M}$ are standard. For example, $+p \rightarrow p$ and $-p \rightarrow \neg p$ are both valid while the other direction is not, i.e., $p \wedge \neg +p$ and $\neg p \wedge \neg -p$ are both satisfiable.

Proposition 2. *The decision problems of DL-PA$^{\pm}$ satisfiability, validity and model checking are all PSPACE complete.*

6 Well-Founded ECA Repairs in DL-PA$^{\pm}$

We now show that well-founded weak ECA repairs and well-founded ECA repairs can be captured in DL-PA$^{\pm}$.

6.1 Well-Founded Weak ECA Repairs

Given a set of ECA constraints $\mathcal{K}$, our translation of well-founded weak ECA repairs uses fresh auxiliary propositional variables $\mathsf{D}(\alpha)$ recording that $\alpha \in \mathbb{A}$ has been executed during the (tentative) repair. For the set of assignments $A = \{\alpha_1, \ldots, \alpha_n\} \subseteq \mathbb{A}$, let the associated set of auxiliary propositional variables be $\mathsf{D}(A) = \{\mathsf{D}(\alpha) \ : \ \alpha \in A\}$. Then the program

$$-\mathsf{D}(A) = \{-\mathsf{D}(\alpha_1), \ldots, -\mathsf{D}(\alpha_n), -\mathsf{D}(\overline{\alpha}_1), \ldots, -\mathsf{D}(\overline{\alpha}_n)\}$$

initialises the auxiliary variables $\mathsf{D}(\alpha_i)$ and $\mathsf{D}(\overline{\alpha}_i)$ to false. Note that the propositional variable $\mathsf{D}(\alpha)$ is to be distinguished from the history atom $\alpha \in \mathbb{A}$ expressing that α was one of the last assignments that brought the database into the violation state.

[8] An alternative to the update of $\mathcal{H}$ would be to erase the history and replace it by A. Altogether, this would make us go from $\langle \mathcal{D}, \mathcal{H} \rangle$ to $\langle \mathcal{D} \circ A, A \rangle$. This would be appropriate in order to model external updates such as the update $\{+\mathtt{mg}_{e,p}\}$ that had occurred in Example 2 and brought the database into an inconsistent state, differing from the kind of updates occurring during the repair process that our definition accounts for.

The next step is to associate repair programs $\mathsf{rep}(\alpha)$ to assignments α. These programs check whether α is triggered by some ECA constraint κ and if so performs it. This involves some bookkeeping by means of the auxiliary variables $\mathsf{D}(\alpha)$, for two reasons: first, to make sure that none of the assignments $\alpha \in A$ is executed twice; second, to make sure that A is consistent w.r.t. α, in the sense that it does not contain both α and its opposite $\overline{\alpha}$.

$$\mathsf{rep}(\alpha) = \neg\mathsf{D}(\alpha) \wedge \neg\mathsf{D}(\overline{\alpha}) \wedge \bigvee_{\kappa \in \mathcal{K} \,:\, \alpha \in \mathbf{A}(\kappa)} (\mathbf{E}(\kappa) \wedge \mathbf{C}(\kappa))?; \{\alpha, +\mathsf{D}(\alpha)\}.$$

The program first performs a test: neither α nor its opposite $\overline{\alpha}$ has been done up to now (this is the conjunct $\neg\mathsf{D}(\alpha) \wedge \neg\mathsf{D}(\overline{\alpha})$) and there is an ECA constraint $\kappa \in \mathcal{K}$ with α in the action part such that the event description $\mathbf{E}(\kappa)$ and the condition $\mathbf{C}(\kappa)$ are both true (this is the conjunct $\bigvee_{\kappa \,:\, \alpha \in \mathbf{A}(\kappa)}(\mathbf{E}(\kappa) \wedge \mathbf{C}(\kappa))$). If that big test program succeeds then α is executed and this is stored by making $\mathsf{D}(\alpha)$ true (this is the update $\{\alpha, +\mathsf{D}(\alpha)\}$).

Theorem 1. *Let Δ be an h-database and $\mathcal{K}$ a set of ECA constraints. There exists a well-founded weak ECA repair of Δ via $\mathcal{K}$ if and only if the program*

$$\mathsf{rep}^{\mathsf{wwf}}_{\mathcal{K}} = -\mathsf{D}(\mathbb{A}); \Big(\bigcup_{\alpha \in \mathbb{A}} \mathsf{rep}(\alpha)\Big)^{*}; \mathsf{OK}(\mathcal{K})?$$

is executable at Δ.

The unbounded iteration $(\bigcup_{\alpha \in \mathbb{A}} \mathsf{rep}(\alpha))^*$ in our repair program $\mathsf{rep}^{\mathsf{wwf}}_{\mathcal{K}}$ can be replaced by $(\bigcup_{\alpha \in \mathbb{A}} \mathsf{rep}(\alpha))^{\leq \mathsf{card}(\mathbb{P})}$, i.e., the number of iterations of $\bigcup_{\alpha \in \mathbb{A}} \mathsf{rep}(\alpha)$ can be bound by the cardinality of $\mathbb{P}$. This is the case because well-founded ECA repairs are consistent update actions: each propositional variable can occur at most once in a well-founded ECA repair. This also holds for the other repair programs that we are going to define below.

Observe that finiteness of $\mathbb{P}$ is necessary for Theorem 1. This is not the case for the next result.

Theorem 2. *Let Δ be an h-database and $\mathcal{K}$ a set of ECA constraints. The set of assignments $A = \{\alpha_1, \ldots, \alpha_n\}$ is a well-founded weak ECA repair of Δ via $\mathcal{K}$ if and only if the program*

$$\mathsf{rep}^{\mathsf{wwf}}_{\mathcal{K}}(A) = -\mathsf{D}(A); \Big(\bigcup_{\alpha \in A} \mathsf{rep}(\alpha)\Big)^{*}; \mathsf{OK}(\mathcal{K})?; \bigwedge_{\alpha \in A} \mathsf{D}(\alpha)?$$

is executable at Δ.

The length of both repair programs is polynomial in the size of the set of ECA constraints and the cardinality of $\mathbb{P}$.

6.2 Well-Founded ECA Repairs

In order to capture well-founded ECA repairs we have to integrate a minimality check into its DL-PA$^{\pm}$ account. We take inspiration from the translation of Winslett's Possible Models Approach for database updates [25,26] into DL-PA of [21]. The translation associates to each propositional variable p a fresh copy p' whose role is to store the truth value of p. This is done before repair programs $\mathsf{rep}(+p)$ or $\mathsf{rep}(-p)$ are executed so that the initial value of p is remembered: once a candidate repair has been computed we check that it is minimal, in the sense that no consistent state can be obtained by modifying less variables.

For α being either $+p$ or $-p$, the programs $\mathtt{copy}(\alpha)$ and $\mathtt{undo}(\alpha)$ respectively copy the truth value of p into p' and, the other way round, copy back the value of p' into p. They are defined as follows

$$\begin{aligned}\mathtt{copy}(\alpha) &= (p?;+p') \cup (\neg p?;-p'),\\ \mathtt{undo}(\alpha) &= (p'?;+p) \cup (\neg p'?;-p).\end{aligned}$$

For example, the formulas $p \to \langle \mathtt{copy}(+p)\rangle(p \land p')$ and $\neg p \to \langle \mathtt{copy}(+p)\rangle(\neg p \land \neg p')$ are both valid, as well as $p' \to \langle \mathtt{undo}(+p)\rangle(p \land p')$ and $\neg p' \to \langle \mathtt{undo}(+p)\rangle(\neg p \land \neg p')$. Then the program

$$\begin{aligned}\mathsf{init}(\{\alpha_1,\dots,\alpha_n\}) = \{&-\mathsf{D}(\alpha_1),\dots,-\mathsf{D}(\alpha_n),\\ &-\mathsf{D}(\overline{\alpha}_1),\dots,-\mathsf{D}(\overline{\alpha}_n)\}; \mathtt{copy}(\alpha_1);\cdots;\mathtt{copy}(\alpha_n)\end{aligned}$$

initialises the values of both kinds of auxiliary variables: it resets all 'done' variables $\mathsf{D}(\alpha_i)$ and $\mathsf{D}(\overline{\alpha}_i)$ to false as before and moreover makes copies of all variables that are going to be assigned by α_i (where the order of the α_i does not matter).

Theorem 3. *Let Δ be an h-database and $\mathcal{K}$ a set of ECA constraints. There exists a well-founded ECA repair of Δ via $\mathcal{K}$ if and only if the program*

$$\begin{aligned}\mathsf{rep}^{\mathsf{wf}}_{\mathcal{K}} = \mathsf{init}(\mathbb{A}); \Big(\bigcup_{\alpha\in\mathbb{A}} \mathsf{rep}(\alpha)\Big)^*; \mathsf{OK}(\mathcal{K})?;\\ \neg\Big\langle\Big(\bigcup_{\alpha\in\mathbb{A}} \mathsf{D}(\alpha)?;\mathtt{undo}(\alpha)\Big)^+\Big\rangle \mathsf{OK}(\mathcal{K})?\end{aligned}$$

is executable at Δ.

Theorem 4. *Let Δ be an h-database and $\mathcal{K}$ a set of ECA constraints. The set of assignments $A = \{\alpha_1,\dots,\alpha_n\}$ is a well-founded ECA repair of Δ via $\mathcal{K}$ if and only if the program*

$$\begin{aligned}\mathsf{rep}^{\mathsf{wf}}_{\mathcal{K}}(A) = \mathsf{init}(A); \Big(\bigcup_{\alpha\in A} \mathsf{rep}(\alpha)\Big)^*; \mathsf{OK}(\mathcal{K})?; \Big(\bigwedge_{\alpha\in A} \mathsf{D}(\alpha)\Big)?;\\ \neg\Big\langle\Big(\bigcup_{\alpha\in A} \mathtt{undo}(\alpha)\Big)^+\Big\rangle \mathsf{OK}(\mathcal{K})?\end{aligned}$$

is executable at Δ.

The length of both repair programs is polynomial in the size of the set of ECA constraints and the cardinality of $\mathbb{P}$.

7 Other Decision Problems

In the present section we discuss how several other interesting decision problems related to well-founded weak ECA repairs and well-founded ECA repairs can be expressed. Let $\mathsf{rep}_{\mathcal{K}}$ stand for either $\mathsf{rep}_{\mathcal{K}}^{\mathsf{wwf}}$ or $\mathsf{rep}_{\mathcal{K}}^{\mathsf{wf}}$, i.e., the program performing well-founded (weak) ECA repairs according to the set of ECA constraints $\mathcal{K}$.

7.1 Properties of a Set of ECA Constraints

Here are three other decision problems about a given set of ECA constraints:

1. Is there a unique repair of Δ via $\mathcal{K}$?
2. Does every Δ have a unique repair via $\mathcal{K}$?
3. Does every Δ have a repair via $\mathcal{K}$?

Each of them can be expressed in DL-PA$^{\pm}$. First, we can verify whether there is a unique repair of Δ via $\mathcal{K}$ by checking whether the program $\mathsf{rep}_{\mathcal{K}}$ is deterministic. This can be done by model checking, for each of the variables $p \in \mathbb{P}$ occurring in $\mathcal{K}$, whether $\Delta \models \langle \mathsf{rep}_{\mathcal{K}} \rangle p \to [\mathsf{rep}_{\mathcal{K}}] p$. Second, global unicity of the repairs (independently of a specific database Δ) can be verified by checking for each of the variables $p \in \mathbb{P}$ whether $\langle \mathsf{rep}_{\mathcal{K}} \rangle p \to [\mathsf{rep}_{\mathcal{K}}] p$ is DL-PA$^{\pm}$ valid. Third, we can verify whether $\mathcal{K}$ can repair every database by checking whether the formula $\langle \mathsf{rep}_{\mathcal{K}} \rangle \top$ is DL-PA$^{\pm}$ valid.

7.2 Comparing Two Sets of ECA Constraints

We start by two definitions that will be useful for our purposes. The program π_1 is *included* in the program π_2 if $||\pi_1|| \subseteq ||\pi_2||$. Two programs π_1 and π_2 are *equivalent* if each is included in the other, that is, if $||\pi_1|| = ||\pi_2||$.

These comparisons can be polynomially reduced into validity checking problems. Our translation makes use of the assignment-recording propositional variables of Sect. 6.1 and of the copies of propositional variables that we have introduced in Sect. 6.2. Hence we suppose that for every variable p there is a fresh variable p' that will store the truth value of p, as well as two fresh variables $\mathsf{D}(+p')$ and $\mathsf{D}(-p')$. For a given set of propositional variables $P \subseteq \mathbb{P}$ we make use of the following three sets of auxiliary variables

$$\begin{aligned} P' &= \{p' \ : \ p \in P\}, \\ \mathsf{D}(+(P')) &= \{\mathsf{D}(+p') \ : \ p \in P\}, \\ \mathsf{D}(-(P')) &= \{\mathsf{D}(-p') \ : \ p \in P\}. \end{aligned}$$

The auxiliary variables are used in a 'generate and test' schema. For a given set of propositional variables $P \subseteq \mathbb{P}$, the program

$$\mathsf{guess}(P) = -(P' \cup \mathsf{D}(+(P')) \cup \mathsf{D}(-(P'))); (\bigcup_{q \in P'} +q)^*$$

guesses nondeterministically which of the auxiliary variables for P are going to be modified: first all are set to false and then some subset is made true. The formula

$$\mathsf{Guessed}(P) = \bigwedge_{p \in P} ((p \leftrightarrow p') \wedge (+p \leftrightarrow \mathsf{D}(+p)) \wedge (-p \leftrightarrow \mathsf{D}(-p)))$$

checks that the guess was correct.

Now we are ready to express inclusion of programs in DL-PA$^{\pm}$: we predict the outcome of π_1 and then check if π_2 produces the same set of changes as π_1.

Proposition 3. *Let π_1 and π_2 be two DL-PA programs.*

1. *π_1 is included in π_2 if and only if*

$$[\mathsf{guess}(\mathbb{P}_{\pi_1} \cup \mathbb{P}_{\pi_2})]\big(\langle \pi_1 \rangle \mathsf{Guessed}(\mathbb{P}_{\pi_1} \cup \mathbb{P}_{\pi_2}) \\ \to \langle \pi_2 \rangle \mathsf{Guessed}(\mathbb{P}_{\pi_1} \cup \mathbb{P}_{\pi_2})\big)$$

   is DL-PA$^{\pm}$ valid.

2. *π_1 and π_2 are equivalent if and only if*

$$[\mathsf{guess}(\mathbb{P}_{\pi_1} \cup \mathbb{P}_{\pi_2})]\big(\langle \pi_1 \rangle \mathsf{Guessed}(\mathbb{P}_{\pi_1} \cup \mathbb{P}_{\pi_2}) \\ \leftrightarrow \langle \pi_2 \rangle \mathsf{Guessed}(\mathbb{P}_{\pi_1} \cup \mathbb{P}_{\pi_2})\big)$$

   is DL-PA$^{\pm}$ valid.

The reduction is polynomial. The following decision problems can be reformulated in terms of program inclusion and equivalence:

1. Is the ECA constraint $\kappa \in \mathcal{K}$ redundant?
2. Are two sets of ECA constraints $\mathcal{K}_1$ and $\mathcal{K}_2$ equivalent?
3. Can all databases that are repaired by $\mathcal{K}_1$ also be repaired by $\mathcal{K}_2$?

Each can be expressed in DL-PA$^{\pm}$ by means of program inclusion or equivalence: the first can be decided by checking whether the programs $\mathsf{rep}_{\mathcal{K}}$ and $\mathsf{rep}_{\mathcal{K} \setminus \{\kappa\}}$ are equivalent; the second can be decided by checking whether the programs $\mathsf{rep}_{\mathcal{K}_1}$ and $\mathsf{rep}_{\mathcal{K}_2}$ are equivalent; the third problem can be decided by checking the validity of $\langle \mathsf{rep}_{\mathcal{K}_1} \rangle \top \to \langle \mathsf{rep}_{\mathcal{K}_2} \rangle \top$.

7.3 Termination

In our DL-PA$^{\pm}$ framework, the taming of termination problems is not only due to the definition of well-founded ECA repairs itself: in dynamic logics, the Kleene star is about unbounded but finite iterations. The modal diamond operator therefore quantifies over terminating executions only, disregarding any infinite executions. We can nevertheless reason about infinite computations by dropping the tests $\neg\mathsf{D}(\alpha) \wedge \neg\mathsf{D}(\overline{\alpha})$ from the definition of $\mathsf{rep}(\alpha)$. Let the resulting 'unbounded' repair program be $\mathsf{urep}(\alpha)$. Let $\mathsf{urep}_{\mathcal{K}}$ stand for either the programs $\mathsf{urep}_{\mathcal{K}}^{\mathsf{wwf}}$ or $\mathsf{urep}_{\mathcal{K}}^{\mathsf{wf}}$ resulting from the replacement of $\mathsf{rep}(\alpha)$ by $\mathsf{urep}(\alpha)$. Then the repair of Δ loops if and only if $\Delta \models [(\mathsf{urep}_{\mathcal{K}})^*]\langle\mathsf{urep}_{\mathcal{K}}\rangle\top$.

8 Conclusion

Our dynamic logic semantics for ECA constraints generalises the well-founded AIC repairs of [7], and several decision problems can be captured in DL-PA$^{\pm}$. Proposition 2 provides a PSPACE upper bound for all these problems. A closer look at the characterisations of Sect. 6 shows that the complexity is actually lower. For the results of Sect. 6.1 (Theorem 1 and Theorem 2), as executability of a program π at Δ is the same as truth of $\langle\pi\rangle\top$ at Δ, our characterisation involves a single existential quantification (a modal diamond operator), with a number of nondeterministic choices that is quadratic in $\mathsf{card}(\mathbb{P})$ (precisely, $1 + 2\mathsf{card}(\mathbb{P})$ nondeterministic choices that are iterated $\mathsf{card}(\mathbb{P})$ times, cf. what we have remarked after the theorem, as well as the definition of $\pi^{\leq n}$ in Sect. 5). Just as the corresponding QBF fragment, this fragment is in NP. For the results of Sect. 6.2 (Theorem 3 and Theorem 4), as executability of $\pi; \neg\langle\pi'\rangle\varphi?$ at Δ is the same as truth of $\langle\pi\rangle[\pi']\varphi$ at Δ, our characterisation involves an existential diamond containing the same program as above that is preceded by $\mathsf{init}(\mathbb{A})$ and that is followed by a universal quantification (a modal box operator). Just as the corresponding QBF fragment, this fragment is in Σ_2^p.

Beyond these decision problems we can express repair algorithms as DL-PA$^{\pm}$ programs, given that the standard programming constructions such as if-then-else and while can all be expressed in dynamic logic. Correctness of such a program π can be verified by checking whether π is included in the program $\mathsf{rep}_{\mathcal{K}}$. The other way round, one can check whether π is able to output *any* well-founded ECA repair by checking whether $\mathsf{rep}_{\mathcal{K}}$ is included in π.

It remains to study further our founded ECA repairs of Sect. 4.2 and their grounded versions. We also plan to check whether the more expressive existential AICs of [9] transfer. Finally, we would like to generalise the history component of h-databases from update actions to event algebra expressions as studied e.g. in [1,22]; dynamic logic should be beneficial here, too.

References

1. Alferes, J.J., Banti, F., Brogi, A.: An event-condition-action logic programming language. In: Fisher, M., van der Hoek, W., Konev, B., Lisitsa, A. (eds.) JELIA 2006. LNCS (LNAI), vol. 4160, pp. 29–42. Springer, Heidelberg (2006). https://doi.org/10.1007/11853886_5
2. Balbiani, P., Herzig, A., Schwarzentruber, F., Troquard, N.: DL-PA and DCL-PC: model checking and satisfiability problem are indeed in PSPACE. CoRR abs/1411.7825 (2014). http://arxiv.org/abs/1411.7825
3. Balbiani, P., Herzig, A., Troquard, N.: Dynamic logic of propositional assignments: a well-behaved variant of PDL. In: 28th Annual ACM/IEEE Symposium on Logic in Computer Science, LICS 2013, New Orleans, LA, USA, 25–28 June 2013, pp. 143–152. IEEE Computer Society (2013). https://doi.org/10.1109/LICS.2013.20
4. Bertossi, L.E.: Database Repairing and Consistent Query Answering. Synthesis Lectures on Data Management. Morgan & Claypool Publishers (2011). https://doi.org/10.2200/S00379ED1V01Y201108DTM020
5. Bertossi, L.E., Pinto, J.: Specifying active rules for database maintenance. In: Saake, G., Schwarz, K., Türker, C. (eds.) Transactions and Database Dynamics, Proceedings of the Eight International Workshop on Foundations of Models and Languages for Data and Objects, Schloß Dagstuhl, Germany, 27–30 September 1999, vol. Preprint Nr. 19, pp. 65–81. Fakultät für Informatik, Otto-von-Guericke-Universität Magdeburg (1999)
6. Bidoit, N., Maabout, S.: A model theoretic approach to update rule programs. In: Afrati, F., Kolaitis, P. (eds.) ICDT 1997. LNCS, vol. 1186, pp. 173–187. Springer, Heidelberg (1997). https://doi.org/10.1007/3-540-62222-5_44
7. Bogaerts, B., Cruz-Filipe, L.: Fixpoint semantics for active integrity constraints. Artif. Intell. **255**, 43–70 (2018)
8. Calautti, M., Caroprese, L., Greco, S., Molinaro, C., Trubitsyna, I., Zumpano, E.: Consistent query answering with prioritized active integrity constraints. In: Desai, B.C., Cho, W. (eds.) IDEAS 2020: 24th International Database Engineering and Applications Symposium, Seoul, Republic of Korea, 12–14 August 2020, pp. 3:1–3:10. ACM (2020). https://dl.acm.org/doi/10.1145/3410566.3410592
9. Calautti, M., Caroprese, L., Greco, S., Molinaro, C., Trubitsyna, I., Zumpano, E.: Existential active integrity constraints. Expert Syst. Appl. **168**, 114297 (2021)
10. Caroprese, L., Greco, S., Sirangelo, C., Zumpano, E.: Declarative semantics of production rules for integrity maintenance. In: Etalle, S., Truszczyński, M. (eds.) ICLP 2006. LNCS, vol. 4079, pp. 26–40. Springer, Heidelberg (2006). https://doi.org/10.1007/11799573_5
11. Caroprese, L., Greco, S., Zumpano, E.: Active integrity constraints for database consistency maintenance. IEEE Trans. Knowl. Data Eng. **21**(7), 1042–1058 (2009). https://doi.org/10.1109/TKDE.2008.226
12. Caroprese, L., Truszczynski, M.: Active integrity constraints and revision programming. TPLP **11**(6), 905–952 (2011)
13. Ceri, S., Fraternali, P., Paraboschi, S., Tanca, L.: Automatic generation of production rules for integrity maintenance. ACM Trans. Database Syst. **19**(3), 367–422 (1994). http://doi.acm.org/10.1145/185827.185828
14. Chomicki, J., Lobo, J., Naqvi, S.A.: Conflict resolution using logic programming. IEEE Trans. Knowl. Data Eng. **15**(1), 244–249 (2003)
15. Chomicki, J., Marcinkowski, J.: Minimal-change integrity maintenance using tuple deletions. Inf. Comput. **197**(1–2), 90–121 (2005). https://doi.org/10.1016/j.ic.2004.04.007

16. Cruz-Filipe, L.: Optimizing computation of repairs from active integrity constraints. In: Beierle, C., Meghini, C. (eds.) FoIKS 2014. LNCS, vol. 8367, pp. 361–380. Springer, Cham (2014). https://doi.org/10.1007/978-3-319-04939-7_18
17. Cruz-Filipe, L., Gaspar, G., Engrácia, P., Nunes, I.: Computing repairs from active integrity constraints. In: Seventh International Symposium on Theoretical Aspects of Software Engineering, TASE 2013, 1–3 July 2013, Birmingham, UK, pp. 183–190. IEEE Computer Society (2013). https://doi.org/10.1109/TASE.2013.32
18. Cruz-Filipe, L., Gaspar, G., Nunes, I., Schneider-Kamp, P.: Active integrity constraints for general-purpose knowledge bases. Ann. Math. Artif. Intell. **83**(3–4), 213–246 (2018)
19. Flesca, S., Greco, S.: Declarative semantics for active rules. Theory Pract. Log. Program. **1**(1), 43–69 (2001). http://journals.cambridge.org/action/displayAbstract?aid=71136
20. Flesca, S., Greco, S., Zumpano, E.: Active integrity constraints. In: Moggi, E., Warren, D.S. (eds.) Proceedings of the 6th International ACM SIGPLAN Conference on Principles and Practice of Declarative Programming, 24–26 August 2004, Verona, Italy, pp. 98–107. ACM (2004). http://doi.acm.org/10.1145/1013963.1013977
21. Herzig, A.: Belief change operations: a short history of nearly everything, told in dynamic logic of propositional assignments. In: Baral, C., Giacomo, G.D., Eiter, T. (eds.) Principles of Knowledge Representation and Reasoning: Proceedings of the Fourteenth International Conference, KR 2014, Vienna, Austria, 20–24 July 2014. AAAI Press (2014). http://www.aaai.org/ocs/index.php/KR/KR14/paper/view/7960
22. Lausen, G., Ludäscher, B., May, W.: On logical foundations of active databases. In: Chomicki, J., Saake, G. (eds.) Logics for Databases and Information Systems (the book grow out of the Dagstuhl Seminar 9529: Role of Logics in Information Systems, 1995), pp. 389–422. Kluwer (1998)
23. Ludäscher, B., May, W., Lausen, G.: Nested transactions in a logical language for active rules. In: Pedreschi, D., Zaniolo, C. (eds.) LID 1996. LNCS, vol. 1154, pp. 197–222. Springer, Heidelberg (1996). https://doi.org/10.1007/BFb0031742
24. Widom, J., Ceri, S.: Active Database Systems: Triggers and Rules for Advanced Database Processing. Morgan Kaufmann, Burlington (1996)
25. Winslett, M.: Reasoning about action using a possible models approach. In: Shrobe, H.E., Mitchell, T.M., Smith, R.G. (eds.) Proceedings of the 7th National Conference on Artificial Intelligence, St. Paul, MN, USA, 21–26 August 1988, pp. 89–93. AAAI Press/The MIT Press (1988). http://www.aaai.org/Library/AAAI/1988/aaai88-016.php
26. Winslett, M.A.: Updating Logical Databases. Cambridge Tracts in Theoretical Computer Science. Cambridge University Press, Cambridge (1990)

Statistics of RDF Store for Querying Knowledge Graphs

Iztok Savnik[1(✉)], Kiyoshi Nitta[2], Riste Skrekovski[3], and Nikolaus Augsten[4]

[1] Faculty of Mathematics, Natural Sciences and Information Technologies, University of Primorska, Koper, Slovenia
iztok.savnik@upr.si
[2] Yahoo Japan Corporation, Tokyo, Japan
knitta@yahoo-corp.jp
[3] Faculty of Information Studies, Novo Mesto, Slovenia
[4] Department of Computer Sciences, University of Salzburg, Salzburg, Austria
nikolaus.augsten@sbg.ac.at

Abstract. Many RDF stores treat graphs as simple sets of vertices and edges without a conceptual schema. The statistics in the schema-less RDF stores are based on the cardinality of the keys representing the constants in triple patterns. In this paper, we explore the effects of storing knowledge graphs in an RDF store on the structure of the space of queries and, consequently, on the definition of the framework for the computation of the statistics. We propose a formal framework for an RDF store with a complete conceptual schema. The poset of schema triples defines the structure of the types of triple patterns and, therefore, the structure of the query space. The set of schema triples, together with the ontology of classes and predicates, form the conceptual schema of a knowledge graph, referred to as a *schema graph*. We present an algorithm for computing the statistics of a schema graph that consists of the schema triples from the stored schema graph and the schema triples that are more general/specific than the stored schema triples up to a user-defined level.

Keywords: RDF stores · graph databases · knowledge graphs · database statistics · statistics of graph databases

1 Introduction

The statistics of RDF stores is an essential tool used in the processes of query optimization [7,19,20,31]. They are used to estimate the size of a query result and the time needed to evaluate a given query. The estimations are required to find the most efficient query evaluation plan for a given input query. Furthermore, the statistics of large graphs are used to solve problems closely related to query evaluation; for example, they can be employed in algorithms for partitioning of large graphs [27].

Many RDF stores treat graphs as simple sets of vertices and edges without a conceptual schema. The statistics in the schema-less RDF stores are based on the cardinality of the keys representing the constants in triple patterns. While some of the initial RDF stores are implemented on top of relational database systems [21,34], most of them are

I. Varzinczak (Ed.): FoIKS 2022, LNCS 13388, pp. 93–110, 2022.
https://doi.org/10.1007/978-3-031-11321-5_6

implemented around a subset of seven indexes corresponding to the keys that are subsets of $\{S, P, O\}$, i.e., the subject, predicate, and object of the triples. The indexes can be either a B+ tree [8,20], a custom-designed index [11,13,33,35], or some other index data structure (e.g., radix trie) [10,12,32]. Statistics at the level of relations (as used in relational systems) are not useful at the level of RDF graphs; therefore, statistics are obtained by sampling the indexes [21], or by creating separate aggregates of indexes [8,20].

There are a few approaches to the computation of statistics of RDF stores that use some form of the semantic structure to which the statistics are attached. Stocker et al. use RDF Schema information to precompute the size of all possible joins where domain/range of some predicate is joined with domain/range component of some other predicate [31]. Neumann and Moetkotte propose the use of the characteristic sets [19], i.e., the sets of predicates with a common S/O component, for the computation of statistics of star-shaped queries. Gubichev and Neumann further organize characteristic sets [7] in a hierarchy to allow precise estimation of joins in star queries.

In this paper, we explore the effects of storing knowledge graphs [15] in an RDF store on the structure of the space of queries and, consequently, on the definition of the framework for the computation of the statistics. The space of SPARQL queries is structured primarily on the basis of the triple patterns. To capture the set of triples addressed by a triple pattern, we define a *type* of a triple pattern to be a triple referred to as a *schema triple*. A schema triple (s, p, o) includes a class s as subject, a predicate p and a class o as object. For example, the type of the triple $(john, livesIn, capodistria)$ is the schema triple $(person, livesIn, location)$ where the domain and the range of predicate $livesIn$ are $person$ and $location$. The set of schema triples is partially ordered by the relationship *more-specific* reflecting sub-class and sub-predicate relationships among classes and predicates.

The space of the sets of triples that are the targets of triple patterns is partially ordered in the same way as are the schema triples that are the types of triple patterns. Namely, the set of triples targeted by a triple pattern depends on the interpretation of the type of a triple pattern. Since the schema triples (as well as their interpretations) are partially ordered, we have the framework to which the statistics of the types of triple patterns are attached. The set of possible types of triple-patters together with the ontology of classes and predicates comprise the conceptual schema of a knowledge graph that we call a *schema graph*.

The selection of the schema triples that form the schema graph is the first step in the computation of the statistics. The *stored schema graph* is the minimal schema graph, including only the schema triples that are stored in an RDF store by means of the definitions of domains and ranges of predicates. The *complete schema graph* includes all legal types (schema triples) of the triples from an RDF store. We propose to use the schema graph that includes the schema triples from a strip around the stored schema graph, i.e., the schema triples from the stored schema graph and some adjacent levels of schema triples. In the second step, we compute the statistics for the selected schema triples that form the schema graph. For each RDF triple, the set of its types (schema triples) that intersects with the schema graph is computed. The statistics are updated for the computed types.

The contributions of the work presented in this paper are the following. First, we propose a formal framework for an RDF store with a complete conceptual schema. The poset of schema triples defines the structure of the types of triple patterns and, therefore, the structure of the query space and the framework for the computation of the statistics. The proposed formalization unifies the representation of the instance and the schema levels of RDF stores. Second, we propose the computation of the statistics of the schema triples from a strip around the stored schema graph. We show that the assignment of types to joining triple patterns can lead to more specific types than those specified in a stored schema. For this reason, we often need the statistics for a schema triple that are more specific/general than some stored schema triple. Finally, we propose two ways of counting the instances and keys of a given schema triple. When adding individual triples to the statistics, one schema triple represents the seven types of keys. The *bound* type of counting a key respects underlying schema triple (the complete type of the key), and the *unbound* type of counting is free from the underlying schema triple.

The paper is organized as follows. Section 2 provides a formalization of a knowledge graph and introduces the denotational semantics of concepts. Section 3 presents three novel algorithms for the computation of statistics. The first algorithm computes the statistics for the stored schema graph, and the second algorithm computes the statistics of all possible schema triples. The third algorithm focuses on the schema triples from a strip around the stored schema graph, i.e., in addition to the stored schema graph, also the schema triples some levels below and above the schema triples in the stored schema graph. Section 3.4 introduces the concepts of a key and a key type that are used for counting the instances of a schema triple. Further, the concepts of bound and unbound counting are discussed. An empirical study of the algorithms for the computation of the statistics is presented in Sect. 4. Two experiments are presented, first, on a simple toy domain and, second, on the core of the Yago2 knowledge graph. Related work is presented in Sect. 5. Finally, concluding remarks are given in Sect. 6.

2 Conceptual Schema of a Knowledge Graph

The formal definition of the schema of a knowledge graph is based on the RDF [25] and RDF-Schema [26] data models. Let I be the set of URI-s, B be the set of blanks and L be the set of literals. Let us also define sets $S = I \cup B$, $P = I$, and $O = I \cup B \cup L$.

A *RDF triple* is a triple $(s, p, o) \in S \times P \times O$. An *RDF graph* $g \subseteq S \times P \times O$ is a set of triples [15]. The set of all graphs is denoted as G. We state that an RDF graph g_1 is a *sub-graph* of g_2, denoted as $g_1 \subseteq g_2$, if all triples of g_1 are also triples of g_2. We call the elements of the set, I, *identifiers* to abstract away the details of the RDF data model. An extended formal definition of the conceptual schema of a knowledge graph is presented in [30].

2.1 Identifiers

We define a set of concepts to be used for a more precise characterization of identifiers I. *Individual identifiers*, denoted as set I_i, are identifiers that have specified their classes using property rdf:type. *Class identifiers* denoted as set I_c, are identifiers that are sub-classes of the top class $\top$ of ontology—Yago, for instance, uses the top class owl:Thing. Further, *predicate identifiers* denoted as set I_p, are individual identifiers (i.e., $I_P \subset I_i$) that represent RDF predicates. The predicate identifiers are, from one perspective, similar to the class identifiers: they have sub-predicates in the same manner as the classes have sub-classes. However, predicates do not have instances. They are the instances of rdf:Property.

We define the partial ordering relationship *more-specific*, formally denoted as $\preceq$, over the complete set of identifiers I. The relationship $\preceq$ subsumes the relationships rdf:type, rdfs:subClassOf and rdfs:subPropertyOf. The top element of partial ordering is the top-class $\top$. The bottom of the partial ordering includes the individual identifiers from I_i. The individual identifiers I_i are related to their classes from I_c using the relationship rdf:type. The class identifiers I_c are related by the relationship rdf:subClassOf, and the predicate identifiers I_P are related by the relationship rdf:subPropertyOf. Relationship "$\preceq$" is reflexive, transitive, and anti-symmetric, i.e., it defines a partial order relationship [30].

Let us now define the two semantic interpretations of the identifiers from I. The *ordinary interpretation* $[\![i]\!]_g \subseteq I_i$ maps individual identifiers to themselves and the class identifiers c to the instances of c or any of c's sub-classes. The *natural interpretation* $[\![i]\!]^*_g$ maps identifiers i to the sets of all identifiers that are more specific than i. The individual identifiers are mapped to themselves, while the class identifiers c are mapped to all instances of c and c's sub-classes, and including all sub-classes of c.

The partial ordering relationship $\preceq$ and the semantic functions $[\![]\!]_g$ and $[\![]\!]^*_g$ are consistent. The relationship $\preceq$ between two identifiers implies the subsumption of the ordinary interpretations as well as natural interpretations.

2.2 Triples, Types and Graphs

The partial ordering relation $\preceq$ is extended to triples. A triple t_1 is more specific or equal to a triple t_2, denoted $t_1 \preceq t_2$, if all components of t_1 are more specific or equal than the related components of t_2. Similar to the set of identifiers I, where we distinguish between I_i, and I_c, the set of triples is divided into *ground triples* and *schema triples*. Ground triples include at least one individual identifier, while the schema triples include the predicate (as a component P) and class identifiers solely.

The schema triples represent the *types* of the ground triples. In order to give a uniform representation of the "schema" and "instance" levels of a datasets we define schema triples as (s, p, o) where $s, o \in I_c$ and $p \in I_P$. For example, the triple (plato,wasBornIn,athens) is a ground triple, and its type, the triple (person,wasBornIn,location), is a schema triple. Further, a schema triple (s, p, o) must represent a type of a nonempty set of ground triples. In terms of the semantic interpretations that are presented shortly: a schema triple has a nonempty interpretation in a graph g.

A special kind of schema triples are the *stored schema triples* (s, p, o) that are represented in a graph g by the triples $(p, \text{rdfs:domain}, s)$ and $(p, \text{rdfs:range}, o)$. Looking from this perspective, a schema triple is any triple $t = (s, p, o)$ (with $s, o \in I_c$ and $p \in I_P$) such that there exist a stored schema triple t_s which is related to t using the relationship *more-specific*, i.e., $t \preceq t_s$ or $t_s \preceq t$. This definition ties the schema triples to the partially ordered set of triples in order to represent a legal type in a given knowledge graph.

We can now define the interpretation of a schema triple t, similarly to the case of identifiers. The interpretation function $[\![]\!]_g$ maps a schema triple t to the set of ground triples that are the instances of a given schema triple t: $[\![t]\!]_g = \{(s', p', o') | (s', p', o') \in g \wedge s' \in [\![s]\!]_g \wedge p' \in [\![p]\!]_g \wedge o' \in [\![o]\!]_g\}$. The natural interpretation function $[\![]\!]_g^*$ maps a schema triple t to the set of triples that are more specific than t, or equal to t: $[\![t]\!]_g^* = \{(s', p', o') | (s', p', o') \in g \wedge s' \preceq s \wedge p' \preceq p \wedge o' \preceq o\}$.

The set of all stored schema triples together with the triples that define the classification hierarchy of the classes and predicates form the *stored schema graph* (abbr. ssg). The ssg defines the structure of ground triples in g; each ground triple is an instance of at least one schema triple $t_s \in ssg$. The ssg represents the minimal schema graph. On the other hand, the *complete schema graph* includes the ssg and *all* possible types that are either more specific or more general than some $t \in ssg$. Therefore, the complete schema graph is a maximal schema graph of g. Finally, the *schema graph* that we use for the computation of the statistics of a knowledge graph contains, besides the ssg, also the schema triples that form a strip around the stored schema triples. These additional schema triples allow us to make a better estimation of the sizes of queries.

2.3 Triple Patterns

We assumed that we have a set of variables V. A *triple pattern* $(s, p, o) \in (S \cup V) \times (P \cup V) \times (O \cup V)$ is a triple that includes at least one variable. The components of a triple pattern $t = (s, p, o)$ can be accessed in, similarly to the elements in an array, as $tp[1] = s$, $tp[2] = p$ and $tp[3] = o$. Let a set $tp_v \subseteq \{1, 2, 3\}$ be a set of indices of components that are variables, i.e., $\forall j \in tp_v : tp[j] \in V$.

The *interpretation* of a triple pattern tp in a graph g is the set of triples $t \in g$ such that t includes any value in place of variables indexed by the elements of tp_v, and, has the values of other components equal to the corresponding tp components. Formally, the interpretation of tp is $[\![tp]\!]_g = \{t \mid t \in g \wedge \forall j \in \{1, 2, 3\} \setminus tp_v : t[j] = tp[j]\}$.

The type of a triple pattern $tp = (s, p, o)$ is a schema triple $t_{tp} = (t_s, t_p, t_o)$ such that the interpretation of the schema triple subsumes the interpretation of the triple pattern, i.e., $[\![tp]\!]_g \subseteq [\![t_{tp}]\!]_g^*$.. Note that we only specify the semantic conditions for the definition of the type of triple patterns.

3 Computing Statistics

The statistical index is implemented as a dictionary where keys represent schema triples, and the values represent the statistical information for the given keys. For instance, the index entry for the schema triple (person,wasBornIn,location) represents the statistical information about the triples that have the instance of a person as the first component, the property wasBornIn as the second component, and the instance of location as the third component.

The main procedure for the computation of the statistics of a knowledge graph is presented as Algorithm 1. The statistic is computed for each triple t from a given knowledge graph. The function STATISTICS-TRIPLE(t) represents one of three functions:

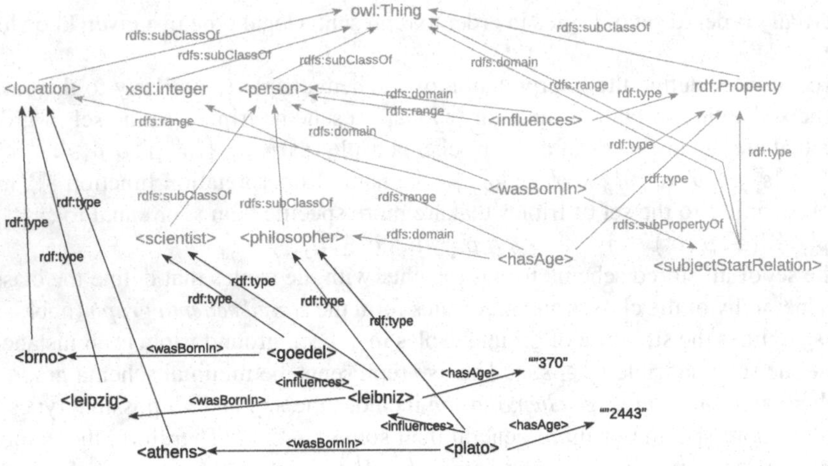

Fig. 1. Triple-store *Simple*.

STATISTICS-STORED(t), STATISTICS-ALL(t) and STATISTICS-LEVELS(t), which are presented in detail in the sequel. Each of the functions computes, in the first phase, the set of schema triples, which include the given triple, t, in their natural interpretations $[\![]\!]^*$. In the second phase, the function STATISTICS-TRIPLE updates the corresponding statistics for the triple t.

Algorithm 1. Procedure COMPUTE-STATISTICS(ts : knowledge-graph)

procedure COMPUTE-STATISTICS(ts : knowlege-graph)
  for all $t \in ts$ **do**
    STATISTICS-TRIPLE(t)

The examples in the following subsections are based on a simple knowledge graph presented in Fig. 1. The knowledge graph referred to in the rest of the text as *Simple* includes 33 triples. Four user-defined classes are included. The class person has subclasses scientist and philosopher. The class location represents the concept of a physical location. The schema part of *Simple* is colored blue, and the data part is colored black. The dataset *Simple* is available from the *epsilon* data repository [28].

Example 1. The stored schema graph of *Simple* is represented by the schema triples (person,wasBornIn, location), (person,influence,person), (philosopher,hasAge,integer) and (owl:Thing,subjectStartRelation,owl:Thing), that are defined by using the predicates rdfs:domain and rdfs:range. Further, the stored schema includes the specialization hierarchy of classes and predicates—the specialization hierarchy is defined with the triples that include the predicates rdfs:subClassOf, rdfs:subPropertyOf, and rdf:type. For example, the hierarchy of a class scientist is represented by triples (scientist,rdfs:-subClassOf,person) and (person,rdfs:subClassOf,owl:Thing).

The interpretation of a schema triple t comprises a set of triples composed of individual identifiers that belong to the classes specified by the schema triple t. The interpretation of the schema triple (person,wasBornIn,location), for example, are the ground triples (goedel,wasBornIn,brno), (leibniz,wasBornIn,leipzig) and (plato,wasBornIn,athens). Therefore, the index key (person,wasBornIn,location) is mapped to the statistical information describing the presented instances. □

In the following Sects. 3.1–3.3, we present the algorithms for the computation of the statistics of the schema graphs. The procedure STATISTICS-STORED that computes the statistics for the stored schema graph is described in Sect. 3.1. Example 2 demonstrates that it is beneficial to store the statistics not only for the schema triples from the stored schema graph but also for more specific schema triples. This allows more precise estimation of the size of triple patterns that form a query.

Example 2. The triple pattern (?x,wasBornIn,?y) has the type (person,wasBornIn,location). Therefore, solely the instances of (person,wasBornIn,location) are addressed by the triple pattern. Further, a join can be defined by triple patterns (?x,wasBornIn,?y) and (?x,hasAge,?z). The types of triple patterns are (person,wasBornIn,-location) and (philosopher,hasAge,integer), respectively. We see that the variable ?x has the type person as well as philosopher. Since the interpretation of a person subsumes the interpretation of philosopher, we can safely use the schema triple (philosopher,wasBornIn,location) as the type of (?x,wasBornIn,?y). □

The second procedure STATISTICS-ALL, presented in Sect. 3.2, computes statistics for all legal schema triples. The problem with storing the statistics for all legal schema triples is the size of such a set. Indeed, knowledge graphs include a large number of classes and predicates [5]. To be able to control the size of the schema graph, we use the stored schema graph as the reference point. The stored schema graph is extended with the schema triples that are up to a given number of levels more general or more specific than the schema triples that form the stored schema graph.

The procedure STATISTICS-LEVELS that computes the statistics for the schema triples included in a strip around the stored schema graph is presented in Sect. 3.3. Finally, the implementation of the procedure UPDATE-STATISTICS for counting the keys for a given schema triple is given in Sect. 3.4.

3.1 Statistics of the Stored Schema Graph

The first procedure for computing the statistics is based on the schema information that is a part of the knowledge graph. As we have already stated in the introduction, we suppose that the knowledge graph would include a complete schema, i.e., the ontology of classes and predicates, as well as the definition of domains and ranges of predicates.

Let us now present Algorithm 2. We assumed that $t = (s, p, o)$ was an arbitrary triple from graph $g \in G$. First, the algorithm initializes set g_p in the line 2 to include the element p, and the transitive closure of $\{p\}$ computed with respect to the relationship rdfs:subPropertyOf to obtain all more general properties of p. Note that '+' denotes one or more applications of the predicate rdfs:subPropertyOf.

Algorithm 2. Function STATISTICS-STORED(t : triple-type)

1: **function** STATISTICS-STORED($t = (s, p, o)$: triple-type)
2: $\quad g_p \leftarrow \{p\} \cup \{c_p | (p, \text{rdfs:subPropertyOf+}, c_p) \in g\}$
3: $\quad$ **for all** $p_g \in g_p$ **do**
4: $\quad\quad d_p \leftarrow \{t_s | (p_g, \text{rdfs:domain}, t_s) \in g\}$
5: $\quad\quad r_p \leftarrow \{t_o | (p_g, \text{rdfs:range}, t_o) \in g\}$
6: $\quad\quad$ **for all** $t_s \in d_p, t_o \in r_p$ **do**
7: $\quad\quad\quad$ UPDATE-STATISTICS($(t_s, p_g, t_o), t$)

After the set g_p is computed, the domains and the ranges of each particular property $p_g \in g_p$ are retrieved from the graph in lines 4–5. After we have computed all the sets, including the types of t's components, the schema triples can be enumerated by taking property p_g and a pair of the domain and range classes of p_g. The statistics are updated using the procedure UPDATE-STATISTICS for each of the generated schema triples in line 7. A detailed description of the procedure UPDATE-STATISTICS is given in Sect. 3.4.

3.2 Statistics of all Schema Triples

Let $t = (s, p, o)$ be arbitrary triple from graph $g \in G$. The set of schema triples for a given triple t is obtained by first computing all possible classes of the components s and o, and all existing super-predicates of p. In this way, we obtain the sets of classes g_s and g_o and, the set of predicates g_p. These sets are then used to generate all possible schema triples, i.e., all possible types of t. The statistics are updated for the generated schema triples.

Algorithm 3. Function STATISTICS-ALL(t : triple-type)

1: **function** STATISTICS-ALL($t = (s, p, o)$: triple-type)
2: $\quad g_s \leftarrow \{t_s | \text{is\_class}(s) \wedge t_s = s \vee \neg\text{is\_class}(s) \wedge (s, \text{rdf:type}, t_s) \in g\}$
3: $\quad g_o \leftarrow \{t_o | \text{is\_class}(o) \wedge t_o = o \vee \neg\text{is\_class}(o) \wedge (o, \text{rdf:type}, t_o) \in g\}$
4: $\quad g_s \leftarrow g_s \cup \{c_s' | c_s \in g_s \wedge (c_s, \text{rdfs:subClassOf+}, c_s') \in g\}$
5: $\quad g_p \leftarrow \{p\} \cup \{c_p | (p, \text{rdfs:subPropertyOf+}, c_p) \in g\}$
6: $\quad g_o \leftarrow g_o \cup \{c_o' | c_o \in g_o \wedge (c_o, \text{rdfs:subClassOf+}, c_o') \in g\}$
7: $\quad$ **for all** $c_s \in g_s, c_p \in g_p, c_o \in g_o$ **do**
8: $\quad\quad$ UPDATE-STATISTICS($(c_s, c_p, c_o), t$)

The procedure for the computation of statistics for a triple t is presented in Algorithm 3. The triple t includes the components s, p and o, i.e., $t = (s, p, o)$. The procedure STATISTICS-ALL computes in lines 1–6 the sets of types g_s, g_p and g_o of the triple elements s, p and o, respectively. The set of types g_s is in line 2 initialized by the set of the types of s, or, by s itself when s is a class. The set g_s is then closed by means of the relationship refs:subClassOf in line 4. Set g_o is computed in the same way as set

g_s. The set of predicates g_p is computed differently. Set g_p obtains the value by closing set $\{p\}$ using the relationship rdfs:subPropertyOf in line 5. Indeed, predicates play a similar role to classes in many knowledge representation systems [24].

The schema triples of t are enumerated in the last part of the procedure STATISTICS-ALL by using sets g_s, g_p and g_o in lines 7-8. For each of the generated schema triples (c_s, c_p, c_o) the interpretation $[\![(c_s, c_p, c_o)]\!]_g^*$ includes the triple $t = (s.p.o)$. Since g_s and g_o include all classes of s and o, and, g_p includes p and all its super-predicates, all possible schema triples in which interpretation includes t are enumerated.

3.3 Statistics of a Strip Around the Stored Schema Graph

In the procedure STATISTICS-STORED, we did not compute the statistics of the schema triples that are either more specific or more general than the schema triples that are included in the knowledge graph for a given predicate. For instance, while we do have the statistics for the schema triple (person,wasBornIn,location) since this schema triple is part of the knowledge graph, we do not have the statistics for the schema triples (scientist,wasBornIn,location) and (philosopher,wasBornIn,location).

The procedure STATISTICS-ALL is in a sense the opposite of the procedure STATISTICS-STORED. Given the parameter triple t, the procedure STATISTICS-ALL updates statistics for *all* possible schema triples which interpretation includes the parameter triple t. The number of all possible schema triples may be much too large for a knowledge graph with a rich conceptual schema. The conceptual schema of Yago knowledge graph, for instance, includes a half-million classes.

Algorithm 4. Function STATISTICS-LEVELS(t : triple-type, k : integer)

1: **function** STATISTICS-LEVELS($t = (s, p, o)$: triple-type, l, u : integer)
2: $\quad g_p \leftarrow \{p\} \cup \{t_p | (p, \text{rdfs:subPropertyOf+}, t_p) \in g\}$
3: $\quad g_s \leftarrow \{t_s | \text{is\_class}(s) \wedge t_s = s \vee \neg\text{is\_class}(s) \wedge (s, \text{rdf:type}, t_s) \in g\}$
4: $\quad g_o \leftarrow \{t_o | \text{is\_class}(o) \wedge t_o = o \vee \neg\text{is\_class}(o) \wedge (o, \text{rdf:type}, t_o) \in g\}$
5: $\quad g_s \leftarrow g_s \cup \{c'_s | c_s \in g_s \wedge (c_s, \text{rdfs:subClassOf+}, c'_s) \in g\}$
6: $\quad g_o \leftarrow g_o \cup \{c'_o | c_o \in g_o \wedge (c_o, \text{rdfs:subClassOf+}, c'_o) \in g\}$
7: $\quad d_p \leftarrow \{c_s | (p, \text{rdfs:domain}, c_s) \in g\}$
8: $\quad r_p \leftarrow \{c_s | (p, \text{rdfs:range}, c_s) \in g\}$
9: $\quad s_s \leftarrow \{c_s | c_s \in g_s \wedge \exists c'_s \in d_p((c_s \succeq c'_s \wedge \text{DIST}(c_s, c'_s) \leq u) \vee$
10: $\qquad (c_s \preceq c'_s \wedge \text{DIST}(c_s, c'_s) \leq l))\}$
11: $\quad s_o \leftarrow \{c_o | c_o \in g_o \wedge \exists c'_o \in r_p((c_o \succeq c'_o \wedge \text{DIST}(c_o, c'_o) \leq u) \vee$
12: $\qquad (c_o \preceq c'_o \wedge \text{DIST}(c_o, c'_o) \leq l))\}$
13: $\quad$ **for all** $c_s \in s_s, p_p \in g_p, c_o \in s_o$ **do**
14: $\qquad$ UPDATE-STATISTICS($(c_s, p_p, c_o), t$)

Let us now present the algorithm STATISTICS-LEVELS. The first part of the algorithm (lines 1–6) is the same as in the algorithm STATISTICS-ALL. Given a triple $t = (s, p, o)$, all types (classes) of s and o and super-predicates of p are computed as the variables g_s, g_o and g_p, respectively.

The domain and range classes of the predicate p are computed in lines 7–8 as the variables d_p and r_p. The sets of classes s_s and s_o that include the set of selected types for the S and P components of t are computed in lines 9–12. The set s_s includes all classes from g_s such that their distance to at least one class defined as the domain of p is less than l for more specific classes, and less than u for more general classes. The function $\text{DIST}(c_1, c_2)$ computes the number of edges on the path from c_1 to c_2 including solely the predicate rdfs:subClassOf, if such a path exists, or, ∞ if there is no such path. Similarly, the set s_o includes all classes from g_o such that their distance to at least one class from r_p is less than l for the classes below the stored schema, and less than u for the classes above the stored schema.

In the last part, the algorithm STATISTICS-LEVELS in lines 13–14 enumerates the types (schema triples) of the triple t by listing the Cartesian product of the sets s_s, g_p and s_o.

Retrieving Statistics of a Schema Triple. We assume that S is a schema graph obtained using the algorithm STATISTICS-LEVELS. When we want to retrieve the statistical data for a schema triple t, we have two possible scenarios. The first is that the schema triple t is an element of S. The statistics are, in this case, directly retrieved from the statistical index. This is the expected scenario, i.e., we expect that S includes all schema triples that represent the types of triple patterns. The second scenario covers cases when the schema triples t are either above or below S. Here, the statistics of t have to be approximated from the statistics computed for the schema graph S. The presentation of the algorithm for retrieving the statistics of a given schema triple is given in [30].

3.4 On Counting Keys

Let $g \in G$ be a knowledge graph stored in a triplestore, and, let $t = (s, p, o)$ be a triple. All algorithms for the computation of the statistics of g presented in the previous section enumerate a set of schema triples S such that for each schema triple $t_s \in S$ the interpretation $[\![t_s]\!]^*_g$ includes the triple t. For each of the computed schema triples $t_s \in S$ the procedure UPDATE-STATISTICS() updates the statistics of t_s as the consequence of the insertion of the triple t into the graph g. The triple $t = (s, p, o)$ includes seven keys: s, p, o, *sp*, *so*, *po* and *spo*. These keys are the targets to be queried in the triple patterns.

Keys and Key Types. Let us now define the concepts of the key and the key type. A *key* is a triple (s_k, p_k, o_k) composed of $s_k \in I \cup \{\_\}$, $p_k \in P \cup \{\_\}$ and $o_k \in O \cup \{\_\}$. Note that we use the notation presented in Sect. 2. The symbol "_" denotes a missing component. A *key type* is a schema triple that includes the types of the key components as well as the *underlined types* of components that are not parts of keys. Formally, a type of a key (s_k, p_k, o_k) is a schema triple (s_t, p_t, o_t), such that $s_k \in [\![s_t]\!]^*_g$, $p_k \in [\![p_t]\!]^*_g$ and $o_k \in [\![o_t]\!]^*_g$ for all the components s_k, p_k and o_k that are not "_". The underlined types of the missing components are computed from the stored schema graph. The *complete* type of the key represents the key type where the underlines are omitted, i.e., the complete type of the key is a bare schema triple.

Example 3. The triple (_,wasBornIn,athens) is a key with the components p=wasBornIn and o=athens, while the component s does not contain a value.

The schema triple (person,wasBornIn,location) is the type of the key (_,wasBornIn,athens). Finally, the complete type of (_,wasBornIn,athens) is the schema triple (person,wasBornIn,location).

Computing Statistics for Keys. Let us now present the procedure for updating the statistics for a given triple $t = (s, p, o)$ and the corresponding schema triple $t_t = (t_s, t_p, t_o)$. In order to store the statistics of a triple t of type t_t, we split t_t into the seven key types: $(t_s, \underline{t_p, t_o})$, $(\underline{t_s}, t_p, \underline{t_o})$, ..., $(\underline{t_s}, t_p, t_o)$, (t_s, t_p, t_o). Furthermore, the triple t is split into the seven keys: $(s, \_, \_)$, $(\_p, \_)$, ..., $(\_, p, o)$, (s, p, o). The statistics is updated for each of the selected *key types*.

There is more than one way of counting the instances of a given *key type*. For the following discussion we select an example of the key type, say $t_k = (\underline{t_s}, t_p, t_o)$, and consider all the ways we can count the key $(\_, p, o)$ for a given key type t_k. The conclusions that we draw in the following paragraphs are general in the sense that they hold for all key types.

1. Firstly, we can either count all keys, including duplicates, or, secondly, we can count only the different keys. We denote these two options by using the parameters *all* or *distinct*, respectively.
2. Secondly, we can either count triples of the type (t_s, t_p, t_o), or, the triples of the type $(\top, t_p, t_o)$. In the first case we call counting *bound* and in the second case we call it *unbound*.

The above stated choices are specified as parameters of the generic procedure UPDATE-KEYTYPE (all|distinct,bound|unbound,t_k,t). The first parameter specifies if we count all or distinct triples. The second parameter defines the domain of counters, which is either restricted by a given complete type t_t of key type t_k, or unrestricted, i.e., the underlined component is not bound by any type. The third parameter is a key type t_k, and, finally, the fourth parameter represents the triple t.

Procedure UPDATE-STATISTICS. Let us now present the procedure UPDATE-STATISTICS(t_t,t) for updating the entry of the statistical index that corresponds to the key schema triple t_t and a triple t. The procedure is presented in Algorithm 5.

Algorithm 5. Procedure UPDATE-STATISTICS(t_t:schema-triple, t : triple)

```
procedure UPDATE-STATISTICS(t_t: schema-triple, t : triple)
    for all key types t_k of t_t do
        UPDATE-KEYTYPE(all,unbound,t_k);
        UPDATE-KEYTYPE(all,bound,t_k);
        UPDATE-KEYTYPE(dist,unbound,t_k, t);
        UPDATE-KEYTYPE(dist,bound,t_k, t);
```

The parameters of the procedure UPDATE-STATISTICS are the schema triple t_t and the triple t such that t_t is a type of t. The FOR statement in line 1 generates all seven key types of the schema triple t_t. The procedure UPDATE-KEYTYPE is applied to each of the generated key types with the different values of the first and the second parameters.

4 Experimental Evaluation

In this section, we present the evaluation of the algorithms for the computation of the statistics for the two example knowledge graphs. First of all, we present the experimental environment used to compute statistics. Secondly, the statistics of a simple knowledge graph introduced in Fig. 1 is presented in Sect. 4.2. Finally, the experiments with the computation of the statistics of the Yago are presented in Sect. 4.3.

4.1 Testbed Description

The algorithms for the computation of the statistics of graphs are implemented in the open-source system for querying and manipulation of graph databases *epsilon* [29]. *epsilon* is a lightweight RDF store based on Berkeley DB [22]. It can execute *basic graph-pattern queries* on datasets that include up to 1G (10^9) triples.

epsilon was primarily used as a tool for browsing ontologies. The operations implemented in *epsilon* are based on the sets of identifiers I, as they are defined in Sect. 2.1. The operations include computing transitive closures based on some relationship (e.g., the relationship rdfs:subClassOf), level-wise computation of transitive closures, and computing the least upper bounds of the set elements with respect to the stored ontology. These operations are used in the procedures for the computation of the triple-store statistics.

4.2 Knowledge Graph *Simple*

Table 1(a) describes the properties of the schema graphs computed by the three algorithms for the computation of the statistics. Each line of the table represents an evaluation of the particular algorithm on the knowledge graph *Simple*. The algorithms denoted by the keywords *ST*, *AL* and *LV* refer to the algorithms STATISTICS-STORED, STATISTICS-ALL and STATISTICS-LEVELS presented in the Sects. 3.1–3.3, respectively. The running time of all the algorithms used in the experiments was below 1ms.

The columns *#ulevel* and *#llevel* represent the number of levels above and below the stored schema graph that are relevant solely for the algorithm *LV*. The columns #bound and #ubound store the number of schema triples included in the schema graph of the computed statistics. The former is computed with the *bound* type of counting, and the latter with the *unbound* type of counting.

Algorithm *ST* computes the statistics solely for the stored schema graph. This algorithm can be compared to the relational approach, where the statistics are computed for each relation. Algorithm *AL* computes the statistics for all possible schema triples. The number of schema triples computed by this algorithm is significant, even for this small instance. In algorithm *LV*, there are only three levels of the schemata, i.e., the maximal #ulevel and #dlevel both equal 2. The statistics of the schema triples that are above the stored schema triples are usually computed since we are interested in having the global statistics based on the most general schema triples from the schema graph. Note that the algorithm *ST* gives the same results as the algorithm *LV* with the parameters #ulevel = 0 and #dlevel = 0.

Table 1. The evaluation of the Algorithms 1–3 on: **a)** *Simple* **b)** Yago-S

algorithm	#ulevel	#dlevel	#bound	#unbound
ST	-	-	63	47
AL	-	-	630	209
LV	0	0	63	47
LV	0	1	336	142
LV	0	2	462	173
LV	1	0	147	72
LV	1	1	476	175
LV	1	2	602	202
LV	2	0	161	76
LV	2	1	490	178
LV	2	2	616	205

algorithm	#ulevel	#dlevel	#bound	time-b	#unbound	time-u
ST	-	-	532	1	405	1
AL	-	-	>1M	>24	>1M	>24
LV	0	0	532	4.9	405	4.9
LV	1	0	3325	6.1	1257	5.4
LV	2	0	8673	8.1	2528	6.2
LV	3	0	12873	9.1	3400	6.8
LV	4	0	15988	10.9	3998	7.1
LV	5	0	18669	11.7	4464	7.5
LV	6	0	20762	12.5	4819	7.8
LV	7	0	21525	13.1	4939	7.9
LV	0	1	116158	5.9	27504	5.4
LV	1	1	148190	7.8	34196	6.3
LV	2	1	183596	10.5	40927	7.4
LV	3	1	207564	12.7	45451	7.8
LV	4	1	225540	14.4	48545	8.3
LV	5	1	239463	15.6	50677	8.7
LV	6	1	250670	17.3	52376	9.1
LV	7	1	255906	19.8	53147	9.2
LV	0	2	969542	7.6	220441	6.6
LV	1	2	>1M	>24	>1M	>24

4.3 Knowledge Graph Yago-S

In this section, we present the evaluation of the algorithms for the computation of the statistics of the Yago-S knowledge graph—we use the core of the Yago 2.0 knowledge base [14], including approximately 25M (10^6) triples. The Yago-S dataset, together with the working environment for the computation of the statistics, is available from the *epsilon* data repository [28].

Yago includes three types of classes: the Wordnet classes, the Yago classes, and the Wikipedia classes. There are approximately 68000 Wordnet classes used in Yago that represent the top hierarchy of the Yago taxonomy. The classes introduced within the Yago dataset are mostly used to link the parts of the datasets. For example, they link the Wikipedia classes to the Wordnet hierarchy. There are less than 30 newly defined Yago classes. The Wikipedia classes are defined to represent group entities, individual entities, or some properties of the entities. There are approximately 500000 Wikipedia classes in Yago 2.0.

We use solely the Wordnet taxonomy and newly defined Yago classes for the computation of the statistics. We do not use the Wikipedia classes since, in most cases, they are very specific. The height of the taxonomy is 18 levels, but there are a small number of branches that are higher than 10. While there are approximately 571000 classes in Yago, it includes only a small number of predicates. There are 133 predicates defined in Yago 2.0 [14], and there are only a few sub-predicates defined; therefore, the structure of the predicates is almost flat.

Table 1(b) presents the execution of the algorithms STATISTICS-STORED, STATISTICS-ALL and STATISTICS-LEVELS on the Yago-S knowledge graph. It includes the same columns as defined for Table 1(a) (see Sect. 4.2), and two additional columns. The two additional columns, named time-b and time-u, represent the time in hours used for the computation of the statistics with either *bound* or *unbound* ways of counting, respectively.

The computation time for updating the statistics for a given triple t depends on the selected algorithm and the complexity of the triple enrollment into the conceptual schemata. In general, the more schema triples that include t in their interpretation, the longer the computation time is. For example, the computation time increases significantly in the case that lower levels of the ontology are used to compute the statistics, simply because there are many schema triples (approx. 450K) on the lower levels of the ontology. Finally, the computation of the statistics for the triples that describe people and their activities takes much more time than the computation of the statistics for the triples that represent links between websites, since the triples describing people have richer relationships to the schema than the triples describing links among URIs.

Let us now give some comments on the results presented in Table 1(b). The algorithm ST for the computation of statistics of a knowledge graph, including 25 Mega triples, takes about 1 h to complete in the case of the bound and unbound types of counting. The computation of the algorithm AL does not complete because it consumes more than 32 GB of main memory after generating more than 2M schema triples.

The algorithm STATISTICS-LEVELS can regulate the amount of the schema triples that serve as the framework for the computation of the statistics. The number of the generated schema triples increases when more levels, either above or below the stored schemata, are taken into account. We executed the algorithm with the parameter #dlevel= $0, 1, 2$. For each fixed value of #dlevel, the second parameter #ulevel value varied from 0 to 7. The number of generated schema triples increases by about a factor of 10 when the value of #dlevel changes from 0 to 1 and from 1 to 2. Note also that varying #ulevel from 0 to 7 for each particular #dlevel results in an almost linear increase of the generated schema triples. This is because the number of schema triples is falling fast as they are closer to the most general schema triples from the knowledge graph.

5 Related Work

In relational systems, the statistics are used for the estimation of the selectivity of simple predicates and join queries in a query optimization process [6]. For each relation, the gathered statistics include the cardinality, the number of pages, and the fraction of the pages that store tuples. Further, for each index, a relational system stores the number of distinct keys and the number of pages of the index. The equidistant histograms [3] were proposed to capture the non-uniform distribution of the attribute values. Piatetsky-Shapiro and Connell show in [23] that the equidistant histograms fail very often to give a precise estimation of the selectivity of simple predicates. To improve the precision of the selectivity estimation, they propose the use of histograms that have equal heights, instead of equal intervals.

Most of the initially designed RDF stores treated stored triples as the edges of the schema-less graphs [9, 10, 13, 18, 21, 33, 34]. To efficiently process triple patterns, these RDF stores use either a subset of the six SPO indexes, or a set of tables (a single triple-table or property tables) stored in a RDBMS. However, the RDF data model [25] was extended with the RDF-Schema [26] that allows for the representation of knowledge bases [2, 17]. Consequently, RDF graphs can separate the conceptual and instance levels of the representation, i.e., between the TBox and ABox [1]. Moreover, RDF-Schema can serve as the means for the definition of taxonomies of classes (or concepts) and predicates, i.e., the roles of a knowledge representation language [2]. Let us now present the existent approaches to collect the statistics of triplestores.

Virtuoso [21] is based on relational technology. Since relational statistics can not capture well the semantics of SPARQL queries, the statistics are computed in real-time by inspecting one of 6 SPO indexes [4]. The size of conjunctive queries that can include one comparison operation ($\geq, >, <$, or $\leq$) is estimated by counting the pointers of index blocks satisfying the conditions on the complete path from the root to the leaf block of the index. In the case there are no conditions in a query, then sampling (1% of triples) is used to estimate the size of the query—we choose pointers of the index blocks at random. The results of query estimation are always stored in the index so that they are available for the following requests.

RDF-3X [20] is a centralized triplestore system. RDF-3X uses six indexes for each ordering of triplestore columns S, P, and O. B+ tree indexes are customized in the following ways. Firstly, triples are stored directly in the leaves of B+ trees. Secondly, each index uses the lexicographic ordering of triples, which provides the opportunity to compress triples in leaves by storing only the differences between the triples. RDF-3X includes additional aggregated indexes where the number of triples is stored for each particular instance (value) of the prefix for each of the six indexes. Aggregate indexes can be used for the selectivity estimation of arbitrary triple patterns. They are converted into selectivity histograms that can be stored in the main memory to improve the performance of selectivity estimation. Furthermore, to provide a more precise estimation of the size of queries in the presence of correlated predicates, frequent paths are determined, and their cardinality is computed and stored.

TriAD [8] is a distributed triplestore implemented on shared-nothing servers running centralized RDF-3X [20]. The triplestore is partitioned utilizing a multilevel graph partitioning algorithm [16] that generates graph summarizations. These are further used for query optimization as well as during the query execution to enable join-ahead pruning. Six distributed indexes are generated for each of the SPO permutations. Partitions are stored and indexed on slave servers, while the summary graph is stored and also indexed at the master server. The statistics of the triplestore are stored locally for the local partitions and, globally, for the summary graph. In both cases, the cardinality is stored for each value of S, P, and O, and, for the pairs of values SP, SO and PO. Furthermore, the selectivity of joins between P_1 and P_2 predicates is also stored in the distributed index on all the slave servers and for the summary graph on the master server.

Stocker et al. [31] was the first to use the statistics based on some form of semantic data. The statistics are used for the computation of the selectivity estimations in the query optimization algorithm that chooses the ordering of joins for basic graph patterns. For the triple patterns, the statistics are gathered for the bound (concrete) S, P, and O components. The number of triples with bound S component is approximated with the total number of triples and the number of distinct S values. Next, the number of triples with a given concrete predicate P is computed. Finally, equidistant histograms are computed for each particular predicate. The histograms represent the O component value distribution. Further, to compute the statistics of joins, RDF schema statements are used to enumerate all pairs of predicates that match in domain/range of the first predicate with the domain/range of the second predicate. The number of triples is computed for each such pair of related predicates and employed as the upper bound of any basic graph pattern that includes a given pair of predicates.

Neumann and Moetkotte propose the use of the characteristic sets [19] (abbr. CS) for the computation of statistics of star-shaped queries. For a given graph g and the subject s, a CS includes the predicates $\{p|(s, p, o) \in g\}$. The statistics are computed for each CS of a given RDF store. Furthermore, the CS-s of data and the CS-s of queries are computed. A star-shaped SPARQL query retrieves instances of all CS-s that are the super-sets of some CS from a query. Besides the statistics of CS-s, the number of triples with a given predicate is computed for each CS and each predicate. The size of joins in star-queries can be accurately estimated in this way. Gubichev and Neumann further extended the work on characteristic sets (abbr. CS) in [7]. CS-s are organized in a hierarchy. The first level includes all CS-s of a given triplestore. The next level includes the cheapest CS-s that are the subsets of CS-s from the previous level. The hierarchical characterization of CS-s allows precise estimation of joins in star queries.

6 Conclusions

This paper presents a new method for the computation of the statistics of knowledge graphs. The statistics are based on the schema graph. We propose a technique to tune the size of the statistics; the user can choose between small, coarse-grained statistics and different levels of larger, finer-grained statistics. The smallest schema graph that can be used as the framework for the computation of the statistics is the stored schema graph, while the largest schema graph corresponds to all possible schema triples that can be induced from the knowledge graph. In between, we can set the number of levels above and below the schema triples from the stored schema graph to be included.

The computation of the statistics of the core of Yago, including around 25M triples, takes from 1–24 h, depending on the size of the resulted schema graph of the statistics. The experiments were run on a low-cost server with a 3.3 GHz Intel Core CPU, 16 GB of RAM, and 7200 RPM 1 TB disk. Therefore, it is feasible to compute statistics even for large knowledge graphs. Note that knowledge graphs are not frequently updated so that the statistics can be computed offline in a batch job.

Finally, the current implementation of algorithms for the computation of the statistics of knowledge graphs is not tuned for performance. While it includes some optimizations, for instance, the transitive closures of specific sets of classes are cached in the main memory index, there is a list of additional tuning options at the implementation level that can speed up the computation of the statistics in practice. For example, the complete schema graph can be stored in the main memory to speed up the operations that involve the schema graph solely.

Acknowledgments. The authors acknowledge the financial support from the Slovenian Research Agency (research core funding No. P1-00383).

References

1. Baader, F., Calvanese, D., McGuinness, D., Nardi, D., Patel-Schneider, P.: Description Logic Handbook. Cambridge University Press, Cambridge (2002)
2. Brachman, R.J., Levesque, H.J.: Knowledge Representation and Reasoning. Elsevier, Amsterdam (2004)
3. Christodoulakis, S.: Estimating block transfers and join sizes. In: Proceedings of SIGMOD 1983, SIGMOD, New York, NY, USA, pp. 40–54. ACM (1903)
4. Erling, O.: Implementing a SPARQL Compliant RDF Triple Store Using a SQL-ORDBMS. OpenLink Software (2009)
5. Färber, M., Bartscherer, F., Menne, C., Rettinger, A.: Linked data quality of DBpedia, Freebase, OpenCyc, Wikidata, and YAGO. Seman. Web J. **1**, 1–53 (2017)
6. Griffiths-Selinger, P., Astrahan, M., Chamberlin, D., Lorie, A., Price, T.: Access path selection in a relational database management system. In: Proceedings of SIGMOD 1979, SIGMOD, New York, NY, USA, pp. 23–34. ACM (1979)
7. Gubichev, A., Neumann, T.: Exploiting the query structure for efficient join ordering in sparql queries. In: Amer-Yahia, S., Christophides, V., Kementsietsidis, A., Garofalakis, M.N., Idreos, S., Leroy, V. (eds.) EDBT, pp. 439–450. OpenProceedings.org (2014)
8. Gurajada, S., Seufert, S., Miliaraki, I., Theobald, M.: TriAD: a distributed shared-nothing RDF engine based on asynchronous message passing. In: Proceedings of the 2014 ACM SIGMOD International Conference on Management of Data, SIGMOD 2014, New York, NY, USA, pp. 289–300. ACM (2014)
9. Harris, S., Gibbins, N.: 3store: efficient bulk RDF storage. In: 1st International Workshop on Practical and Scalable Semantic Systems (PSSS 2003), pp. 1–15, 20 October 2003
10. Harris, S., Lamb, N., Shadbolt, N.: 4store: the design and implementation of a clustered Rdf store. In: Proceedings of the 5th International Workshop on Scalable Semantic Web Knowledge Base Systems (2009)
11. Harth, A., Decker, S.: Optimized index structures for querying RDF from the web. In: Third Latin American Web Congress (LA-WEB'2005), vol. 2005, pp. 71–80, January 2005
12. Harth, A., Hose, K., Schenkel, R.: Database techniques for linked data management. In: Proceedings of the 2012 ACM SIGMOD International Conference on Management of Data, SIGMOD 2012, pp. 597–600, New York, NY, USA. ACM (2012)
13. Harth, A., Umbrich, J., Hogan, A., Decker, S.: YARS2: a federated repository for querying graph structured data from the web. In: Aberer, K., et al. (eds.) ASWC/ISWC -2007. LNCS, vol. 4825, pp. 211–224. Springer, Heidelberg (2007). https://doi.org/10.1007/978-3-540-76298-0_16
14. Hoffart, J., Suchanek, F.M., Berberich, K., Weikum, G.: YAGO2: a spatially and temporally enhanced knowledge base from Wikipedia. Artif. Intell. **194**, 28–61 (2013). Artificial Intelligence, Wikipedia and Semi-Structured Resources
15. Hogan, A., et al.: Knowledge graphs. ACM Comput. Surv. **54**(4) (2021)
16. Karypis, G., Kumar, V.: A fast and high quality multilevel scheme for partitioning irregular graphs. SIAM J. Sci. Comput. **20**(1), 359–392 (1999)
17. Lenat, D.B.: Cyc: a large-scale investment in knowledge infrastructure. Commun. ACM **38**(11), 33–38 (1995)
18. McBride, B.: Jena: a semantic web toolkit. IEEE Internet Comput. **6**(6), 55–59 (2002)
19. Neumann, T., Moerkotte, G.: Characteristic sets: accurate cardinality estimation for RDF queries with multiple joins. In: Proceedings of the 2011 IEEE 27th International Conference on Data Engineering, ICDE 2011, pp. 984–994, Washington, DC, USA. IEEE Computer Society (2011)

20. Neumann, T., Weikum, G.: The RDF3X engine for scalable management of RDF data. VLDB J. **19**(1), 91–113 (2010)
21. OpenLink Software Documentation Team. OpenLink Virtuoso Universal Server: Documentation (2009)
22. Oracle Corporation. Oracle Berkeley DB 11g Release 2 (2011)
23. Piatetsky-Shapiro, G., Connell, C.: Accurate estimation of number of tuples satisfying condition. In: Proceedings of SIGMOD 1984, SIGMOD, pp. 256–276, New York, NY, USA. ACM (1984)
24. Ramachandran, D., Reagan, P., Goolsbey, K.: First-orderized ResearchCyc: expressivity and efficiency in a common-sense ontology. In: AAAI Reports, AAAI (2005)
25. Resource description framework (RDF) (2004). http://www.w3.org/RDF/
26. RDF schema (2004). http://www.w3.org/TR/rdf-schema/
27. Savnik, I., Nitta, K.: Method of Big-graph partitioning using a skeleton graph. In: Milutinovic, V., Kotlar, M. (eds.) Exploring the DataFlow Supercomputing Paradigm. CCN, pp. 3–39. Springer, Cham (2019). https://doi.org/10.1007/978-3-030-13803-5_1
28. Savnik, I., Nitta, K.: Datasets: Simple and YAGO-s (2021). http://osebje.famnit.upr.si/~savnik/epsilon/datasets/
29. Savnik, I., Nitta, K.: Epsilon: data and knowledge graph database system (2021). http://osebje.famnit.upr.si/~savnik/epsilon/epsilon/
30. Savnik, I., Nitta, K., Skrekovski, R., Augsten, N.: Statistics of knowledge graphs based on the conceptual schema. Technical Report, Cornell University (2021). arXiv:2109.09391
31. Stocker, M., Seaborne, A., Bernstein, A., Kiefer, C., Reynolds, D.: Sparql basic graph pattern optimization using selectivity estimation. In: WWW 2008, Semantic Web II, WWW 2008, pp. 595–604, New York, NY, USA. ACM (2008)
32. Webber, J.: A programmatic introduction to Neo4j. In: Proceedings of the 3rd Annual Conference on Systems, Programming, and Applications: Software for Humanity, SPLASH 2012, pp. 217–218, New York, NY, USA. ACM (2012)
33. Weiss, C., Karras, P., Bernstein, A.: Hexastore: sextuple indexing for semantic web data management. Proc. VLDB Endow. **1**(1), 1008–1019 (2008)
34. Wilkinson, K., Sayers, C., Kuno, H., Reynolds, D.: Efficient RDF storage and retrieval in Jena2 (2003)
35. Zou, L., Özsu, M.T., Chen, L., Shen, X., Huang, R., Zhao, D.: gStore: a graph-based SPARQL query engine. VLDB J. **23**(4), 565–590 (2014)

Can You Answer While You Wait?

Luís Cruz-Filipe[1(✉)], Graça Gaspar[2], and Isabel Nunes[2]

[1] Department of Mathematics and Computer Science, University of Southern Denmark, Odense, Denmark
lcfilipe@gmail.com

[2] LASIGE, Department of Informatics, Faculty of Sciences, University of Lisbon, Lisbon, Portugal
{mdgaspar,minunes}@fc.ul.pt

Abstract. Continuous query answering is a challenging problem faced by systems that need to reason over data as it arrives. Recently, a logic-based approach to this problem has been proposed that advocates generating hypothetical query answers – potential answers that are consistent with the available data, but still require confirmation by future input.

The current work studies hypothetical query answering in realistic settings, where data may arrive out of order. This requires revising its semantics and reanalysing the intuitions that led to the design of the existing algorithms, in order to develop a novel incremental online algorithm that takes into account that past data may yet arrive. We also discuss how our methods may be extended to channels with losses.

1 Introduction

Reasoning systems used in today's world must react in real time to information that they receive from e.g. sensors, continuously producing results in an online fashion. Conceptually, one of the ways to model this influx of information is as a data stream, and the reasoning tasks that these systems have to perform are usually called *continuous queries*.

One important line of work [9,17,39,42,46] views continuous queries as logic programs whose facts arrive through a data stream, and applies logic-based methods to compute answers to those queries. In practice, answers may take a long time to compute, and it may be useful to know that some answers are likely to be produced in the future, based on available information that might lead to their generation. This motivated the introduction of *hypothetical answers* [15]: answers supported by the information provided so far by the input stream, but that depend on other facts in the future being true.

Depending on the underlying communication system, information may be delayed, and may arrive unordered. The reasoning system must therefore allow for the possibility that absent data may still arrive, and if necessary delay the production of answers [6,18,43]. The approaches mentioned above abstract from communication delays by assuming that the data from the data stream is ordered: information with timestamp greater than t is "put on hold" until all

I. Varzinczak (Ed.): FoIKS 2022, LNCS 13388, pp. 111–129, 2022.
https://doi.org/10.1007/978-3-031-11321-5_7

data about timestamp t is produced. This assumption simplifies the theoretical development greatly, but is not easily implementable.

As soon as we remove the assumption that the data stream is ordered, the typical strategies for continuous query answering start failing, since they rely on the possibility of processing information grouped by time point. In the present work, we propose an alternative approach to computing and updating hypothetical answers that treats communication delays explicitly. We define a procedure for incrementally computing hypothetical answers in the presence of delays, by means of flexible strategies that deal with information as it arrives while acknowledging the possibility that older data may arrive later on. The only requirement is that a limit is known to how delayed the information may be, which we argue is reasonable in many practical applications. In the conclusions, we discuss how to remove this requirement.

Structure. Section 2 revisits the syntax of Temporal Datalog [42] and the main ideas behind the formalism of hypothetical answers [15], and introduces the running example that we use throughout this article. Section 3 updates the declarative and operational semantics of hypothetical answers to allow for communication delays. The new online algorithm for maintaining and updating hypothetical answers, given in Sect. 4, relies on the novel concept of *local mgu*. This section also includes proofs of soundness and completeness of the algorithm. Section 5 discusses alternative approaches to communication delays, and Sect. 6 includes some conclusions and directions for future work.

2 Background

2.1 Continuous Queries in Temporal Datalog

We work in *Temporal Datalog*, the fragment of negation-free Datalog extended with the special temporal sort from [13]. The formalism for writing continuous queries over datastreams follows [42] with slight adaptations.

Syntax of Temporal Datalog. Temporal Datalog is an extension of Datalog [12] where constants and variables can have two sorts: *object* or *temporal*. Terms are also sorted: an *object term* is either an object constant or an object variable, and a *time term* is either a natural number, a time variable, or an expression of the form $T + \mathsf{k}$ where T is a time variable and k is an integer. Time constants are also called *timestamps*.

Predicates take exactly one temporal parameter, which is the last one. (Some works also allow predicates with no temporal parameter, but this adds no expressive power to the language.) Atoms, rules, facts and programs are defined as usual. Rules are assumed to be safe: each variable in the head must occur in the body.

A term, atom, rule, or program is *ground* if it contains no variables. We write $\mathsf{var}(\alpha)$ for the set of variables occurring in an atom α, and extend this function homomorphically to rules and sets. A *fact* is a function-free ground atom.

A predicate symbol is said to be *intensional* or IDB if it occurs in an atom in the head of a rule with non-empty body, and *extensional* or EDB if it is defined only through facts. This classification extends to atoms in the natural way.

Substitutions are functions mapping a finite set of variables to terms of the expected sort. Given a rule r and a substitution θ, the corresponding *instance* $r' = r\theta$ of r is obtained by simultaneously replacing every variable X in r by $\theta(X)$ and computing any additions of temporal constants.

A *temporal query* is a pair $Q = \langle P, \Pi \rangle$ where Π is a program and P is an IDB atom in the language underlying Π. We do not require P to be ground, and typically the temporal parameter is uninstantiated.[1]

A *dataset* is a family $D = \{D|_\tau \mid \tau \in \mathbb{N}\}$, where $D|_\tau$ represents the set of EDB facts delivered by a data stream at time point τ. Note that every fact in $D|_\tau$ has timestamp at most τ; facts with timestamp lower than τ correspond to communication delays. We call $D|_\tau$ the τ*-slice* of D, and define also the τ-history $D_\tau = \bigcup\{D|_{\tau'} \mid \tau' \leq \tau\}$. It follows that $D|_\tau = D_\tau \setminus D_{\tau-1}$ for every τ, and that D_τ also contains only facts whose temporal argument is at most τ. By convention, $D_{-1} = \emptyset$.

Semantics. The semantics of Temporal Datalog is defined in the usual way over Herbrand models, evaluating time terms to a natural number in the obvious way.

An *answer* to a query $Q = \langle P, \Pi \rangle$ over a set of ground facts S is a ground substitution θ over the set of variables in P such that $\Pi \cup S \models P\theta$. In this work, S is typically a τ-history of some dataset D. We denote the set of all answers to Q over D_τ as $\mathcal{A}(Q, D, \tau)$.

We illustrate these concepts with our running example, which is a variant of Example 1 in [42].

Example 1. The following program Π_E tracks activation of cooling measures in a set of wind turbines equipped with sensors, recording malfunctions and shutdowns, based on temperature readings $\mathsf{Temp}(\mathit{Device}, \mathit{Level}, \mathit{Time})$.

$$\begin{aligned}
\mathsf{Temp}(X, \mathsf{high}, T) &\rightarrow \mathsf{Flag}(X, T)\\
\mathsf{Flag}(X, T) \wedge \mathsf{Flag}(X, T+1) &\rightarrow \mathsf{Cool}(X, T+1)\\
\mathsf{Cool}(X, T) \wedge \mathsf{Flag}(X, T+1) &\rightarrow \mathsf{Shdn}(X, T+1)\\
\mathsf{Shdn}(X, T) &\rightarrow \mathsf{Malf}(X, T-2)
\end{aligned}$$

Malfunctions correspond to answers to the query $Q_E = \langle \mathsf{Malf}(X, T), \Pi_E \rangle$. If we assume $D_0 = \{\mathsf{Temp}(\mathsf{wt2}, \mathsf{high}, 0)\}$, then at time point 0 there is no answer to Q_E. If $\mathsf{Temp}(\mathsf{wt2}, \mathsf{high}, 1)$ arrives to D at time point 1, then $D_1 = D_0 \cup \{\mathsf{Temp}(\mathsf{wt2}, \mathsf{high}, 1)\}$, and there still is no answer to Q_E. Finally, the arrival of $\mathsf{Temp}(\mathsf{wt2}, \mathsf{high}, 2)$ to D at time point 2 yields $D_2 = D_1 \cup \{\mathsf{Temp}(\mathsf{wt2}, \mathsf{high}, 2)\}$, allowing us to infer $\mathsf{Malf}(\mathsf{wt2}, 0)$. Then $\{X := \mathsf{wt2}, T := 0\} \in \mathcal{A}(Q_E, D, 2)$. ◁

[1] The most common exception is if P represents a property that does not depend on time, where by convention the temporal parameter is instantiated to 0.

In this example, all facts were delivered instantly by the data stream. In the presence of communication delays, the answer $\{X := \mathsf{wt2}, T := 0\}$ might not be produced at time point 2, but only later.

2.2 SLD-resolution

Our development is based on SLD-resolution. We summarize the key relevant concepts and results, following [10,35].

A *definite clause* is a disjunction of literals containing at most one positive literal. A definite clause with only negative literals is called a *goal* and written $\neg \bigwedge_j \beta_j$, where the β_j are atoms. Definite clauses with one positive literal are written in the standard rule notation $\bigwedge_i \alpha_i \rightarrow \alpha$, where the α_i and α are atoms.

A *most general unifier (mgu)* of two atomic formulas $P(\vec{X})$ and $P(\vec{Y})$ is a substitution θ such that: (i) $P(\vec{X})\theta = P(\vec{Y})\theta$ and (ii) for all σ, if $P(\vec{X})\sigma = P(\vec{Y})\sigma$ then $\sigma = \theta\gamma$ for some substitution γ. If C is a rule $\bigwedge_i \alpha_i \rightarrow \alpha$, G is a goal $\neg \bigwedge_j \beta_j$ with $\mathsf{var}(G) \cap \mathsf{var}(C) = \emptyset$, and θ is an mgu of α and β_k, then the *resolvent* of G and C using θ is the goal $\neg \left(\bigwedge_{j<k} \beta_j \wedge \bigwedge_i \alpha_i \wedge \bigwedge_{j>k} \beta_j \right) \theta$.

Let P be a program and G be a goal. An *SLD-derivation* of $P \cup \{G\}$ is a sequence $G_0, G_1, \ldots$ of goals, a sequence $C_1, C_2, \ldots$ of α-renamings of program clauses of P and a sequence $\theta_1, \theta_2, \ldots$ of substitutions such that $G_0 = G$ and G_{i+1} is the resolvent of G_i and C_{i+1} using θ_{i+1}. An *SLD-refutation* of $P \cup \{G\}$ is a finite SLD-derivation of $P \cup \{G\}$ ending in the empty clause ($\Box$), and its *computed answer* is obtained by restricting the composition of $\theta_1, \ldots, \theta_n$ to the variables occurring in G.

SLD-resolution is sound and complete. If θ is a computed answer for $P \cup \{G\}$, then $P \models \neg(\forall G\theta)$. Conversely, if $P \models \neg(\forall G\theta)$, then there exist σ and γ such that $\theta = \sigma\gamma$ and σ is a computed answer for $P \cup \{G\}$. The *independence of the computation rule* states that the order in which the literals in the goal are resolved with rules from the program does not affect the existence of an SLD-refutation.

2.3 Hypothetical Answers

In Example 1, the answer to the query Q_E could only be determined when $\tau = 2$. However, we could already infer that this answer might arise from the fact that $D_0 = \{\mathsf{Temp}(\mathsf{wt2}, \mathsf{high}, 0)\}$. The later delivery of $\mathsf{Temp}(\mathsf{wt2}, \mathsf{high}, 1)$ supports this possibility, while its absence from the data stream would have discarded it.

The formalism of *hypothetical answers* [15] builds on this idea. A hypothetical answer to a query $Q = \langle P, \Pi \rangle$ given a dataset D and a time instant τ is a pair $\langle \theta, H \rangle$ where θ is a substitution over the variables in P and H is a finite set of ground EDB atoms (the hypotheses) such that all atoms in H have a timestamp larger than τ, $\Pi \cup D_\tau \cup H \models P\theta$, and H is minimal with respect to set inclusion. Furthermore, if the minimal subset E of D_τ such that $\Pi \cup E \cup H \models P\theta$ is non-empty, then $\langle \theta, H \rangle$ is said to be *supported* by E (the *evidence*).

Example 2. In Example 1, $\langle [X := \mathsf{wt2}, T := 0], \{\mathsf{Temp}(\mathsf{wt2}, \mathsf{high}, 2)\}\rangle$ is a hypothetical answer for Q_E and D at time 1. This answer is supported by the evidence $\{\mathsf{Temp}(\mathsf{wt2}, \mathsf{high}, 0), \mathsf{Temp}(\mathsf{wt2}, \mathsf{high}, 1)\}$.

The pair $\langle [X := \mathsf{wt2}, T := 3], \{\mathsf{Temp}(\mathsf{wt2}, \mathsf{high}, t) \mid t = 3, 4, 5\}\rangle$ is also a hypothetical answer for Q_E and D at time 1, but it is not supported. ◁

Hypothetical answers can be computed dynamically as information is delivered by the data stream. Given a query $Q = \langle P, \Pi\rangle$, an offline pre-processing step applies SLD-resolution to Π and $\leftarrow P$ until only EDB atoms remain. All leaves and corresponding substitutions are collected in a set $\mathcal{P}_Q$ of pairs $\langle \theta, H\rangle$.

For each time point τ, a set of schematic hypothetical answers $\mathcal{S}_\tau$ of the form $\langle \theta, E, H\rangle$ is computed online as follows.

1. If $\langle \theta, H\rangle \in \mathcal{P}_Q$ and σ is such that $H\sigma \setminus D|_\tau$ only contains atoms with timestamp higher than τ, then $\langle \theta\sigma, H\sigma \cap D|_\tau, H\sigma \setminus D|_\tau\rangle$ is added to $\mathcal{S}_\tau$.
2. If $\langle \theta, E, H\rangle \in \mathcal{S}_{\tau-1}$ and σ is such that $H\sigma \setminus D|_\tau$ only contains atoms whose timestamp is higher than τ, then the triple $\langle \theta\sigma, E \cup (H\sigma \cap D|_\tau), H\sigma \setminus D|_\tau\rangle$ is added to $\mathcal{S}_\tau$.

The candidate substitutions σ are easily obtained by unifying the elements of each set H that have minimal temporal argument with the elements of $D|_\tau$.

Example 3. In the context of our running example, the pre-processing step yields the singleton set

$$\mathcal{P}_{Q_E} = \{\langle \emptyset, \{\mathsf{Temp}(X, \mathsf{high}, T), \mathsf{Temp}(X, \mathsf{high}, T+1), \mathsf{Temp}(X, \mathsf{high}, T+2)\}\rangle\}.$$

Taking D_0, D_1 and D_2 as in Example 1, the previous algorithm yields

$$\begin{aligned}
\mathcal{S}_0 &= \{\langle [X := \mathsf{wt2}, T := 0], \{\mathsf{Temp}(\mathsf{wt2}, \mathsf{high}, 0)\}, \{\mathsf{Temp}(\mathsf{wt2}, \mathsf{high}, i) \mid i = 1, 2\}\rangle\}\\
\mathcal{S}_1 &= \{\langle [X := \mathsf{wt2}, T := 0], \{\mathsf{Temp}(\mathsf{wt2}, \mathsf{high}, i) \mid i = 0, 1\}, \{\mathsf{Temp}(\mathsf{wt2}, \mathsf{high}, 2)\}\rangle,\\
&\qquad \langle [X := \mathsf{wt2}, T := 1], \{\mathsf{Temp}(\mathsf{wt2}, \mathsf{high}, 1)\}, \{\mathsf{Temp}(\mathsf{wt2}, \mathsf{high}, i) \mid i = 2, 3\}\rangle\}\\
\mathcal{S}_2 &= \{\langle [X := \mathsf{wt2}, T := 0], \{\mathsf{Temp}(\mathsf{wt2}, \mathsf{high}, i) \mid i = 0, 1, 2\}, \emptyset\rangle,\\
&\qquad \{\langle [X := \mathsf{wt2}, T := 1], \{\mathsf{Temp}(\mathsf{wt2}, \mathsf{high}, i) \mid i = 1, 2\}, \{\mathsf{Temp}(\mathsf{wt2}, \mathsf{high}, 3)\}\rangle,\\
&\qquad \langle [X := \mathsf{wt2}, T := 2], \{\mathsf{Temp}(\mathsf{wt2}, \mathsf{high}, 2)\}, \{\mathsf{Temp}(\mathsf{wt2}, \mathsf{high}, i) \mid i = 3, 4\}\rangle\}
\end{aligned}$$

This algorithm is sound and complete: the supported hypothetical answers at each time point τ are the ground instantiations of the schematic answers computed at the same time point.

3 Introducing Communication Delays

So far, facts are assumed to be received instantaneously at the data stream: any fact with timestamp τ must be in $D|_\tau$, if present. However, communications take time, so a fact with timestamp τ may arrive later.

In this section, we extend the formalism of hypothetical answers to deal with these communication delays. We assume that communications may be delayed (but not lost) and that there is a known upper bound on the delay. In practice, there are communication protocols that ensure this property (with high enough probability).

The bound on the delay may be different for different predicates, and for different instantiations of the same predicate; the only restriction is that it may not depend on the timestamp.[2] We model this as a function δ mapping each ground EDB atom in the language of Π to a natural number, with the restriction that: if a, a' differ only in their temporal argument, then $\delta(a) = \delta(a')$. We extend δ to non-ground atoms by defining $\delta(P(t_1, \ldots, t_n))$ as the maximum of all $\delta(P(t'_1, \ldots, t'_n))$ such that $P(t'_1 \ldots, t'_n)$ is a ground instance of $P(t_1, \ldots, t_n)$, and to predicate symbols by $\delta(P) = \delta(P(X_1, \ldots, X_n))$.

3.1 Declarative Semantics

The first ingredient in our extension is the following notion.

Definition 1. *A ground atom* $P(t_1, \ldots, t_n)$ *is* future-possible for τ *if* $\tau < t_n + \delta(P(t_1, \ldots, t_n))$.

In other words, an atom is future-possible for τ if it still may be delivered by the data stream. Future-possible atoms generalize the notion of "atom with timestamp greater than τ" – in particular, any such atom is future-possible for τ.

Hypothetical answers can now be generalized to include future-possible atoms.

Definition 2. *A* hypothetical answer *to query* Q *over* D_τ *is a pair* $\langle \theta, H \rangle$*, where* θ *is a substitution and* H *is a finite set of ground EDB atoms (the hypotheses) such that:*

- $\mathsf{supp}(\theta) = \mathsf{var}(()P)$, *i.e.,* θ *only changes variables that occur in* P*;*
- H *only contains atoms future-possible for* τ*;*
- $\Pi \cup D_\tau \cup H \models P\theta$*;*
- H *is minimal with respect to set inclusion.*

$\mathcal{H}(Q, D, \tau)$ *is the set of hypothetical answers to* Q *over* D_τ*.*

As before, if the minimal subset E of D_τ such that $\Pi \cup E \cup H \models P\theta$ is non-empty, then $\langle \theta, H, E \rangle$ is a supported answer to Q over D_τ. We denote the set of all supported answers to Q over D_τ by $\mathcal{E}(Q, D, \tau)$.

The key properties of hypothetical and supported answers still hold for this generalized notion. The proofs are straightforward adaptations of those in [14].

[2] This is a reasonable assumption e.g. in the case of information originating from sensors, where the delay may depend on the distance and infrastructure and therefore be different for each sensor, but typically does not change over time. At the end of this article, we briefly discuss how to extend our framework to deal with unbounded communication delays and possible loss of information.

Proposition 1. *Let $Q = \langle P, \Pi \rangle$ be a query, D be a dataset and τ be a time instant. If $\langle \theta, \emptyset \rangle \in \mathcal{H}(Q, D, \tau)$, then $\theta \in \mathcal{A}(Q, D, \tau)$.*

Proposition 2. *Let $Q = \langle P, \Pi \rangle$ be a query, D be a dataset and τ be a time instant. If $\langle \theta, H \rangle \in \mathcal{H}(Q, D, \tau)$, then there exist a time point $\tau' \geq \tau$ and a dataset D' such that $D_\tau = D'_\tau$ and $\theta \in \mathcal{A}(Q, D', \tau')$.*

Proposition 3. *Let $Q = \langle P, \Pi \rangle$ be a query, D be a dataset and τ be a time instant. If $\langle \theta, H \rangle \in \mathcal{H}(Q, D, \tau)$, then there exists H^0 such that $\langle \theta, H^0 \rangle \in \mathcal{H}(Q, D, -1)$ and $H = H^0 \setminus D_\tau$. Furthermore, if $H \neq H^0$, then $\langle \theta, H, H^0 \setminus H \rangle \in \mathcal{E}(Q, D, \tau)$.*

Proposition 4. *Let $Q = \langle P, \Pi \rangle$ be a query, D be a dataset and τ be a time instant. If $\langle \theta, H \rangle \in \mathcal{H}(Q, D, \tau)$ and $\tau' < \tau$, then there exists $\langle \theta, H' \rangle \in \mathcal{H}(Q, D, \tau')$ such that $H = H' \setminus (D_\tau \setminus D_{\tau'})$.*

Example 4. We illustrate these concepts with the program from Example 1. We assume that $\delta(\mathsf{Temp}) = 1$, that $D|_0 = \{\mathsf{Temp}(\mathsf{wt4}, \mathsf{high}, 0)\}$, and that $D|_1 = \emptyset$. Let $\theta = [X := \mathsf{wt4}, T := 0]$. Then

$$\langle \theta, \{\mathsf{Temp}(\mathsf{wt4}, \mathsf{high}, 1), \mathsf{Temp}(\mathsf{wt4}, \mathsf{high}, 2)\} \rangle \in \mathcal{H}(Q_E, D, 1),$$

reflecting the intuition that $\mathsf{Temp}(\mathsf{wt4}, \mathsf{high}, 1)$ may still arrive in $D|_2$. This answer is supported by $\mathsf{Temp}(\mathsf{wt4}, \mathsf{high}, 0)$. ◁

3.2 Operational Semantics

The operational semantics based on SLD-resolution, which forms the basis for the online algorithm in [15], also extends to a scenario with communication delays: the key change is definining an operational counterpart to the notion of future-possible atom.

Definition 3. *An atom $P(t_1, \ldots, t_n)$ is a* potentially future atom wrt τ *if either t_n contains a temporal variable or t_n is ground and $\tau < t_n + \delta(P(t_1, \ldots, t_n))$.*

This concept generalizes the notion of future-possible atom for τ to possibly non-ground atoms. In particular, any atom whose temporal parameter contains a variable is potentially future, since it may be instantiated to a future timestamp.

Without loss of generality, we assume that all SLD-derivations have the following property: when unifying a goal $\neg \bigwedge_i \alpha_i$ with a clause $\bigwedge_j \beta_j \to \beta$, the chosen mgu $\theta = [X_1 := t_1, \ldots, X_n := t_n]$ is such that all variables occurring in $t_1, \ldots, t_n$ also occur in the chosen atom α_k.

Definition 4. *An* SLD-refutation with future premises *of Q over D_τ is a finite SLD-derivation of $\Pi \cup D_\tau \cup \{\neg P\}$ whose last goal only contains potentially future EDB atoms wrt τ.*

If $\mathcal{R}$ is an SLD-refutation with future premises of Q over D_τ with last goal $G = \neg \bigwedge_i \alpha_i$ and θ is the substitution obtained by restricting the composition of the mgus in $\mathcal{R}$ to $\mathsf{var}(P)$, then $\langle \theta, \bigwedge_i \alpha_i \rangle$ is a computed answer with premises *to Q over D_τ, denoted $\langle Q, D_\tau \rangle \vdash_{\mathsf{SLD}} \langle \theta, \bigwedge_i \alpha_i \rangle$.*

Example 5. Consider the setting of Example 4 and $\tau = 1$. There is an SLD-derivation of $\Pi \cup D_1 \cup \{\neg\mathsf{Malf}(X,T)\}$ ending with

$$\leftarrow \mathsf{Temp}(\mathsf{wt4},\mathsf{high},1), \mathsf{Temp}(\mathsf{wt4},\mathsf{high},2),$$

which contains two potentially future EDB atoms with respect to 1. Thus,

$$\langle Q_E, D_1\rangle \vdash_{\mathsf{SLD}} \langle \theta, \mathsf{Temp}(\mathsf{wt4},\mathsf{high},1) \wedge \mathsf{Temp}(\mathsf{wt4},\mathsf{high},2)\rangle$$

with $\theta = [X := \mathsf{wt4}, T := 0]$. ◁

As before, computed answers with premises are the operational counterpart to hypothetical answers, with two caveats: a computed answer with premises need not be ground, as the corresponding SLD-derivation may include some universally quantified variables in the last goal; and $\wedge_i \alpha_i$ may contain redundant conjuncts, in the sense that they might not be needed to establish the goal (see the examples in [15]).

With the new definitions, computed answers with premises and hypothetical answers are related as before.

Proposition 5 (Soundness). *Let $Q = \langle P, \Pi\rangle$ be a query, D be a dataset and τ be a time instant. Assume that $\langle Q, D_\tau\rangle \vdash_{\mathsf{SLD}} \langle \theta, \wedge_i \alpha_i\rangle$. Let σ be a ground substitution such that: (i) $\mathsf{supp}(\sigma) = \mathsf{var}(\wedge_i \alpha_i) \cup (\mathsf{var}(P) \setminus \mathsf{supp}(\theta))$ and (ii) if α_i is $P(t_1,\ldots,t_n)$, then $t_n\sigma + \delta(P(t_1,\ldots,t_n)) > \tau$. Then there is a set $H \subseteq \{\alpha_i\sigma\}_i$ such that $\langle (\theta\sigma)|_{\mathsf{var}(P)}, H\rangle \in \mathcal{H}(Q, D, \tau)$.*

Proposition 6 (Completeness). *Let $Q = \langle P, \Pi\rangle$ be a query, D be a dataset and τ be a time instant. If $\langle \theta, H\rangle \in \mathcal{H}(Q, D, \tau)$, then there exist substitutions ρ and σ and a finite set of atoms $\{\alpha_i\}_i$ such that $\theta = \rho\sigma$, $H = \{\alpha_i\sigma\}_i$ and $\langle Q, D_\tau\rangle \vdash_{\mathsf{SLD}} \langle \rho, \wedge_i \alpha_i\rangle$.*

The incremental algorithm in [15] is based on the idea of "organizing" SLD-derivations adequately to pre-process Π independently of D_τ, so that the computation of (hypothetical) answers can be split into an offline part and a less expensive online part. The following result asserts that this can be done.

Proposition 7. *Let $Q = \langle P, \Pi\rangle$ be a query and D be a dataset. For any time constant τ, if $\langle Q, D_\tau\rangle \vdash_{\mathsf{SLD}} \langle \theta, \wedge_i \alpha_i\rangle$, then there exist an SLD-refutation with future premises of Q over D_τ computing $\langle \theta, \wedge_i \alpha_i\rangle$ and a sequence $k_{-1} \leq k_0 \leq \ldots \leq k_\tau$ such that:*

- *goals $G_1, \ldots, G_{k_{-1}}$ are obtained by resolving with clauses from Π;*
- *for $0 \leq i \leq \tau$, goals $G_{k_{i-1}+1}, \ldots, G_{k_i}$ are obtained by resolving with clauses from $D|_i$.*

Proof. Immediate consequence of the independence of the computation rule. □

An SLD-refutation with the structure described in Proposition 7, which we call *stratified*, also has the property that all goals after $G_{k_{-1}}$ are always resolved with EDB atoms. Furthermore, each goal G_{k_i} contains only potentially future EDB atoms with respect to i. Let θ_i be the restriction of the composition of all substitutions in the SLD-derivation up to step k_i to $\mathsf{var}(P)$. Then $G_{k_i} = \neg \bigwedge_j \alpha_j$ represents all hypothetical answers to Q over D_i of the form $\langle (\theta_i\sigma)|_{\mathsf{var}(P)}, \bigwedge_j \alpha_j \rangle$ for some ground substitution σ (cf. Proposition 5).

This yields an online procedure to compute supported answers to continuous queries over data streams. In a pre-processing step, we calculate all computed answers with premises to Q over D_{-1}, and keep the ones with minimal set of formulas. (Note that Proposition 6 guarantees that all minimal sets are generated by this procedure, although some non-minimal sets may also appear.) The online part of the procedure then resolves each of these sets with the facts delivered by the data stream, adding the resulting resolvents to a set of schemata of supported answers (i.e. where variables may still occur). By Proposition 7, if there is at least one resolution step at this stage, then the hypothetical answers represented by these schemata all have evidence, so they are indeed supported.

The pre-processing step coincides exactly with that described in [15], which also includes examples and sufficient conditions for termination. The online part, though, needs to be reworked in order to account for communication delays. This is the topic of the next section.

4 Incremental Computation of Hypothetical Answers

Pre-processing a query returns a finite set $\mathcal{P}_Q$ that represents $\mathcal{H}(Q, D, -1)$: for each computed answer $\langle \theta, \wedge_i \alpha_i \rangle$ with premises to Q over D_{-1} where $\{\alpha_i\}_i$ is minimal, $\mathcal{P}_Q$ contains an entry $\langle \theta, \{\alpha_i\}_i \rangle$. Each tuple $\langle \theta, H \rangle \in \mathcal{P}_Q$ represents the set of all hypothetical answers $\langle \theta\sigma, H\sigma \rangle$ as in Proposition 5. We now show how to compute and update the set $\mathcal{E}(Q, D, \tau)$ schematically.

Definition 5. *Let Γ and Δ be sets of atoms such that all atoms in Δ are ground. A substitution σ is a* local mgu *for Γ and Δ if: for every substitution θ such that $\Gamma\theta \cap \Delta = \Gamma\sigma \cap \Delta$ there exists another substitution ρ such that $\theta = \sigma\rho$.*

Intuitively, a local mgu for Γ and Δ is a substitution that behaves as an mgu of some subsets of Γ and Δ. Local mgus allow us to postpone instantiating some atoms, in order to cope with delayed information that needs to be processed later – see Example 6 below.

Lemma 1. *If Γ and Δ are finite, then the set of local mgus for Γ and Δ is computable.*

Proof. We claim that the local mgus for Γ and Δ are precisely the computed substitutions at some node of an SLD-tree for $\neg \bigwedge \Gamma$ and Δ.

Suppose σ is a local mgu for Γ and Δ, and let $\Psi = \{a \in \Gamma | a\sigma \in \Delta\}$. Build an SLD-derivation by unifying at each stage an element of Ψ with an element of

Δ. This is trivially a valid SLD-derivation, and the substitution θ it computes must be σ: (i) by soundness of SLD-resolution, θ is an mgu of Ψ and a subset of Δ; but $\Psi\theta \cap \Delta = \Delta$ is a set of ground atoms, so θ must coincide with σ on all variables occurring in Ψ, and be the identity on all other variables; and (ii) by construction $\Gamma\theta \cap \Delta = \Gamma\sigma \cap \Delta$, so θ cannot instantiate fewer variables than σ.

Consider now an arbitrary SLD-derivation for $\neg \bigwedge \Gamma$ and Δ with computed substitution σ, and let Ψ be the set of elements of Γ that were unified in this derivation. Then σ is an mgu of Ψ and a subset of Δ, and as before this means that it maps every variable in Ψ to a ground term and every other variable to itself. Suppose that θ is such that $\Gamma\theta \cap \Delta = \Psi\theta \cap \Delta$. In particular, θ also unifies Ψ and a subset of Δ, so it must coincide with σ in all variables that occur in Ψ. Taking ρ as the restriction of θ to the variables outside Ψ shows that σ is a local mgu for Γ and Δ. □

Example 6. Consider the program Π consisting of the rule

$$q(X,T) \leftarrow p(X,T), r(Y,T),$$

where p is an EDB with $\delta(p) = 2$ and r is an EDB with $\delta(r) = 0$, and the query $\langle q(X,T), \Pi\rangle$. The pre-processing step for this query yields the singleton set $\mathcal{P}_Q = \{\langle\emptyset, \{p(X,T), r(Y,T)\}\rangle\}$.

Suppose that $D|_0 = \{p(\mathsf{a},0), r(\mathsf{b},0)\}$. There are two SLD-trees for the goal $\leftarrow p(X,T), r(Y,T)$ and $D|_0$:

$$\begin{array}{c} \leftarrow p(X,T), r(Y,T) \\ \downarrow [X{:=}\mathsf{a}, T{:=}0] \\ \leftarrow r(Y,0) \\ \downarrow [Y{:=}\mathsf{b}] \\ \Box \end{array} \qquad\qquad \begin{array}{c} \leftarrow p(X,T), r(Y,T) \\ \downarrow [Y{:=}\mathsf{b}, T{:=}0] \\ \leftarrow p(X,0) \\ \downarrow [X{:=}\mathsf{a}] \\ \Box \end{array}$$

These trees include the computed substitutions $\sigma_0 = \emptyset$, $\sigma_1 = [X := \mathsf{a}, T := 0]$, $\sigma_2 = [Y := \mathsf{b}, T := 0]$ and $\sigma_3 = [X := \mathsf{a}, Y := \mathsf{b}, T := 0]$, which are easily seen to be local mgus.

Although σ_3 is an answer to the query, we also need to consider substitution σ_2: since $\delta(p) = 2$, it may be the case that additional answers are produced (e.g. if $p(\mathsf{c}, 0) \in D|_2$). However, since $\delta(r) = 0$, substitution σ_1 can be safely discarded. Substitution σ_0 does not need to be considered: since it does not unify any element of H with $D|_0$, any potential answers it might produce would be generated by step 1 of the algorithm in future time points. ◁

Definition 6. *The set $\mathcal{S}_\tau$ of* schematic supported answers *for query Q at time τ is defined as follows.*

- $\mathcal{S}_{-1} = \{\langle\theta, \emptyset, H\rangle \mid \langle\theta, H\rangle \in \mathcal{P}_Q\}$.
- *If $\langle\theta, E, H\rangle \in \mathcal{S}_{\tau-1}$ and σ is a local mgu for H and $D|_\tau$ such that $H\sigma \backslash D|_\tau$ only contains potentially future atoms wrt τ, then $\langle\theta\sigma, E \cup E', H\sigma \setminus D|_\tau\rangle \in \mathcal{S}_\tau$, where $E' = H\sigma \cap D|_\tau$.*

Example 7. We illustrate this mechanism in the setting of Example 4, where

$$\mathcal{P}_Q = \{\langle\emptyset, \underbrace{\{\mathsf{Temp}(X,\mathsf{high},T),\mathsf{Temp}(X,\mathsf{high},T+1),\mathsf{Temp}(X,\mathsf{high},T+2)\}}_{H}\rangle\}$$

and we assume that $\delta(\mathsf{Temp}) = 1$. We start by setting $\mathcal{S}_{-1} = \{\langle\emptyset,\emptyset,H\rangle\}$.

Since $D|_0 = \{\mathsf{Temp}(\mathsf{wt2},\mathsf{high},0)\}$, SLD-resolution between H and $D|_0$ yields the local mgus $\emptyset$ and $[X := \mathsf{wt2}, T := 0]$. Therefore,

$$\begin{aligned}\mathcal{S}_0 = \{&\langle\emptyset,\emptyset,\{\mathsf{Temp}(X,\mathsf{high},T),\mathsf{Temp}(X,\mathsf{high},T+1),\mathsf{Temp}(X,\mathsf{high},T+2)\}\rangle,\\ &\langle[X := \mathsf{wt2}, T := 0], \{\mathsf{Temp}(\mathsf{wt2},\mathsf{high},0)\}, \underbrace{\{\mathsf{Temp}(\mathsf{wt2},\mathsf{high},i) \mid i = 1,2\}}_{H_0}\rangle\}.\end{aligned}$$

Next, $D|_1 = \emptyset$, so trivially $D|_1 \subseteq H_0$ and therefore the empty substitution is a local mgu of H_0 and $D|_1$. Furthermore, H_0 only contains potentially future atoms wrt 1 because $\delta(\mathsf{Temp}) = 1$. The same argument applies to $D|_1$ and H. So $\mathcal{S}_1 = \mathcal{S}_0$.

We now consider several possibilities for what happens to the schematic supported answer $\langle[X := \mathsf{wt2}, T := 0], \{\mathsf{Temp}(\mathsf{wt2},\mathsf{high},0)\}, H_0\rangle$ at time instant 2. Since H_0 is ground, the only local mgu of H_0 and $D|_2$ will always be $\emptyset$.

- If $\mathsf{Temp}(\mathsf{wt2},\mathsf{high},1) \notin D|_2$, then $H_0 \setminus D|_2$ contains $\mathsf{Temp}(\mathsf{wt2},\mathsf{high},1)$, which is not a potentially future atom wrt 2, and therefore this schematic supported answer is discarded.
- If $\mathsf{Temp}(\mathsf{wt2},\mathsf{high},1) \in D|_2$ but $\mathsf{Temp}(\mathsf{wt2},\mathsf{high},2) \notin D|_2$, then $H_0 \setminus D|_2 = \{\mathsf{Temp}(\mathsf{wt2},\mathsf{high},2)\}$, which only contains potentially future atoms wrt 2, and therefore $\mathcal{S}_2$ contains the schematic supported answer

$$\langle[X := \mathsf{wt2}, T := 0], \{\mathsf{Temp}(\mathsf{wt2},\mathsf{high},i) \mid i = 0,1,2\}\rangle.$$

- Finally, if $\{\mathsf{Temp}(\mathsf{wt2},\mathsf{high},1),\mathsf{Temp}(\mathsf{wt2},\mathsf{high},2)\} \subseteq D|_2$, then $H_2 \setminus D|_2 = \emptyset$, and the system can output the answer $[X := \mathsf{wt2}, T := 0]$ to the original query. In this case, this answer would be added to $\mathcal{S}_2$, and then trivially copied to all subsequent $\mathcal{S}_\tau$. ◁

As in this example, answers are always propagated from $\mathcal{S}_\tau$ to $\mathcal{S}_{\tau+1}$. In an actual implementation of this algorithm, we would likely expect these answers to be output when generated, and discarded afterwards.

Proposition 8 (Soundness). *If $\langle\theta, E, H\rangle \in \mathcal{S}_\tau$, $E \neq \emptyset$, and σ instantiates all free variables in $E \cup H$, then $\langle\theta\sigma, H', E\sigma\rangle \in \mathcal{E}(Q, D, \tau)$ for some $H' \subseteq H\sigma$.*

Proof. By induction on τ, we show that $\langle Q, D_\tau\rangle \vdash_{\mathsf{SLD}} \langle\theta, \wedge_i\alpha_i\rangle$ with $H = \{\alpha_i\}_i$. For $\mathcal{S}_{-1}$ this is trivially the case, due to the way that $\mathcal{P}_Q$ is computed.

Assume now that there exists $\langle\theta_0, E_0, H_0\rangle \in \mathcal{S}_{\tau-1}$ such that σ_0 is a local mgu of H_0 and $D|_\tau$, $\theta = \theta_0\sigma_0$, $E = E_0 \cup (H_0\sigma_0 \cap D|_\tau)$ and $H = H_0\sigma_0 \setminus D|_\tau$. If $H_0\sigma_0 \cap D_\tau = \emptyset$, then necessarily $\sigma_0 = \emptyset$ (no steps of SLD-resolution were performed), and the thesis holds by induction hypothesis. Otherwise, we can build the required derivation by extending the derivation obtained by induction hypothesis with unification steps between the relevant elements of H_0 and $D|_\tau$. By Lemma 1, this derivation computes the substitution θ.

Applying Proposition 5 to this SLD-derivation yields an $H' \subseteq H\sigma$ such that $\langle \theta\sigma, H' \rangle \in \mathcal{H}(Q, D, \tau)$. By construction, $E\sigma \neq \emptyset$ is evidence for this answer. □

It may be the case that $\mathcal{S}_\tau$ contains some elements that do not correspond to hypothetical answers because of the minimality requirement. Consider a simple case of a query Q where

$$\mathcal{P}_Q = \{\langle \emptyset, \{p(a,0), p(b,1), p(c,2)\}\rangle, \langle \emptyset, \{p(a,0), p(c,2), p(d,3)\}\rangle\},$$

and suppose that $D|_0 = \{p(a,0)\}$ and $D|_1 = \{p(b,1)\}$. Then

$$\mathcal{S}_1 = \{\langle \emptyset, \{p(a,0), p(b,1)\}, \{p(c,2)\}\rangle, \langle \emptyset, \{p(a,0)\}, \{p(c,2), p(d,3)\}\rangle\},$$

and the second element of this set has a non-minimal set of hypotheses. For simplicity, we do not to include a test for set inclusion in the definition of $\mathcal{S}_\tau$.

Proposition 9 (Completeness). *If $\langle \sigma, H, E \rangle \in \mathcal{E}(Q, D, \tau)$, then there exist a substitution ρ and a triple $\langle \theta, E', H' \rangle \in \mathcal{S}_\tau$ such that $\sigma = \theta\rho$, $H = H'\rho$ and $E = E'\rho$.*

Proof. By Proposition 6, $\langle Q, D_\tau \rangle \vdash_{\mathsf{SLD}} \langle \theta, \wedge_i \alpha_i \rangle$ for some substitution ρ and set of atoms $H' = \{\alpha_i\}_i$ with $H = \{\alpha_i\rho\}_i$ and $\sigma = \theta\rho$ for some θ. By Proposition 7, there is a stratified SLD-derivation computing this answer. The sets of atoms from $D|_\tau$ unified in each stratum of this derivation define the set of elements from H that need to be unified to construct the corresponding element of $\mathcal{S}_\tau$. □

The use of local mgus gives a worst-case exponential complexity to the computation of $\mathcal{S}_\tau$ from $\mathcal{S}_{\tau-1}$: since every node of the SLD-tree can give a local mgu as in the proof of Lemma 1, there can be as many local mgus as subsets of H that can be unified with $D|_\tau$. In practice, the number of substitutions that effectively needs to be considered is at most $(k+1)^v$, where v is the number of variables in H and k is the maximum number of possible instantiations for a variable in $D|_\tau$. In practice, v only depends on the program Π, and is likely to be small, while k is 0 if the elements of H are already instantiated.

If there are no communication delays and the program satisfies an additional property, we can regain the polynomial complexity from [15].

Definition 7. *A query $Q = \langle P, \Pi \rangle$ is* connected *if each rule in P contains at most one temporal variable, which occurs in the head if it occurs in the body.*

Proposition 10. *If the query Q is connected and $\delta(P) = 0$ for every predicate symbol P, then $\mathcal{S}_\tau$ can be computed from $\mathcal{P}_Q$ and $\mathcal{S}_{\tau-1}$ in time polynomial in the size of $\mathcal{P}_Q$, $\mathcal{S}_{\tau-1}$ and $D|_\tau$.*

Proof. Suppose that $\delta(P) = 0$ for every P.

If $\langle \theta, \emptyset, H \rangle \in \mathcal{S}_{\tau-1}$, then (i) $\emptyset$ is always a local mgu of H and $D|_\tau$, so this schematic answer can be added to $\mathcal{S}_\tau$ (in constant time); and (ii) we can compute in polynomial time the set $M \subseteq H$ of elements that need to be in $D|_\tau$ in order to yield a non-empty local mgu σ of H and $D|_\tau$ such that $H\sigma \setminus D|_\tau$ only contains potentially future atoms wrt τ – these are the elements whose timestamps are less or equal to all other elements' timestamps. (Since there are no delays, they must all arrive simultaneously at the data stream.) To decide which substitutions make M a subset of $D|_\tau$, we can perform classical SLD-resolution between M and $D|_\tau$. For each such element of $\mathcal{S}_{\tau-1}$, the size of every SLD-derivation that needs to be constructed is bound by the number of atoms in the initial goal, since $D|_\tau$ only contains facts. Furthermore, all unifiers can be constructed in time linear in the size of the formulas involved, since the only function symbol available is addition of temporal terms. Finally, the total number of SLD-derivations that needs to be considered is bound by the size of $\mathcal{S}_{\tau-1} \times D|_\tau$.

If $\langle \theta, E, H \rangle \in \mathcal{S}_{\tau-1}$ and $E \neq \emptyset$, then we observe that the temporal argument is instantiated in all elements of E and H – this follows from connectedness and the fact that the elements of E are ground by construction. Therefore, we know exactly which facts in H must unify with $D|_\tau$ – these are the elements of H whose timestamp is exactly τ. As above, the elements that must be added to $\mathcal{S}_\tau$ can then be computed in polynomial time by SLD-resolution. □

The key step of this proof amounts to showing that the algorithm from [15] computes all local mgus that can lead to schematic supported answers.

Both conditions stated above are necessary. If Q is not connected, then $\mathcal{S}_\tau$ may contain elements $\langle \theta, E, H \rangle$ where $E \neq \emptyset$ and H contains elements with uninstantiated timestamps; and if for some predicate $\delta(P) \neq 0$, then we must account for the possibility that facts about P arrive with some delay to the data stream. In either case, this means that we do not know the set of elements that need to unify with the data stream, and need to consider all possibilities.

Example 8. In the context of our running example, we say that a turbine has a manufacturing defect if it exhibits two specific failures during its lifetime: at some time it overheats, and at some (different) time it does not send a temperature reading. Since this is a manufacturing defect, we set it to hold at timepoint 0, regardless of when the failures occur. We model this property by the rule

$$\mathsf{Temp}(X, \mathsf{high}, T_1), \mathsf{Temp}(X, \mathsf{n/a}, T_2) \rightarrow \mathsf{Defective}(X, 0)\,.$$

Let Π'_E be the program obtained from Π_E by adding this rule, and consider now the query $Q' = \langle \mathsf{Defective}(X, T), \Pi'_E \rangle$. Performing SLD-resolution between Π'_E and $\mathsf{Defective}(X, 0)$ yields the goal $\neg(\mathsf{Temp}(X, \mathsf{high}, T_1) \wedge \mathsf{Temp}(X, \mathsf{n/a}, T_2))$, which only contains potentially future atoms with respect to -1.

Assume that $\mathsf{Temp}(\mathsf{wt2}, \mathsf{high}, 0) \in D_0$. Then

$$\langle \theta' = [X := \mathsf{wt2}, T := 0]\,, \{\mathsf{Temp}(\mathsf{wt2}, \mathsf{high}, 0)\}, \{\mathsf{Temp}(\mathsf{wt2}, \mathsf{n/a}, T_2)\} \rangle \in \mathcal{S}_0\,.$$

We do not know if or when θ' will become an answer to the original query, but our algorithm is still able to output relevant information to the user. ◁

Under some assumptions, we can also bound the interval in which a schematic hypothetical answer can be present in $\mathcal{S}$. Assume that the query Q has a temporal variable T and that $\mathcal{S}_\tau$ contains a triple $\langle\theta, E, H\rangle$ with $T\theta \leq \tau$ and $H \neq \emptyset$.

- Suppose that there is a number d such that for every substitution σ and every $\tau \geq T\sigma + d$, $\sigma \in \mathcal{A}(Q, D, \tau)$ iff $\sigma \in \mathcal{A}(Q, D, T\sigma + d)$. Then the timestamp of each element of H is at most $\tau + d$.
- Similarly, suppose that there is a natural number w such that each σ is an answer to Q over D_τ iff σ is an answer to Q over $D_\tau \setminus D_{\tau-w}$. Then all elements in E must have timestamp at least $\tau - w$.

The values d and w are known in the literature as *(query) delay* and *window size*, respectively, see e.g. [42].

5 Related Work

This work contributes to the field of stream reasoning, the task of conjunctively reasoning over streaming data and background knowledge [44].

Research advances on Complex Event Processors and Data Stream Management Systems [16], together with Knowledge Representation and the Semantic Web, all contributed to the several stream reasoning languages, systems and mechanisms proposed during the last decade [18].

Computing answers to a query over a data source that is continuously producing information, be it at slow or very fast rates, asks for techniques that allow for some kind of *incremental evaluation*, in order to avoid reevaluating the query from scratch each time a new tuple of information arrives. Several efforts have been made in that direction, capitalising on incremental algorithms based on seminaive evaluation [1,7,24,25,37], on truth maintenance systems [8], window oriented [21] among others. The framework we build upon [15] fits naturally in the first class, as it is an incremental variant of SLD-resolution.

Hypothetical query answering over streams is broadly related to works that study abduction in logic programming [19,26], namely those that view negation in logic programs as hypotheses and relate it to contradiction avoidance [3,20]. Furthermore, our framework can be characterized as applying an incremental form of data-driven abductive inference. Such forms of abduction have been used in other works [36], but with a rather different approach and in the plan interpretation context. To our knowledge, hypothetical or abductive reasoning has not been previously used in the context of stream reasoning, to answer continuous queries, although it has been applied to annotation of stream data [4].

A similar problem also arises in the area of incomplete databases, where the notions of possible and certain answers have been developed [28]. In this context, possible answers are answers to a complete database D' that the incomplete database D can represent, while certain answers consist of the tuples that belong to all complete databases that D can represent (the intersection of all possible answers). Libkin [34] the way those notions were defined, exploring an alternative way of looking at incomplete databases that dates back to Reiter [41], and that views a database as a logical theory. Libkin's approach was to explore the semantics of incompleteness further in a setting that is independent of a particular data model, and appealing to orderings to describe the degree of incompleteness of a data model and relying on those orderings to define certain answers. Other authors [22,30,31,40] have also investigated ways to assign confidence levels to the information output to the user.

Most theoretical approaches to stream-processing systems commonly require input streams to be ordered. However, some approaches, namely from the area of databases and event processing, have developed techniques to deal with out-of-order data. An example of such a technique requires inserting special marks in the input stream (punctuations) to guide window processing [32]. Punctuations assert a timestamp that is a lower bound for all future incoming values of a attribute. Another technique starts by the potential generation of out-of-order results, which are then ordered by using stackbased data structures and associated purge algorithms [33].

Commercial systems [27] for continuous query answering use other techniques to deal with communication delays, not always with a solid foundational background. One such technique is the use of watermarks [2], which specify how long the system should wait for information to arrive. In practice, watermarks serve the same purpose as the function δ in our framework. However, they do not come with guarantees of correctness: in fact, information may arrive after the time point specified by the watermark, in which case it is typically discarded.

Memory consumption is a concern in [45], where a sound and complete stream reasoning algorithm is presented for a fragment of datalogMTL – forward-propagating datalogMTL – that also disallows propagation of derived information towards the past. DatalogMTL is a Datalog extension of the Horn fragment of MTL [5,29], which was proposed in [11] for ontology-based access to temporal log data. DatalogMTL rules allow propagation of derived information to both past and future time points. Concerned with the possibly huge or unbounded set of input facts that have to be kept in memory, the authors of [45] restrict the set of operators of DatalogMTL to one that generates only so-called forward-propagating rules. They present a sound and complete algorithm for stream reasoning with forward-propagating datalogMTL that limits memory usage through the use of a sliding window.

6 Discussion, Conclusions and Future Work

In the current work, we expanded the formalism of hypothetical answers to continuous queries with the possibility of communication delays, and showed how we could define a declarative semantics and a corresponding operational semantics by suitably adapting and enriching the definitions from [15].

Our work deals with communication delays without requiring any previous processing of the input stream, as long as bounds for delays are known. To the best of our knowledge, this is the first work providing this flexibility in the context of a logical approach to stream query answering. This motivated us to develop a resolution search strategy driven by the arrival of data, instead of a static one. It would be interesting to try to combine our resolution strategy with dynamic strategies [23], or techniques to produce compact representations of search trees [38].

The algorithm we propose needs to store significant amounts of information. This was already a problem in [15], and the addition of delays enhances it (since invalid answers are in general discarded later). One possible way to overcome this limitation would be to assign confidence levels to schematic supported answers, and either discard answers when their confidence level is below a given threshold, or discard the answers with lowest confidence when there are too many. This requires having information about (i) the probability that a given premise is true, and (ii) the probability that premise arrives with a given delay. These probabilities can be used to infer the likelihood that a given schematic supported answer will be confirmed. The relevant probabilities could be either evaluated by experts, or inferred using machine-learning tools.

Our approach extends naturally extends to the treatment of lossy channels. If we have a sub-probability distribution for communication delays (i.e. where the sum of all probabilities is strictly smaller than 1), then we are also allowing for the chance that information is lost in transit. Thus, the same analysis would also be able to estimate the likelihood that a given substitution is an answer, even though some of the relevant premises are missing. We plan to extend our framework along these lines, in order to have enough ingredients to develop a prototype of our system and conduct a full practical evaluation.

The presentation in [15] also discussed briefly how to add negation to the language. Those ideas are incompatible with the way we deal with communication delays, since they relied heavily on the fact that the exact time at which facts arrive is known. Adding negation to the current framework is also a direction for future research.

Acknowledgements. This work was partially supported by FCT through the LASIGE Research Unit, ref. UIDB/00408/2020 and ref. UIDP/00408/2020.

References

1. Abiteboul, S., Hull, R., Vianu, V.: Foundations of Databases. Addison-Wesley, Boston (1995)
2. Akidau, T., et al.: Millwheel: fault-tolerant stream processing at internet scale. Proc. VLDB Endow. **6**(11), 1033–1044 (2013)
3. Alferes, J.J., Moniz Pereira, L. (eds.): Reasoning with Logic Programming. LNCS, vol. 1111. Springer, Heidelberg (1996). https://doi.org/10.1007/3-540-61488-5
4. Alirezaie, M., Loutfi, A.: Automated reasoning using abduction for interpretation of medical signals. J. Biomed. Semant. **5**, 35 (2014)
5. Alur, R., Henzinger, T.A.: Real-time logics: complexity and expressiveness. Inf. Comput. **104**(1), 35–77 (1993)
6. Babcock, B., Babu, S., Datar, M., Motwani, R., Widom, J.: Models and issues in data stream systems. In: Popa, L., Abiteboul, S., Kolaitis, P.G. (eds.) Proceedings of PODS, pp. 1–16. ACM (2002)

7. Barbieri, D.F., Braga, D., Ceri, S., Della Valle, E., Grossniklaus, M.: Incremental reasoning on streams and rich background knowledge. In: Aroyo, L., et al. (eds.) ESWC 2010. LNCS, vol. 6088, pp. 1–15. Springer, Heidelberg (2010). https://doi.org/10.1007/978-3-642-13486-9_1
8. Beck, H., Dao-Tran, M., Eiter, T.: Answer update for rule-based stream reasoning. In: Yang, Q., Wooldridge, M.J. (eds.) Proceedings of IJCAI, pp. 2741–2747. AAAI Press (2015)
9. Beck, H., Dao-Tran, M., Eiter, T., Fink, M.: LARS: a logic-based framework for analyzing reasoning over streams. In: Bonet, B., Koenig, S. (eds.) 29th AAAI Conference on Artificial Intelligence, AAAI 2015, Austin, TX, USA, pp. 1431–1438. AAAI Press (2015)
10. Ben-Ari, M.: Mathematical Logic for Computer Science, 3rd edn. Springer, London (2012). https://doi.org/10.1007/978-1-4471-4129-7
11. Brandt, S., Kalayci, E.G., Ryzhikov, V., Xiao, G., Zakharyaschev, M.: Querying log data with metric temporal logic. J. Artif. Intell. Res. **62**, 829–877 (2018)
12. Ceri, S., Gottlob, G., Tanca, L.: What you always wanted to know about datalog (and never dared to ask). IEEE Trans. Knowl. Data Eng. **1**(1), 146–166 (1989)
13. Chomicki, J., Imielinski, T.: Temporal deductive databases and infinite objects. In: Edmondson-Yurkanan, C., Yannakakis, M. (eds.) Proceedings of PODS, pp. 61–73. ACM (1988)
14. Cruz-Filipe, L., Gaspar, G., Nunes, I.: Hypothetical answers to continuous queries over data streams. CoRR abs/1905.09610 (2019)
15. Cruz-Filipe, L., Gaspar, G., Nunes, I.: Hypothetical answers to continuous queries over data streams. In: Proceedings of AAAI, pp. 2798–2805. AAAI Press (2020)
16. Cugola, G., Margara, A.: Processing flows of information: from data stream to complex event processing. ACM Comput. Surv. **44**(3), 15:1–15:62 (2012)
17. Dao-Tran, M., Eiter, T.: Streaming multi-context systems. In: Sierra, C. (ed.) Proceedings of IJCAI, pp. 1000–1007. ijcai.org (2017)
18. Dell'Aglio, D., Valle, E.D., van Harmelen, F., Bernstein, A.: Stream reasoning: a survey and outlook. Data Sci. **1**(1–2), 59–83 (2017)
19. Denecker, M., Kakas, A.: Abduction in logic programming. In: Kakas, A.C., Sadri, F. (eds.) Computational Logic: Logic Programming and Beyond. LNCS (LNAI), vol. 2407, pp. 402–436. Springer, Heidelberg (2002). https://doi.org/10.1007/3-540-45628-7_16
20. Dung, P.M.: Negations as hypotheses: an abductive foundation for logic programming. In: Furukawa, K. (ed.) Proceedings of ICLP, pp. 3–17. MIT Press (1991)
21. Ghanem, T.M., Hammad, M.A., Mokbel, M.F., Aref, W.G., Elmagarmid, A.K.: Incremental evaluation of sliding-window queries over data streams. IEEE Trans. Knowl. Data Eng. **19**(1), 57–72 (2007)
22. Gray, A.J., Nutt, W., Williams, M.H.: Answering queries over incomplete data stream histories. IJWIS **3**(1/2), 41–60 (2007)
23. Guo, H., Gupta, G.: Dynamic reordering of alternatives for definite logic programs. Comput. Lang. Syst. Struct. **35**(3), 252–265 (2009)
24. Gupta, A., Mumick, I.S., Subrahmanian, V.: Maintaining views incrementally. In: Buneman, P., Jajodia, S. (eds.) Proceedings of SIGMOD, pp. 157–166. ACM Press (1993)
25. Hu, P., Motik, B., Horrocks, I.: Optimised maintenance of datalog materialisations. In: McIlraith, S.A., Weinberger, K.Q. (eds.) 32nd AAAI Conference on Artificial Intelligence, AAAI 2018, New Orleans, LA, USA, pp. 1871–1879. AAAI Press (2018)

26. Inoue, K.: Hypothetical reasoning in logic programs. J. Log. Program. **18**(3), 191–227 (1994)
27. Isah, H., Abughofa, T., Mahfuz, S., Ajerla, D., Zulkernine, F.H., Khan, S.: A survey of distributed data stream processing frameworks. IEEE Access **7**, 154300–154316 (2019)
28. Jr., W.L.: On semantic issues connected with incomplete information databases. ACM Trans. Database Syst. **4**(3), 262–296 (1979)
29. Koymans, R.: Specifying real-time properties with metric temporal logic. Real-Time Syst. **2**(4), 255–299 (1990)
30. Lang, W., Nehme, R.V., Robinson, E., Naughton, J.F.: Partial results in database systems. In: Dyreson, C.E., Li, F., Özsu, M.T. (eds.) Proceedings of SIGMOD, pp. 1275–1286. ACM (2014)
31. de Leng, D., Heintz, F.: Approximate stream reasoning with metric temporal logic under uncertainty. In: 33rd AAAI Conference on Artificial Intelligence, AAAI 2019, Honolulu, Hawaii, USA, pp. 2760–2767. AAAI Press (2019)
32. Li, J., Tufte, K., Shkapenyuk, V., Papadimos, V., Johnson, T., Maier, D.: Out-of-order processing: a new architecture for high-performance stream systems. Proc. VLDB Endow. **1**(1), 274–288 (2008)
33. Li, M., Liu, M., Ding, L., Rundensteiner, E.A., Mani, M.: Event stream processing with out-of-order data arrival. In: Proceedings of ICDCS, p. 67. IEEE Computer Society (2007)
34. Libkin, L.: Incomplete data: what went wrong, and how to fix it. In: Hull, R., Grohe, M. (eds.) Proceedings of PODS, pp. 1–13. ACM (2014)
35. Lloyd, J.W.: Foundations of Logic Programming. Springer, Heidelberg (1984). https://doi.org/10.1007/978-3-642-96826-6
36. Meadows, B.L., Langley, P., Emery, M.J.: Seeing beyond shadows: incremental abductive reasoning for plan understanding. In: Plan, Activity, and Intent Recognition. AAAI Workshops, vol. WS-13-13. AAAI (2013)
37. Motik, B., Nenov, Y., Piro, R.E.F., Horrocks, I.: Incremental update of datalog materialisation: the backward/forward algorithm. In: Bonet, B., Koenig, S. (eds.) 29th AAAI Conference on Artificial Intelligence, AAAI 2015, Austin, TX, USA, pp. 1560–1568. AAAI Press (2015)
38. Nishida, N., Vidal, G.: A framework for computing finite SLD trees. J. Log. Algebraic Methods Program. **84**(2), 197–217 (2015)
39. Özçep, Ö.L., Möller, R., Neuenstadt, C.: A stream-temporal query language for ontology based data access. In: Lutz, C., Thielscher, M. (eds.) KI 2014. LNCS (LNAI), vol. 8736, pp. 183–194. Springer, Cham (2014). https://doi.org/10.1007/978-3-319-11206-0_18
40. Razniewski, S., Korn, F., Nutt, W., Srivastava, D.: Identifying the extent of completeness of query answers over partially complete databases. In: Sellis, T.K., Davidson, S.B., Ives, Z.G. (eds.) Proceedings of SIGMOD, pp. 561–576. ACM (2015)
41. Reiter, R.: Towards a logical reconstruction of relational database theory. In: Brodie, M.L., Mylopoulos, J., Schmidt, J.W. (eds.) On Conceptual Modelling. Topics in Information Systems, pp. 191–233. Springer, New York (1984). https://doi.org/10.1007/978-1-4612-5196-5_8
42. Ronca, A., Kaminski, M., Grau, B.C., Motik, B., Horrocks, I.: Stream reasoning in temporal datalog. In: McIlraith, S.A., Weinberger, K.Q. (eds.) 32nd AAAI Conference on Artificial Intelligence, AAAI 2018, New Orleans, LA, USA, pp. 1941–1948. AAAI Press (2018)

43. Stonebraker, M., Çetintemel, U., Zdonik, S.B.: The 8 requirements of real-time stream processing. SIGMOD Rec. **34**(4), 42–47 (2005)
44. Valle, E.D., Ceri, S., van Harmelen, F., Fensel, D.: It's a streaming world! Reasoning upon rapidly changing information. IEEE Intell. Syst. **24**(6), 83–89 (2009)
45. Walega, P.A., Kaminski, M., Grau, B.C.: Reasoning over streaming data in metric temporal datalog. In: 33rd AAAI Conference on Artificial Intelligence, AAAI 2019, Honolulu, Hawaii, USA, pp. 3092–3099. AAAI Press (2019)
46. Zaniolo, C.: Logical foundations of continuous query languages for data streams. In: Barceló, P., Pichler, R. (eds.) Datalog 2.0 2012. LNCS, vol. 7494, pp. 177–189. Springer, Heidelberg (2012). https://doi.org/10.1007/978-3-642-32925-8_18

The Implication Problem for Functional Dependencies and Variants of Marginal Distribution Equivalences

Minna Hirvonen(✉)

Department of Mathematics and Statistics, University of Helsinki, Helsinki, Finland
minna.hirvonen@helsinki.fi

Abstract. We study functional dependencies together with two different probabilistic dependency notions: unary marginal identity and unary marginal distribution equivalence. A unary marginal identity states that two variables x and y are identically distributed. A unary marginal distribution equivalence states that the multiset consisting of the marginal probabilities of all the values for variable x is the same as the corresponding multiset for y. We present a sound and complete axiomatization and a polynomial-time algorithm for the implication problem for the class of these dependencies, and show that this class has Armstrong relations.

Keywords: Armstrong relations · Complete axiomatization · Functional dependence · Marginal distribution equivalence · Polynomial-time algorithm · Probabilistic team semantics

1 Introduction

Notions of dependence and independence are central to many areas of science. In database theory, the study of dependencies (or logical integrity constraints) is a central topic because it has applications to database design and many other data management tasks. For example, functional dependencies and inclusion dependencies are commonly used in practice as primary key and foreign key constraints. In this paper, we study functional dependencies together with unary marginal identity and unary marginal distribution equivalence. The latter two are probabilistic dependency notions that compare distributions of two variables. A unary marginal identity states that two variables x and y are identically distributed. A unary marginal distribution equivalence states that the multiset consisting of the marginal probabilities of all the values for variable x is the same as the corresponding multiset for y. Marginal identity can actually be viewed as a probabilistic version of inclusion dependency; it is sometimes called "probabilistic inclusion" [8].

The author was supported by the Finnish Academy of Science and Letters (the Vilho, Yrjö and Kalle Väisälä Foundation) and by grant 345634 of the Academy of Finland.

I. Varzinczak (Ed.): FoIKS 2022, LNCS 13388, pp. 130–146, 2022.
https://doi.org/10.1007/978-3-031-11321-5_8

We consider the so-called implication problem for the class of these dependencies. The *implication problem* for a class of dependencies is the problem of deciding whether for a given finite set $\Sigma \cup \{\sigma\}$ of dependencies from the class, any database that satisfies every dependency from Σ also satisfies σ. If the databases are required to be finite, the problem is called the *finite implication problem.* Otherwise, we speak of the *unrestricted implication problem.* We axiomatize the finite implication problem for functional dependence, unary marginal identity, and unary marginal distribution equivalence over uni-relational databases that are complemented with a probability distribution over the set of tuples appearing in the relation.

The implication problem that we axiomatize contains one qualitative class (functional dependence) and two probabilistic classes (marginal identity and marginal distribution equivalence) of dependencies, so we have to consider how these different kinds of dependencies interact. Some probabilistic dependencies have already been studied separately. The implication problem for probabilistic independence has been axiomatized over 30 years ago in [12]. More recently, the implication problem for marginal identity (over finite probability distributions) was axiomatized in [16]. The study of joint implication problems for different probabilistic and relational atoms have potential for practical applications because various probabilistic dependency notions appear in many situations. An example of an interesting class of probabilistic dependencies is probabilistic independencies together with marginal identities. In probability theory and statistics, random variables are often assumed to be independent and identically distributed (IID), a property which can be expressed with probabilistic independencies and marginal identities. Another example comes from the foundations of quantum mechanics where functional dependence and probabilistic conditional independence can be used to express certain properties of hidden-variable models [1,2].

For practical purposes, it is important to consider implication problems also from a computational point of view: the usability of a class of dependencies, e.g. in database design, depends on the computational properties of its implication problem. We show that the implication problem for functional dependencies, unary marginal identities, and unary marginal distribution equivalences has a polynomial-time algorithm.

A class of dependencies is said to have *Armstrong relations* if for any finite set Σ of dependencies from the class, there is a relation that satisfies a dependency σ in the class if and only if σ is true for every relation that satisfies Σ. An Armstrong relation can be used as a convenient representation of a dependency set. If a class of dependecies has Armstrong relations, then the implication problem for a fixed set Σ and an arbitrary dependency σ is reduced to checking whether the Armstrong relation for Σ satisfies σ. When Armstrong axiomatized functional dependence in [3], he also implicitly proved that the class has Armstrong relations [9]. Unfortunately, there are sets of functional dependecies for which the size of a minimal Armstrong relation is exponential in the number of variables (attributes) of the set [5]. Armstrong relations can still be useful in practice [18,19].

Sometimes integrity constraints, e.g. inclusion dependencies, are considered on multi-relational databases. In this case one looks for *Armstrong databases* [10] instead of single relations. In this terminology, an Armstrong relation is simply a uni-relational Armstrong database. Not all classes of dependencies enjoy Armstrong databases: functional dependence together with unary inclusion dependence (over multi-relational databases where empty relations are allowed) does not have Armstrong databases [10]. However, standard functional dependencies (i.e. functional dependencies with a nonempty left-side) and inclusion dependencies do have Armstrong databases. It is known that probabilistic independence has Armstrong relations [12]. We show that the class of functional dependencies, unary marginal identities, and unary marginal distribution equivalences enjoys Armstrong relations.

Instead of working with notions and conventions from database theory, we have chosen to formulate the axiomatization in *team semantics* which is a semantical framework for logics. This is because the dependency notions that we consider can be viewed as certain kinds of atomic formulas in logics developed for expressing probabilistic dependencies (see the logics studied e.g. in [14]), and we want to keep this connection[1] to these logics explicit. A "team" in team semantics is basically a uni-relational database, so the proofs that we present could easily also be stated in terms of databases.

Team semantics was originally introduced by Hodges [17]. The systematic development of the framework began with *dependence logic*, a logic for functional dependence introduced by Väänänen [24], and the setting has turned out to be useful for formulating logics for other dependency notions as well. These logics include, e.g., inclusion logic [11] and (conditional) independence logic [13]. In team semantics, logical formulas are evaluated over sets of assignments (called *teams*) instead of single assignments as, for example, in first-order logic. This allows us to define atomic formulas that express dependencies. For example, the dependency atom $=\!(\bar{x}, \bar{y})$ expresses the functional dependency stating that the values of $\bar{x}$ determine the values of $\bar{y}$. As mentioned above, a team of assignments can be thought of as a relation (or a table) in a database: each assignment corresponds to a tuple in the relation (or a row in the table).

Since we want to study functional dependencies together with probabilistic dependency notions, we turn to *probabilistic team semantics* which is a generalization of the relational team semantics. A *probabilistic team* is a set of assignments with an additional function that maps each assignment to some numerical value, a *weight*. The function is usually a probability distribution. As probabilistic team semantics is currently defined only for discrete distributions that have a finite number of possible values for variables, we consider the implication problem only for finite teams.

Although some probabilistic dependencies might seem similar to their qualitative variants, their implication problems are different: probabilistic dependencies refer to the weights of the rows rather than the rows themselves. Probabilistic dependencies can be tricky, especially if one considers two or more different vari-

[1] Our axiomatization is obviously only for the atomic level of these logics.

ants together. For example, consider marginal identity together with probabilistic independence. The chase algorithm that was used for proving the completeness of the axiomatization of marginal identity in [16] uses inclusion dependencies that contain index variables for counting multiplicities of certain tuples. There does not seem to be a simple way of extending this procedure to also include probabilistic independencies. This is because adding a new row affects the probability distribution and often breaks existing probabilistic independencies. On the other hand, the approach that was used for probabilistic independencies in [12] cannot easily be generalized to also cover inclusion dependencies either.

Luckily, in our case we can utilize the implication problem for functional dependencies and unary inclusion dependencies which has been axiomatized in [7] for both finite and unrestricted implication. The axiomatization for finite implication is proved to be complete by constructing relations with the help of multigraphs that depict the dependecies between variables. This approach has also been applied to unary functional dependence, unary inclusion dependence, and independence [15]. Our approach is similar, but since we are working in the probabilistic setting, we have to be careful that our construction also works with the two kinds of variants of unary marginal distribution equivalences.

2 Preliminaries

Let D be a finite set of variables and A a finite set of values. We usually denote variables by x, y, z and values by a, b, c. Tuples of variables and tuples of values are denoted by $\bar{x}, \bar{y}, \bar{z}$ and $\bar{a}, \bar{b}, \bar{c}$, respectively. The notation $|\bar{x}|$ means the length of the tuple $\bar{x}$, and $\mathsf{var}(\bar{x})$ means the set of variables that appear in the tuple $\bar{x}$.

An assignment of values from A for the set D is a function $s\colon D \to A$. A team X of A over D is a finite set of assignments $s\colon D \to A$. When the variables of D are ordered in some way, e.g. $D = \{x_1, \ldots, x_n\}$, we identify the assignment s with the tuple $s(x_1, \ldots, x_n) = s(x_1) \ldots s(x_n) \in A^n$, and also explicitly call s a tuple. A team X over $D = \{x_1, \ldots, x_n\}$ can then be viewed as a table whose columns are the variables $x_1, \ldots, x_n$, and rows are the tuples $s \in X$. For any tuple of variables $\bar{x}$ from D, we let

$$X(\bar{x}) := \{s(\bar{x}) \in A^{|\bar{x}|} \mid s \in X\}.$$

A probabilistic team $\mathbb{X}$ is a function $\mathbb{X}\colon X \to (0, 1]$ such that $\sum_{s \in X}(s) = 1$. For a tuple of variables $\bar{x}$ from D and a tuple of values $\bar{a}$ from A, we let

$$|\mathbb{X}_{\bar{x}=\bar{a}}| := \sum_{\substack{s(\bar{x})=\bar{a} \\ s \in X}} \mathbb{X}(s).$$

Let $\bar{x}, \bar{y}$ be tuples of variables from D. Then $=\!\!(\bar{x}, \bar{y})$ is a *(functional) dependency atom*. If the tuples $\bar{x}, \bar{y}$ are of the same length, then $\bar{x} \approx \bar{y}$ and $\bar{x} \approx^* \bar{y}$ are *marginal identity* and *marginal distribution equivalence atoms*, respectively. We also use the abbreviations FD (functional dependency), MI (marginal identity), and MDE (marginal distribution equivalence) for the atoms. If $|\bar{x}| = |\bar{y}| = 1$, an atom is called *unary* and abbreviated by UFD, UMI, or UMDE.

Before defining the semantics for the atoms, we need to introduce the notion of a *multiset*. A multiset is a pair (B, m) where B is a set, and $m\colon B \to \mathbb{N}$ is a multiplicity function. The function m determines for each element $b \in B$ how many multiplicities of b the multiset (B, m) contains. We often denote multisets using double wave brackets, e.g., $(B, m) = \{\{0, 1, 1\}\}$ when $B = \{0, 1\}$ and $m(n) = n + 1$ for all $n \in B$.

Let σ be either a dependency, a marginal identity, or a marginal distribution equivalence atom. The notation $\mathbb{X} \models \sigma$ means that a probabilistic team $\mathbb{X}$ satisfies σ, which is defined as follows:

(i) $\mathbb{X} \models =\!\!(\bar{x}, \bar{y})$ iff for all $s, s' \in X$, if $s(\bar{x}) = s'(\bar{x})$, then $s(\bar{y}) = s'(\bar{y})$.
(ii) $\mathbb{X} \models \bar{x} \approx \bar{y}$ iff $|\mathbb{X}_{\bar{x}=\bar{a}}| = |\mathbb{X}_{\bar{y}=\bar{a}}|$ for all $\bar{a} \in A^{|\bar{x}|}$.
(iii) $\mathbb{X} \models \bar{x} \approx^* \bar{y}$ iff $\{\{|\mathbb{X}_{\bar{x}=\bar{a}}| \mid \bar{a} \in X(\bar{x})\}\} = \{\{|\mathbb{X}_{\bar{y}=\bar{a}}| \mid \bar{a} \in X(\bar{y})\}\}$.

An atom $=\!\!(\bar{x}, \bar{y})$ is called a functional dependency, because $\mathbb{X} \models =\!\!(\bar{x}, \bar{y})$ iff there is a function $f\colon X(\bar{x}) \to X(\bar{y})$ such that $f(s(\bar{x})) = s(\bar{y})$ for all $s \in X$. An FD of the form $=\!\!(\lambda, x)$, where λ is the empty tuple, is called a constant atom and denoted by $=\!\!(x)$. Intuitively, the constant atom $=\!\!(x)$ states that variable x is constant in the team. The atom $\bar{x} \approx \bar{y}$ states that the tuples $\bar{x}$ and $\bar{y}$ give rise to identical distributions. The meaning of the atom $\bar{x} \approx^* \bar{y}$ is similar but allows the marginal probabilities to be attached to different tuples of values for $\bar{x}$ and $\bar{y}$.

Let $\Sigma \cup \{\sigma\}$ be a set of dependency, marginal identity, and marginal distribution equivalence atoms. We write $\mathbb{X} \models \Sigma$ iff $\mathbb{X} \models \sigma'$ for all $\sigma' \in \Sigma$, and we write $\Sigma \models \sigma$ iff $\mathbb{X} \models \Sigma$ implies $\mathbb{X} \models \sigma$ for all $\mathbb{X}$.

3 Axiomatization for FDs+UMIs+UMDEs

In this section, we present an axiomatization for dependency, unary marginal identity, and unary marginal distribution equivalence atoms. The axiomatization is infinite.

3.1 The Axioms

The axioms for unary marginal identity and unary marginal distribution equivalence are the equivalence axioms of reflexivity, symmetry, and transitivity:

UMI1: $x \approx x$
UMI2: If $x \approx y$, then $y \approx x$.
UMI3: If $x \approx y$ and $y \approx z$, then $x \approx z$.

UMDE1: $x \approx^* x$
UMDE2: If $x \approx^* y$, then $y \approx^* x$.
UMDE3: If $x \approx^* y$ and $y \approx^* z$, then $x \approx^* z$.

For functional dependencies, we take the Armstrong axiomatization [3] which consists of reflexivity, transitivity, and augmentation:

FD1: $=\!\!(\bar{x}, \bar{y})$ when $\mathsf{var}(\bar{y}) \subseteq \mathsf{var}(\bar{x})$.
FD2: If $=\!\!(\bar{x}, \bar{y})$ and $=\!\!(\bar{y}, \bar{z})$, then $=\!\!(\bar{x}, \bar{z})$.
FD3: If $=\!\!(\bar{x}, \bar{y})$, then $=\!\!(\bar{x}\bar{z}, \bar{y}\bar{z})$.

Since marginal identity is a special case of marginal distribution equivalence, we have the following axiom:

UMI & UMDE: If $x \approx y$, then $x \approx^* y$.

For the unary functional dependencies and unary marginal distribution equivalencies, we have the *cycle rule*:

If $=\!\!(x_0, x_1)$ and $x_1 \approx^* x_2$ and $\ldots$ and $=\!\!(x_{k-1}, x_k)$ and $x_k \approx^* x_0$,
then $=\!\!(x_1, x_0)$ and $\ldots$ and $=\!\!(x_k, x_{k-1})$ and $x_0 \approx^* x_1$ and $\ldots$ and $x_{k-1} \approx^* x_k$

for all $k \in \{1, 3, 5, \ldots\}$. Note that the cycle rule is not a single axiom but an infinite axiom schema, making our axiomatization infinite. The following inference rule is as a useful special case of the cycle rule

UMDE & FD: If $=\!\!(x, y)$ and $=\!\!(y, x)$, then $x \approx^* y$.

This can be seen by noticing that by the cycle rule

$$=\!\!(x, y) \text{ and } y \approx^* y \text{ and } =\!\!(y, x) \text{ and } x \approx^* x$$

implies $x \approx^* y$.

From here on, the "inference rules" refers to the axioms defined above. Let $\Sigma \cup \{\sigma\}$ be a set of dependency, marginal identity, and a marginal distribution equivalence atoms. We write $\Sigma \vdash \sigma$ iff σ can be derived from Σ by using the inference rules. We denote by $\mathsf{cl}(\Sigma)$ the set of atoms obtained by closing Σ under the inference rules, i.e., for all σ with variables from D, $\sigma \in \mathsf{cl}(\Sigma)$ iff $\Sigma \vdash \sigma$.

We say that an axiomatization is *sound* if for any set $\Sigma \cup \{\sigma\}$ of dependencies, $\Sigma \vdash \sigma$ implies $\Sigma \models \sigma$. Correspondingly, an axiomatization is *complete* if for any set $\Sigma \cup \{\sigma\}$ of dependencies, $\Sigma \models \sigma$ implies $\Sigma \vdash \sigma$.

3.2 Soundness of the Axioms

In this section, we show that our axiomatization is sound. We only show that the cycle rule is sound; since it is straightforward to check the soundness of other the axioms, we leave out their proofs.

We notice that for all $x, y \in D$, $\mathbb{X} \models =\!\!(x, y)$ implies $|X(x)| \geq |X(y)|$ and $\mathbb{X} \models x \approx^* y$ implies $|X(x)| = |X(y)|$. Suppose now that the precedent of the cycle rule holds for $\mathbb{X}$. It then follows that

$$|X(x_0)| \geq |X(x_1)| = |X(x_2)| \geq \cdots = |X(x_{k-1})| \geq |X(x_k)| = |X(x_0)|,$$

which implies that $|X(x_i)| = |X(x_j)|$ for all $i, j \in \{0, \dots k\}$. If i, j are additionally such that $\mathbb{X} \models =\!\!(x_i, x_j)$, then there is a surjective function $f\colon X(x_i) \to X(x_j)$ for which $f(s(x_i)) = s(x_j)$ for all $s \in X$. Since $X(x_i)$ and $X(x_j)$ are both finite and have the same number of elements, the function f is also one-to-one. Therefore the inverse of f is also a function, and we have $\mathbb{X} \models =\!\!(x_j, x_i)$, as wanted. Since f is bijective, there is a one-to-one correspondence between $X(x_i)$ and $X(x_j)$. Thus $|\mathbb{X}_{x_i=a}| = |\mathbb{X}_{x_j=f(a)}|$ for all $a \in X(x_i)$, and we have $\{\{|\mathbb{X}_{x_i=a}| \mid a \in X(x_i)\}\} = \{\{|\mathbb{X}_{x_j=a}| \mid a \in X(x_j)\}\}$, which implies that $\mathbb{X} \models x_i \approx^* x_j$.

4 Completeness and Armstrong Relations

In this section, we show that our axiomatization is complete. We do this by showing that for any set Σ of FDs, UMIs, and UMDEs, there is a probabilistic team $\mathbb{X}$ such that $\mathbb{X} \models \sigma$ iff $\sigma \in \mathsf{cl}(\Sigma)$. Note that proving this implies completeness: if $\sigma \notin \mathsf{cl}(\Sigma)$ (i.e. $\Sigma \nvdash \sigma$), then $\mathbb{X} \not\models \sigma$. Since $\mathbb{X} \models \Sigma$, we have $\Sigma \not\models \sigma$. By doing the proof this way, we obtain Armstrong relations because the constructed team $\mathbb{X}$ is an Armstrong relation[2] for Σ.

We first define a multigraph in which different-colored edges correspond to different types of dependencies between variables. By using the properties of the graph, we can construct a suitable team, which can then be made into a probabilistic team by taking the uniform distribution over the assignments.

Definition 1. *Let Σ be a set of FDs, UMIs, and UMDEs. We define a multigraph $G(\Sigma)$ as follows:*

(i) the set of vertices consists of the variables appearing in Σ,
(ii) for each marginal identity $x \approx y \in \Sigma$, there is an undirected black edge between x and y,
(iii) for each marginal distribution equivalence $x \approx^* y \in \Sigma$, there is an undirected blue edge between x and y, and
(iv) for each functional dependency $=\!\!(x, y) \in \Sigma$, there is a directed red edge from x to y.

If there are red directed edges both from x to y and from y to x, they can be thought of as a single red undirected edge between x and y.

In order to use the above graph construction to show the existence of Armstrong relations, we need to consider certain properties of the graph.

Lemma 2. *Let Σ be a set of FDs, UMIs and UMDEs that is closed under the inference rules, i.e. $\mathsf{cl}(\Sigma) = \Sigma$. Then the graph $G(\Sigma)$ has the following properties:*

[2] Note that a probabilistic team is actually not a relation but a probability distribution. Therefore, to be exact, instead of Armstrong relations, we should speak of *Armstrong models*, which is a more general notion introduced in [9]. In our setting, the Armstrong models we construct are uniform distributions over a relation, so each model is determined by a relation, and it suffices to speak of Armstrong relations.

(i) *Every vertex has a black, blue, and red self-loop.*
(ii) *The black, blue, and red subgraphs of $G(\Sigma)$ are all transitively closed.*
(iii) *The black subgraph of $G(\Sigma)$ is a subgraph of the blue subgraph of $G(\Sigma)$.*
(iv) *The subgraphs induced by the strongly connected components of $G(\Sigma)$ are undirected. Each such component contains a black, blue, and red undirected subgraph. In the subgraph of each color, the vertices of the component can be partitioned into a collection of disjoint cliques. All the vertices of the component belong to a single blue clique.*
(v) *If $=\!\!(\bar{x}, y) \in \Sigma$ and the vertices $\bar{x}$ have a common ancestor z in the red subgraph of $G(\Sigma)$, then there is a red edge from z to y.*

Proof. (i) By the reflexivity rules UMI1, UMDE1, and FD1, atoms $x \approx x$, $x \approx^* x$, and $=\!\!(x, x)$ are in Σ for every vertex x.
(ii) The transitivity rules UMI3, UMDE3, and FD2 ensure the transitivity of the black, blue, and red subgraphs, respectively.
(iii) By the UMI & UMDE rule, we have that if there is a black edge between x and y, then there is also a blue edge between x and y.
(iv) In a strongly connected component every black or blue edge is undirected by the definition of $G(\Sigma)$. Consider then a red edge from x to y. Since we are in a strongly connected component, there is also a path from y to x. By (iii), we may assume that this path only consist of blue and red edges. By adding the red edge from x to y to the beginning of the path, we obtain a cycle from x to x. (Note that in this cycle, some vertices might be visited more than once.) By using the part (i), we can add blue and red self-loops, if necessary, to make sure that the cycle is constructed from alternating blue and red edges as in the cycle rule. From the corresponding k-cycle rule, it then follows that there is also a red edge from y to x.
The existence of a partition follows from the fact that in each strongly connected component the black/blue/red edges define an equivalence relation for the vertices. (The reflexivity and transitivity follow from parts (i) and (ii), and the fact that each strongly connected component is undirected implies symmetry.) Lastly, pick any vertices x and y in a strongly connected component. We claim that there is a blue edge between x and y. Since we are in a strongly connected component, there is a path from x to y. The path consists of undirected black, blue and red edges. By (iii), each black edge of the path can be replaced with a blue one. Similarly, by the rule UMDE & FD, each undirected red edge can be replaced with a blue one. Thus, there is a blue path from x to y, and, by transitivity, also a blue edge.
(v) Let $\bar{x} = x_0 \dots x_n$. Suppose that we have $=\!\!(z, x_0 \dots x_k)$ and $=\!\!(z, x_{k+1})$ for some $0 \leq k \leq n-1$. By FD1 and FD3, we have $=\!\!(z, zz)$, $=\!\!(zz, x_{k+1}z)$, and $=\!\!(x_{k+1}z, zx_{k+1})$. Then by using FD2 twice, we obtain $=\!\!(z, zx_{k+1})$. Next, by FD3, we also have $=\!\!(zx_{k+1}, x_0 \dots x_k x_{k+1})$, and then by using FD2, we obtain $=\!\!(z, x_0 \dots x_{k+1})$. Since we have $=\!\!(z, x_j)$ for all $0 \leq j \leq n$, this shows that $=\!\!(z, \bar{x})$. By using FD2 to $=\!\!(z, \bar{x})$ and $=\!\!(\bar{x}, y)$, we obtain $=\!\!(z, y)$.

We next give a unique number to each strongly connected component of $G(\Sigma)$. The numbers are assigned such that the number of a descendant component is always greater that the number of its ancestor component. We call these numbers scc-numbers and denote by $\mathsf{scc}(x)$ the scc-number of the strongly connected component that the vertex x belongs to.

Definition 3. *Let Σ and $G(\Sigma)$ be as in Lemma 2 and assign an scc-numbering to $G(\Sigma)$. Let D be the set of the variables appearing in Σ. We construct a team X over the variables D as follows:*

(i) Add a tuple of all zeroes, i.e., an assignment s such that $s(x) = 0$ for all $x \in D$.

(ii) Process each strongly connected component in turn, starting with the one with the smallest scc-number and then proceeding in the ascending order of the numbers. For each strongly connected component, handle each of its maximal red cliques in turn.

(a) For each maximal red clique k, add a tuple with zeroes in the columns corresponding to the variables in k and to the variables that are in any red clique that is a red descendant of k. Leave all the other positions in the tuple empty for now.

(b) Choose a variable in k and count the number of zeroes in a column corresponding to the chosen variable. It suffices to consider only one variable, because the construction ensures that all the columns corresponding to variables in k have the same number of zeroes. Denote this number by $count(k)$.

After adding one tuple for each maximal red clique, check that the $count(k)$ is equal for every clique k in the current component and strictly greater than the count of each clique in the previous component. If it is not, repeat some of the tuples added to make it so. This can be done, because a red clique k can be a red descendant of another red clique j only if j is in a strongly connected component with a strictly smaller scc-number than the one of the component that k is in, and thus j's component is already processed. Note that the counts of the cliques in a component do not change after the component has been processed.

(iii) The last component is a single red clique consisting of those variables x for which $=\!(x) \in \Sigma$, if any. Each variable in this clique functionally depends on all the other variables in the graph, so we do not leave any empty positions in its column. Therefore the columns corresponding to these variables contain only zeroes. If there are no variables x for which $=\!(x) \in \Sigma$, we finish processing the last component by adding one tuple with all positions empty.

(iv) After all strongly connected components have been processed, we start filling the empty positions. Process again each strongly connected component in turn, starting with the one with the smallest scc-number. For each strongly connected component, count the number of maximal black cliques. If there are n such cliques, number them from 0 to $n-1$. Then handle each maximal black clique k, $0 \leq k \leq n-1$ in turn.

(a) For each column in clique k, count the number of empty positions. If the column has $d > 0$ empty positions, fill them with numbers $1, \ldots d-1, d+k$ without repetitions. (Note that each column in k has the same number of empty positions.)

(v) If there are variables x for which $=\!\!(x) \in \Sigma$, they are all in the last component. The corresponding columns contain only zeroes and have no empty positions. As before, count the number of maximal black cliques. If there are n such cliques, number them from 0 to $n-1$. Then handle each maximal black clique k, $0 \le k \le n-1$ in turn.
(a) For each column in clique k, change all the zeroes into k's.
After handling all the components (including the one consisting of constant columns, if there are any), there are no empty positions anymore, and the construction is finished.

The next lemma shows that team we constructed above already has many of the wanted properties:

Lemma 4. *Let Σ, $G(\Sigma)$, and X be as in Definition 3. Define a probabilistic team $\mathbb{X}$ as the uniform distribution over X. Then the following statements hold:*

(i) For any nonunary functional dependency σ, if $\sigma \in \Sigma$, then $\mathbb{X} \models \sigma$.
(ii) For any unary functional dependency or constancy atom σ, $\mathbb{X} \models \sigma$ if and only if $\sigma \in \Sigma$.
(iii) For any unary marginal distribution equivalence atom σ, $\mathbb{X} \models \sigma$ if and only if $\sigma \in \Sigma$.
(iv) For any unary marginal identity atom σ, $\mathbb{X} \models \sigma$ if and only if $\sigma \in \Sigma$.

Proof. (i) Suppose that $\sigma = \,=\!\!(\bar{x}, \bar{y}) \in \Sigma$. We may assume that $\bar{x}$,$\bar{y}$ do not contain variables z for which $=\!\!(z) \in \Sigma$, as it is easy to show that $\mathbb{X} \models \,=\!\!(\bar{x}, \bar{y})$ implies $\mathbb{X} \models \,=\!\!(\bar{x}z, \bar{y})$ and $\mathbb{X} \models \,=\!\!(\bar{x}, \bar{y}z)$ for any such variable z. For a contradiction suppose that $\mathbb{X} \not\models \sigma$. Then there are tuples s, s' that violate σ. By our assumption, the only repeated number in each column of variables $\bar{x}$,$\bar{y}$ is 0. Thus $s(\bar{x}) = s'(\bar{x}) = \bar{0}$, and either $s(\bar{y}) \neq \bar{0}$ or $s'(\bar{y}) \neq \bar{0}$. Assume that $s(\bar{y}) \neq \bar{0}$. Then there is $y_0 \in \mathsf{var}(\bar{y})$ such that $s(y_0) \neq 0$. By our construction, tuple s corresponds to a red clique. This means that in s, each variable that is 0, is functionally determined by every variable of the corresponding red clique. Since Σ is closed under FD rules, there is a variable z such that $s(z) = 0$ and $=\!\!(z, \bar{x}) \in \Sigma$. Then also $=\!\!(z, y_0) \in \Sigma$, and thus, by the construction, $s(y_0) = 0$, which is a contradiction.

(ii) Suppose that $\sigma = \,=\!\!(x)$. If $\sigma \in \Sigma$, then x is determined by all the other variables in Σ. This means that x is in the last component (the one with the highest scc-number), and thus the only value appearing in column x is k, where k is the number assigned to the maximal black clique of x. Hence, we have $\mathbb{X} \models \sigma$. If $\sigma \notin \Sigma$, then either x is not in the last component or there are no variables y for which $=\!\!(y) \in \Sigma$. In either case, we have at some point added a tuple in which the position of x is empty, and therefore there are more then one value appearing in column x, i.e., $\mathbb{X} \not\models \sigma$.
Suppose then that $\sigma = \,=\!\!(x, y)$. We may assume that neither $=\!\!(x)$ nor $=\!\!(y)$ are in Σ. If $=\!\!(y) \in \Sigma$, then y is constant, and we have $\sigma \in \Sigma$ and $\mathbb{X} \models \sigma$. If $=\!\!(x) \in \Sigma$ and $=\!\!(y) \notin \Sigma$, then $\sigma \notin \Sigma$ and $\mathbb{X} \not\models \sigma$. If $\sigma \in \Sigma$, we are done since this is a special case of (i). If $\sigma \notin \Sigma$, then the first tuple and the tuple that was added for the red clique of x violate σ.

(iii) Since $\mathbb{X}$ is obtained by taking the uniform distribution over X, $\mathbb{X} \models x \approx^* y$ holds if and only if $|X(x)| = |X(y)|$. The latter happens if and only if x and y are in the same strongly connected component. Since any strongly connected component is itself the maximal blue clique, this is equivalent to x and y being in the same maximal blue clique in a strongly connected component, i.e., $x \approx^* y \in \Sigma$.

(iv) Since $\mathbb{X}$ is obtained by taking the uniform distribution over X, $\mathbb{X} \models x \approx y$ holds if and only if $X(x) = X(y)$. The latter happens if and only if x and y are in the same maximal black clique in a strongly connected component. This happens if and only if $x \approx y \in \Sigma$.

Example 1. Let $\Delta = \{=\!(x_0, x_1), x_1 \approx x_2, =\!(x_2, x_3), x_3 \approx^* x_0, =\!(x_4), =\!(x_5)\}$, and $\Sigma = \mathsf{cl}(\Delta)$. Then the strongly connected components of $G(\Sigma)$ are $\{x_0, x_1, x_2, x_3\}$ and $\{x_4, x_5\}$, the maximal red cliques are $\{x_0, x_1\}$, $\{x_2, x_3\}$, and $\{x_4, x_5\}$, the maximal blue cliques are $\{x_0, x_1, x_2, x_3\}$ and $\{x_4, x_5\}$ and the maximal black cliques are $\{x_0\}$, $\{x_1, x_2\}$, $\{x_3\}$, $\{x_4\}$, and $\{x_5\}$. The probabilistic team $\mathbb{X}$ for Σ (as in Lemma 4) and the graph $G(\Sigma)$ are depicted in Fig. 1.

The construction of Lemma 4 does not yet give us Armstrong relations. This is because there might be nonunary functional dependencies σ such that $\mathbb{X} \models \sigma$ even though $\sigma \notin \Sigma$. In the lemma below, we show how to construct a probabilistic team for which $\mathbb{X} \models \sigma$ implies $\sigma \in \Sigma$ also for nonunary functional dependencies σ.

Lemma 5. *Let Σ and $G(\Sigma)$ be as in Lemma 2. Then there exists a probabilistic team $\mathbb{X}$ such that $\mathbb{X} \models \sigma$ if and only if $\sigma \in \Sigma$.*

Proof. Let D be the set of the variables appearing in Σ. We again begin by constructing a relational team X over D, but this time we modify the construction from Definition 3 by adding a new step (0), which will be processed first, before continuing with step (i). From step (i), we will proceed as in Definition 3. Let Σ' be a set of all nonunary functional dependencies $=\!(\bar{x}, \bar{y}) \notin \Sigma$ where $\bar{x}, \bar{y} \in D$. The step (0) is then defined as follows:

x_0	x_1	x_2	x_3	x_4	x_5	$\mathbb{X}$
0	0	0	0	0	1	1/4
0	0	1	1	0	1	1/4
1	1	0	0	0	1	1/4
2	3	3	4	0	1	1/4

(a)

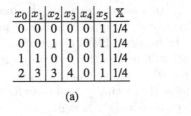

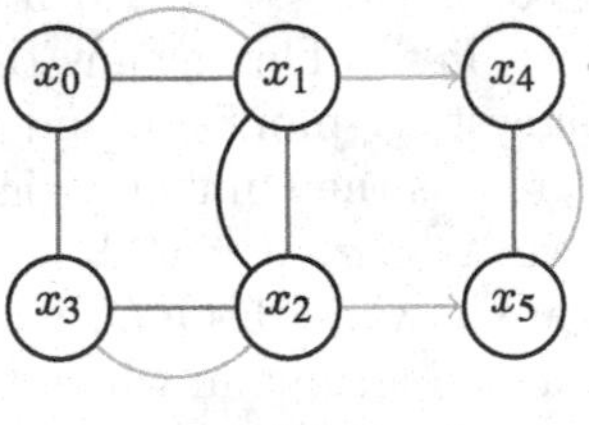

(b)

Fig. 1. (a) the probabilistic team $\mathbb{X}$ and (b) the graph $G(\Sigma)$ of Example 1. For the sake of clarity, we have removed from $G(\Sigma)$ all self-loops and some edges that are implied by transitivity.

(0) For each $C \subseteq D$ with $|C| \geq 2$, check if there is $=\!\!(\bar{x}, \bar{y}) \in \Sigma'$ such that $\mathsf{var}(\bar{x}) = C$. If yes, add a tuple with zeroes in exactly those positions z that are functionally determined by $\bar{x}$, i.e., those z for which $=\!\!(\bar{x}, z) \in \Sigma$. Leave all the other positions in the tuple empty for now.

As before, we define the probabilistic team $\mathbb{X}$ as the uniform distribution over X. We first show that $\mathbb{X}$ violates all atoms from Σ'. Let $\sigma = \,=\!\!(\bar{x}, \bar{y}) \in \Sigma'$. We may assume that $\bar{x}$,$\bar{y}$ do not contain variables z for which $=\!\!(z) \in \Sigma$ because $\mathbb{X} \not\models \,=\!\!(\bar{x}, \bar{y})$ implies $\mathbb{X} \not\models \,=\!\!(\bar{x}z, \bar{y})$ and $\mathbb{X} \not\models \,=\!\!(\bar{x}, \bar{y}z)$ for any such variable z. Since $\sigma \notin \Sigma$, there is $y_0 \in \mathsf{var}(\bar{y})$ such that y_0 is not determined by $\bar{x}$, i.e., $=\!\!(\bar{x}, y_0) \notin \Sigma$. Thus, the tuple that was added for the set $C = \mathsf{var}(\bar{x})$ in step (0), and the tuple added in step (i), agree on variables $\bar{x}$ (they are all zeroes), but in the first tuple, variable y_0 is nonzero and in the other tuple, y_0 is zero. Hence, we have $\mathbb{X} \not\models \sigma$.

We still need to show that $\mathbb{X} \models \Sigma$, and that $\sigma \notin \Sigma$ implies $\mathbb{X} \not\models \sigma$, when σ is a UMI, UMDE, UFD or a constancy atom. Note that since the construction proceeds from step (i) as in Definition 3, it ensures that the counts of the empty positions are the same for all the columns that correspond to the variables in the same strongly connected component. Therefore, for any unary marginal identity or distribution equivalence atom σ, $\mathbb{X} \models \sigma$ if and only if $\sigma \in \Sigma$ as shown in Lemma 4, parts (iii) & (iv). For a unary functional dependency, or a constancy atom $\sigma \notin \Sigma$, we obtain, by Lemma 4, part (ii), that already the part of the team $\mathbb{X}$ that was constructed as in Definition 3 contains tuples that violate σ.

If $=\!\!(z) \in \Sigma$, then clearly $=\!\!(\bar{x}, z) \in \Sigma$ for any $\bar{x}$, and thus adding the tuples in step (0) does not end up violating $=\!\!(z)$. Suppose then that $=\!\!(\bar{z}, \bar{z}') \in \Sigma$, where $\bar{z}$ and $\bar{z}'$ can also be single variables. For a contradiction, suppose there are two tuples that together violate $=\!\!(\bar{z}, \bar{z}')$. We may again assume that $\bar{z}$,$\bar{z}'$ do not contain variables u for which $=\!\!(u) \in \Sigma$. By our assumption, the only repeated numbers in columns $\bar{z}$ are zeroes, so it has to be the case that the variables $\bar{z}$ are all zeroes in the two tuples that violate $=\!\!(\bar{z}, \bar{z}')$. We now show that in any such tuple, the variables $\bar{z}'$ are also all zeroes, contradicting the assumption that there are two tuples that agree on $\bar{z}$ but not on $\bar{z}'$.

Suppose that the tuple was added on step (0). By the construction, we then have some variables $\bar{x}$, $\bar{y}$ such that $|\bar{x}| \geq 2$, $=\!\!(\bar{x}, \bar{y}) \in \Sigma'$, and $=\!\!(\bar{x}, \bar{z}) \in \Sigma$. Since $=\!\!(\bar{z}, \bar{z}') \in \Sigma$, by transitivity, we have $=\!\!(\bar{x}, \bar{z}') \in \Sigma$, and thus the variables $\bar{z}'$ are also all zeroes in the tuple, as wanted. Suppose then that the tuple was not added in step (0), i.e., it was added for some maximal red clique in a strongly connected component. Suppose then that there is $z_0 \in \mathsf{var}(\bar{z}')$ which is not zero in the tuple. This leads to a contradiction as shown in Lemma 4 part (i). Hence, the tuple has to have all zeroes for variables $\bar{z}'$.

5 Complexity of the Implication Problem for FDs+UMIs+UMDEs

Let $\Sigma \cup \{\sigma\}$ be a set of functional dependencies, unary marginal identity, and unary marginal distribution equivalence atoms. The decision problem of checking whether $\Sigma \models \sigma$ is called the *implication problem* for FDs+UMIs+UMDEs. We show that this implication problem has a polynomial-time algorithm. Denote by D the set of the variables appearing in $\Sigma \cup \{\sigma\}$, and suppose that Σ is partitioned into the sets Σ_{FD}, Σ_{UMI}, and Σ_{UMDE} that consist of the FDs, UMIs, and UMDEs from Σ, respectively.

First, note that if σ is a UMI, it suffices to check whether $\sigma \in \mathsf{cl}(\Sigma_{\mathrm{UMI}})$ as the inference rules for FDs and UMDEs do not produce new UMIs. Notice that Σ_{UMI} can be viewed as an undirected graph $G = (D, \approx)$ such that each $x \approx y \in \Sigma_{\mathrm{UMI}}$ means there is an undirected edge between x and y. (We may assume that each vertex has a self-loop and all the edges are undirected, even though the corresponding UMIs might not be explicitly listed in Σ_{UMI}.) Suppose that $\sigma = x \approx y$. Then $\sigma \in \mathsf{cl}(\Sigma_{\mathrm{UMI}})$ iff y is reachable from x in G. This is the problem of *undirected s-t connectivity*, which is known to be L-complete [20].

Thus, we can assume that σ is an FD or a UMDE. The idea behind the following part of the algorithm is similar to the algorithm (introduced in [7]) for the implication problem for FDs and unary inclusion dependencies.

The implication problem for FDs is known to be linear-time computable by the *Beeri-Bernstein algorithm* [4]. Given a set $\Delta \cup \{=\!(\bar{x}, \bar{y})\}$ of FDs, the Beeri-Bernstein algorithm computes the set $\mathsf{fdclosure}(\bar{x}, \Delta) = \{z \mid \Delta \models =\!(\bar{x}, z)\}$. Then $\Delta \models =\!(\bar{x}, \bar{y})$ holds iff $\mathsf{var}(\bar{y}) \subseteq \mathsf{fdclosure}(\bar{x}, \Delta)$. The Beeri-Bernstein algorithm keeps two lists: $\mathsf{fdlist}(\bar{x})$ and $\mathsf{attrlist}(\bar{x})$. The set $\mathsf{fdlist}(\bar{x})$ consists of FDs and is updated in order to keep the algorithm in linear-time (for more details, see [4]). The set $\mathsf{attrlist}(\bar{x})$ lists the variables that are found to be functionally dependent on $\bar{x}$. The algorithm[3] is as follows:

(i) Initialization:
   Let $\mathsf{fdlist}(\bar{x}) = \Delta$ and $\mathsf{attrlist}(\bar{x}) = \mathsf{var}(\bar{x})$.
(ii) Repeat the following until no new variables are added to $\mathsf{attrlist}(\bar{x})$:
   (a) For all $z \in \mathsf{attrlist}(\bar{x})$, if there is an FD $=\!(u_0 \dots u_k, \bar{u}') \in \mathsf{fdlist}(\bar{x})$ such that $z = u_i$ for some $i = 0, \dots, k$, replace $=\!(u_0 \dots u_k, \bar{u}')$ with $=\!(u_0 \dots u_{i-1} u_{i+1} \dots u_k, \bar{u}')$.
   (b) For each constant atom $=\!(\bar{u}) \in \mathsf{fdlist}(\bar{x})$, add $\mathsf{var}(\bar{u})$ to $\mathsf{attrlist}(\bar{x})$.
(ii) Return $\mathsf{attrlist}(\bar{x})$.

Let $x \in D$, and define $\mathsf{desclist}(x) = \{y \in D \mid \Sigma \models =\!(x, y)\}$. We will construct an algorithm, called $\mathsf{Closure}(x, \Sigma)$, that computes the set $\mathsf{desclist}(x)$ by utilizing the Beeri-Bernstein algorithm. If we view the variables D and the atoms from $\mathsf{cl}(\Sigma)$ as a multigraph of Definition 1, the set $\mathsf{desclist}(x)$ consists of the variables that are red descendants of x. Since FDs and UMIs interact only via UMDEs, there is no need to construct the whole multigraph. We only consider the UMIs in the initialization step, as they imply UMDEs. The algorithm $\mathsf{Closure}(x, \Sigma)$ keeps and updates $\mathsf{fdlist}(y)$ and $\mathsf{attrlist}(y)$ for each $y \in D$. When the algorithm halts, $\mathsf{attrlist}(x) = \mathsf{desclist}(x)$. The algorithm is as follows:

(i) Initialization:

[3] This presentation of the algorithm is based on [7].

(a) For each variable $y \in D$, $\mathsf{attrlist}(y)$ is initialized to $\{y\}$ and $\mathsf{fdlist}(y)$ to Σ_{FD}. We begin by running the Beeri-Bernstein algorithm for each $y \in D$. The algorithm adds to $\mathsf{attrlist}(y)$ all variables $z \in D$ that are functionally dependent on y under Σ_{FD}. Note that each $\mathsf{fdlist}(y)$ is also modified.

(b) We then construct a multigraph G similarly as before, but the black edges are replaced with blue ones, i.e. the set of vertices is D, and there is a directed red edge from y to z iff $z \in \mathsf{attrlist}(y)$, and there is an undirected blue edge between y and z iff at least one of the following $y \approx z, z \approx y, y \approx^* z, z \approx^* y$ is in $\Sigma_{\mathrm{UMI}} \cup \Sigma_{\mathrm{UMDE}}$.

(ii) Repeat the following iteration until halting and then return $\mathsf{attrlist}(x)$:

(1) Ignore the colors of the edges, and find the strongly connected components of G.

(2) For any y and z that are in the same strongly connected component, if there is a red directed edge from y to z, check whether there is a red directed edge from z to y and a blue undirected edge between y and z. If either or both are missing, add the missing edges.

(3) If no new red edges were added in step (2), then halt. Otherwise, add all the new red edges to $\mathsf{fdlist}(y)$.

(4) For each y, continue the Beeri-Bernstein algorithm.

(5) If for all y, no new variables were added to $\mathsf{attrlist}(y)$ in step (4), then halt. Otherwise, for each new $z \in \mathsf{attrlist}(y)$, add to G a directed red edge from y to z.

Let $n = |D|$. Since we can add at most n^2 new red edges to the graph, $|\mathsf{fdlist}(y)| \leq |\Sigma_{\mathrm{FD}}| + O(n^2)$ for each $y \in D$. Thus the Beeri-Bernstein algorithm takes $O(|\Sigma_{\mathrm{FD}}| + n^2)$ time for each $y \in D$. Because we run the algorithm for all $y \in D$, the total time is $O(n(n^2 + |\Sigma_{\mathrm{FD}}|))$.

In each iteration, before running the Beeri-Bernstein algorithm (step (4)), we find the strongly connected components (step (1)) and check all pairs of vertices for red edges and add red and blue edges if needed (step (2)). The time required for finding the strongly connected components is linear in the number of vertices and edges [23], so it takes $O(|D| + |\Sigma_{\mathrm{FD}}| + n^2 + |\Sigma_{\mathrm{UMI}}| + |\Sigma_{\mathrm{UMDE}}|)$, i.e. $O(n^2)$ time. Going through the pairs of vertices in step (2) also takes $O(n^2)$ time. Thus each iteration takes $O(n^3 + n|\Sigma_{\mathrm{FD}}|)$ time in total.

An iteration proceeds to step (4) only if new red edges are added in step (2). In each iteration, if new red edges are added, they merge together two or more red cliques in the same strongly connected component. In the beginning we have at most n such cliques, so the algorithm cannot take more than $O(n)$ iterations. The total running time of $\mathsf{Closure}(x, \Sigma)$ is then $O(n^4 + n^2|\Sigma_{\mathrm{FD}}|)$.

If $\sigma = {=\!\!(x, y)}$, then $\Sigma \models \sigma$ iff $y \in \mathsf{desclist}(x)$. If σ is a non-unary FD, in order to check whether $\Sigma \models \sigma$, it suffices to check whether $\Sigma_{\mathrm{FD}} \cup \Delta \models \sigma$, where $\Delta = \{{=\!\!(y, \bar{z})} \mid y \in D, \mathsf{var}(\bar{z}) = \mathsf{desclist}(y)\}$. This, again, can be done using the Beeri-Bernstein algorithm. Since the order of the variables $\mathsf{var}(\bar{z})$ in the tuple $\bar{z}$ does not matter for each FD ${=\!\!(y, \bar{z})}$, we may assume that the set Δ contains only one FD for each $y \in D$. Thus this step only adds $O(|\Sigma_{\mathrm{FD}}| + n)$ extra time. (Note that constructing the set Δ does not affect the time-bound because $\mathsf{Closure}(x, \Sigma)$ actually computes $\mathsf{desclist}(y)$ for all $y \in D$.)

If $\sigma = x \approx^* y$, we can use the blue (undirected) subgraph of G to check whether $\Sigma \models \sigma$. After the algorithm $\mathsf{Closure}(x, \Sigma)$ has halted, we only need to check whether y is reachable from x in the subgraph, which is an L-complete problem, as noted before.

6 Conclusion

We have presented a sound and complete axiomatization for functional dependence, unary marginal identity, and unary marginal distribution equivalence. The class of these dependencies was also shown to have Armstrong relations because in the completeness proof, we constructed an Armstrong relation for any set of FDs, UMIs, and UMDEs. We also showed that the implication problem for these dependencies is in polynomial-time.

The following questions remain open:

(i) Can we extend the axiomatization to nonunary marginal identity and marginal distribution equivalence?
(ii) What is the connection between the implication problem considered in this paper and the finite implication problem for functional dependence and unary inclusion dependence?
(iii) Can we find an axiomatization for probabilistic independence and marginal identity (and marginal distribution equivalence)?

Concerning the first question, one should not rush to conjectures based on the results in the qualitative (non-probabilistic) case. Our results for functional dependencies and unary variants of marginal identities and marginal distribution equivalences were obtained by methods that resemble the ones used in the case of functional dependence and unary inclusion dependence in [7]. The implication problem for functional and inclusion dependencies (where also nonunary inclusions are allowed) is undecidable for both finite and unrestricted implication [6], so one might think that allowing nonunary marginal identities with functional dependencies makes the corresponding (finite) implication problem undecidable. However, it is not clear whether this is the case, as the reduction proof for the undecidability result does not directly generalize to the probabilistic case.

The second question also relates to how similar the qualitative version (inclusion dependence) is to the probabilistic version (marginal identity). It seems that the implication problem for functional dependencies and unary marginal identities (without unary marginal distribution equivalencies) can be simulated with the implication problem for functional dependencies and unary inclusion dependecies, by simulating each marginal identity $x \approx y$ with two inclusion atoms ($x \subseteq y$ and $y \subseteq x$). It would be interesting to see whether the full implication problem (i.e. the one with unary marginal equivalencies as well) can also be simulated with the implication problem for functional and unary inclusion dependence.

As for the third question, the implication problems for probabilistic independence and marginal identity have already been studied separately: the problem for probabilistic independence has been axiomatized in [12], and the problem for marginal identity (over finite probability distributions) has recently been axiomatized in [16]. Obtaining an axiomatization for both of these dependencies together is appealing, given how commonly IID assumption is used in probability theory and statistics.

In addition to probabilistic independence, we can consider probabilistic conditional independence $\bar{y} \perp\!\!\!\perp_{\bar{x}} \bar{z}$ which states that $\bar{y}$ and $\bar{z}$ are conditionally independent given $\bar{x}$. A probabilistic conditional independency of the form $\bar{y} \perp\!\!\!\perp_{\bar{x}} \bar{y}$[4] is equivalent with the functional dependency $=\!\!(\bar{x}, \bar{y})$, so functional dependence can be seen as a special case of probabilistic conditional independence. Probabilistic conditional independence has no finite complete axiomatization [21], although sound (but not complete) axiomatizations, the so-called graphoid and semigraphoid axioms, exist and are often used in practice. Every sound inference rule for probabilistic conditional independence that has at most two antecedents can be derived from the semigraphoid axioms [22]. It would be interesting to find axiomatizations for marginal identity and some subclasses of probabilistic conditional independence. The results of this paper imply that if we consider a subclass that contains probabilistic conditional independencies of the form $\bar{y} \perp\!\!\!\perp_{\bar{x}} \bar{y}$, we effectively include functional dependencies and thus we will also have the cycle rules.

Acknowledgements. I would like to thank the anonymous referees for valuable comments, and Miika Hannula and Juha Kontinen for useful discussions and advice.

References

1. Abramsky, S., Puljujärvi, J., Väänänen, J.: Team semantics and independence notions in quantum physics (2021)
2. Albert, R., Grädel, E.: Unifying hidden-variable problems from quantum mechanics by logics of dependence and independence. Ann. Pure Appl. Logic 103088 (2022)
3. Armstrong, W.W.: Dependency structures of data base relationships. In: Proceedings of IFIP World Computer Congress, pp. 580–583 (1974)
4. Beeri, C., Bernstein, P.A.: Computational problems related to the design of normal form relational schemas. ACM Trans. Database Syst. **4**(1), 30–59 (1979)
5. Beeri, C., Dowd, M., Fagin, R., Statman, R.: On the structure of armstrong relations for functional dependencies. J. ACM **31**(1), 30–46 (1984)
6. Chandra, A.K., Vardi, M.Y.: The implication problem for functional and inclusion dependencies is undecidable. SIAM J. Comput. **14**(3), 671–677 (1985)
7. Cosmadakis, S.S., Kanellakis, P.C., Vardi, M.Y.: Polynomial-time implication problems for unary inclusion dependencies. J. ACM **37**(1), 15–46 (1990)

[4] Sometimes the tuples of variables are assumed to be disjoint which disallows the probabilistic conditional independencies of this form. In the context of team semantics, such an assuption is usually not made.

8. Durand, A., Hannula, M., Kontinen, J., Meier, A., Virtema, J.: Approximation and dependence via multiteam semantics. Ann. Math. Artif. Intell. **83**(3–4), 297–320 (2018)
9. Fagin, R.: Horn clauses and database dependencies. J. ACM **29**, 952–985 (1982)
10. Fagin, R., Vardi, M.Y.: Armstrong databases for functional and inclusion dependencies. Inf. Process. Lett. **16**(1), 13–19 (1983)
11. Galliani, P.: Inclusion and exclusion dependencies in team semantics: on some logics of imperfect information. Ann. Pure Appl. Logic **163**(1), 68–84 (2012)
12. Geiger, D., Paz, A., Pearl, J.: Axioms and algorithms for inferences involving probabilistic independence. Inf. Comput. **91**(1), 128–141 (1991)
13. Grädel, E., Väänänen, J.: Dependence and independence. Stud. Logica. **101**(2), 399–410 (2013)
14. Hannula, M., Hirvonen, Å., Kontinen, J., Kulikov, V., Virtema, J.: Facets of distribution identities in probabilistic team semantics. In: Calimeri, F., Leone, N., Manna, M. (eds.) JELIA 2019. LNCS (LNAI), vol. 11468, pp. 304–320. Springer, Cham (2019). https://doi.org/10.1007/978-3-030-19570-0_20
15. Hannula, M., Link, S.: On the interaction of functional and inclusion dependencies with independence atoms. In: Pei, J., Manolopoulos, Y., Sadiq, S., Li, J. (eds.) DASFAA 2018. LNCS, vol. 10828, pp. 353–369. Springer, Cham (2018). https://doi.org/10.1007/978-3-319-91458-9_21
16. Hannula, M., Virtema, J.: Tractability frontiers in probabilistic team semantics and existential second-order logic over the reals (2021)
17. Hodges, W.: Compositional Semantics for a Language of Imperfect Information. J. Interest Group Pure Appl. Logics **5**(4), 539–563 (1997)
18. Langeveldt, W.-D., Link, S.: Empirical evidence for the usefulness of armstrong relations in the acquisition of meaningful functional dependencies. Inf. Syst. **35**(3), 352–374 (2010)
19. Mannila, H., Räihä, K.-J.: Design by example: an application of armstrong relations. J. Comput. Syst. Sci. **33**(2), 126–141 (1986)
20. Reingold, O.: Undirected connectivity in log-space. J. ACM **55**(4), 1–24 (2008)
21. Studený, M.: Conditional independence relations have no finite complete characterization, pp. 377–396. Kluwer (1992)
22. Studený, M.: Semigraphoids are two-antecedental approximations of stochastic conditional independence models. In: Uncertainty in Artificial Intelligence, Proceedings of the Tenth Conference, pp. 546–552 (1994)
23. Tarjan, R.: Depth first search and linear graph algorithms. SIAM J. Comput. **1**(2), 146–160 (1972)
24. Väänänen, J.: Dependence Logic. Cambridge University Press, Cambridge (2007)

Approximate Keys and Functional Dependencies in Incomplete Databases with Limited Domains

Munqath Al-atar[1,2] and Attila Sali[1,3(✉)]

[1] Department of Computer Science and Information Theory, Budapest University of Technology and Economics, Budapest, Hungary
m.attar@cs.bme.hu

[2] ITRDC, University of Kufa, Kufa, Iraq
munqith.alattar@uokufa.edu.iq

[3] Alfréd Rényi Institute of Mathematics, Budapest, Hungary
sali.attila@renyi.hu

Abstract. A possible world of an incomplete database table is obtained by imputing values from the attributes (infinite) domain to the place of NULLs. A table satisfies a possible key or possible functional dependency constraint if there exists a possible world of the table that satisfies the given key or functional dependency constraint. A certain key or functional dependency is satisfied by a table if all of its possible worlds satisfy the constraint. Recently, an intermediate concept was introduced. A strongly possible key or functional dependency is satisfied by a table if there exists a strongly possible world that satisfies the key or functional dependency. A strongly possible world is obtained by imputing values from the active domain of the attributes, that is from the values appearing in the table. In the present paper, we study approximation measures of strongly possible keys and FDs. Measure g_3 is the ratio of the minimum number of tuples to be removed in order that the remaining table satisfies the constraint. We introduce a new measure g_5, the ratio of the minimum number of tuples to be added to the table so the result satisfies the constraint. g_5 is meaningful because the addition of tuples may extend the active domains. We prove that if g_5 can be defined for a table and a constraint, then the g_3 value is always an upper bound of the g_5 value. However, the two measures are independent of each other in the sense that for any rational number $0 \leq \frac{p}{q} < 1$ there are tables of an arbitrarily large number of rows and a constant number of columns that satisfy $g_3 - g_5 = \frac{p}{q}$. A possible world is obtained usually by adding many new values not occurring in the table before. The measure g_5 measures the smallest possible distortion of the active domains.

Research of the second author was partially supported by the National Research, Development and Innovation Office (NKFIH) grants K–116769 and SNN-135643. This work was also supported by the BME- Artificial Intelligence FIKP grant of EMMI (BME FIKP-MI/SC) and by the Ministry of Innovation and Technology and the National Research, Development and Innovation Office within the Artificial Intelligence National Laboratory of Hungary.

I. Varzinczak (Ed.): FoIKS 2022, LNCS 13388, pp. 147–167, 2022.
https://doi.org/10.1007/978-3-031-11321-5_9

Keywords: Strongly possible functional dependencies · Strongly possible keys · incomplete databases · data Imputation · Approximate functional dependencies · approximate keys

1 Introduction

The information in many industrial and research databases may usually be incomplete due to many reasons. For example, databases related to instrument maintenance, medical applications, and surveys [10]. This makes it necessary to handle the cases when some information missing from a database and are required by the user. Imputation (filling in) is one of the common ways to handle the missing values [20].

A new approach for imputing values in place of the missing information was introduced in [2], to achieve complete data tables, using only information already contained in the SQL table attributes (which are called the active domain of an attribute). Any total table obtained in this way is called a strongly possible world. We use only the data shown on the table to replace the missing information because in many cases there is no proper reason to consider any other attribute values than the ones that already exist in the table. Using this concept, new key and functional dependency constraints called strongly possible keys (spKeys) and strongly possible functional dependencies (spFDs) were defined in [3,5] that are satisfied after replacing any missing value (`NULL`) with a value that is already shown in the corresponding attribute. In Sect. 2, we provide the formal definitions of spKeys and spFDs.

The present paper continues the work started in [5], where an approximation notion was introduced to calculate how close any given set of attributes can be considered as a key, even when it does not satisfy the conditions of spKeys. This is done by calculating the minimum number of tuples that need to be removed from the table so that the spKey constraint holds.

Tuple removal may be necessary because the active domains do not contain enough values to be able to replace the `NULL` values so that the tuples are pairwise distinct on a candidate key set of attributes K. In the present paper, we introduce approximation measures of spKeys and spFDs by adding tuples. Adding a tuple with new unique values will add more values to the attributes' active domains, thus some unsatisfied constraints may get satisfied. For example, Car_Model and $DoorNo$ is designed to form a key in the Cars Types table shown in Table 1 but the table does not satisfy the spKey $sp\langle Car\_Model, DoorNo\rangle$. Two tuples would need to be removed, but adding a new tuple with distinct door number value to satisfy $sp\langle Car\_Model, DoorNo\rangle$ is better than removing two tuples. In addition to that, we know that the car model and door number determines the engine type, then the added tuple can also have a new value in the $DoorNo$ attribute so that the table satisfy $(Car\_Model, DoorNo) \rightarrow_{sp} Engine\_Type$ rather than removing other two tuples.

Table 1. Cars Types Incomplete Table

Car_Model	Door No	Engine_Type
BMW I3	4 doors	⊥
BMW I3	⊥	electric
Ford explorer	⊥	V8
Ford explorer	⊥	V6

Adding tuples with new values provides more values in the active domains used to satisfy the spKey. But if the total part of the table does not satisfy the key, then it is useless to add more values to the active domain. Thus, we assume throughout this paper that the K-total part of the table satisfies the spKey $sp\langle K\rangle$ constraint, and that the X-total part satisfies the spFD constraint $X \rightarrow_{sp} Y$ (for exact definitions see Sect. 2). The interaction between spFDs and spKeys is studied in [1]. We also assume that every attribute has at least one non-null value (so that the active domain is not the empty set) and we have at least 2 attributes in the key set K since it was observed in [5] that a single attribute can only be an spKey if the table does not contain `NULL` in it.

The main objectives of this paper are:

- Extend the g_3 measure defined for spKeys in [5] to spFDs.
- Propose a new approximation measure for spKeys and spFDs called g_5, that adopt adding tuples with new values to the tables that violate the constraints.
- Compare the newly proposed measure g_5 with the earlier introduced measure g_3 and show that adding new tuples is more effective than removing violating ones.
- Nevertheless, g_3 and g_5 are independent of each other.

It is important to observe the difference between possible worlds and strongly possible worlds. The former one was defined and studied by several sets of authors, for example in [9,18,28]. In possible worlds, any value from the usually countably infinite domain of the attribute can be imputed in place of `NULL`s. This allows an infinite number of worlds to be considered. By taking the newly introduced active domain values given by the added tuples and minimizing the number of the tuples added, we sort of determine a minimum world that satisfies the constraints and contains an spWorld allowed by the original table given.

The paper is organized as follows. Section 2 gives the basic definitions and notations. Some related work and research results are discussed in Sect. 3. The approximation measures for spKeys and spFDs are provided in Sects. 4 and 5 respectively. And finally, the conclusions and the future directions are explained in Sect. 6.

2 Basic Definitions

Let $R = \{A_1, A_2, \dots A_n\}$ be a relation schema. The set of all the possible values for each attribute $A_i \in R$ is called the domain of A_i and denoted as $D_i = dom(A_i)$ for $i = 1,2,\dots n$. Then, for $X \subseteq R$, then $D_X = \prod_{\forall A_i \in K} D_i$.

An instance $T = (t_1,t_2,\dots t_s)$ over R is a list of tuples such that each tuple is a function $t : R \rightarrow \bigcup_{A_i \in R} dom(A_i)$ and $t[A_i] \in dom(A_i)$ for all A_i in R. By taking a list of tuples we use the bag semantics that allows several occurrences of the same tuple. Usage of the bag semantics is justified by that SQL allows multiple occurrences of tuples. Of course, the order of the tuples in an instance is irrelevant, so mathematically speaking we consider a multiset of tuples as an instance. For a tuple $t_r \in T$ and $X \subset R$, let $t_r[X]$ be the restriction of t_r to X.

It is assumed that $\perp$ is an element of each attribute's domain that denotes missing information. t_r is called V-total for a set V of attributes if $\forall A \in V$, $t_r[A] \neq \perp$. Also, t_r is a total tuple if it is R-total. t_1 and t_2 are weakly similar on $X \subseteq R$ denoted as $t_1[X] \sim_w t_2[X]$ defined by Köhler et.al. [17] if

$$\forall A \in X \quad (t_1[A] = t_2[A] \text{ or } t_1[A] = \perp \text{ or } t_2[A] = \perp).$$

Furthermore, t_1 and t_2 are strongly similar on $X \subseteq R$ denoted by $t_1[X] \sim_s t_2[X]$ if

$$\forall A \in X \quad (t_1[A] = t_2[A] \neq \perp).$$

For the sake of convenience we write $t_1 \sim_w t_2$ if t_1 and t_2 are weakly similar on R and use the same convenience for strong similarity. Let $T = (t_1, t_2, \dots t_s)$ be a table instance over R. Then, $T' = (t'_1, t'_2, \dots t'_s)$ is a possible world of T, if $t_i \sim_w t'_i$ for all $i = 1, 2, \dots s$ and T' is completely NULL-free. That is, we replace the occurrences of $\perp$ with a value from the domain D_i different from $\perp$ for all tuples and all attributes. A active domain of an attribute is the set of all the distinct values shown under the attribute except the NULL. Note that this was called the visible domain of the attribute in papers [1–3,5].

Definition 2.1. *The active domain of an attribute A_i (VD_i^T) is the set of all distinct values except $\perp$ that are already used by tuples in T:*

$$VD_i^T = \{t[A_i] : t \in T\} \setminus \{\perp\} \text{ for } A_i \in R.$$

To simplify notation, we omit the upper index T if it is clear from the context what instance is considered.

Then the VD_1 in Table 2 is {Mathematics, Datamining}. The term active domain refers to the data that already exist in a given dataset. For example, if we have a dataset with no information about the definitions of the attributes' domains, then we use the data itself to define their own structure and domains. This may provide more realistic results when extracting the relationship between data so it is more reliable to consider only what information we have in a given dataset.

While a possible world is obtained by using the domain values instead of the occurrence of NULL, a strongly possible world is obtained by using the active domain values.

Definition 2.2. *A possible world T' of T is called a strongly possible world (spWorld) if $t'[A_i] \in VD_i^T$ for all $t' \in T'$ and $A_i \in R$.*

The concept of strongly possible world was introduced in [2]. A strongly possible worlds allow us to define strongly possible keys (spKeys) and strongly possible functional dependencies (spFDs).

Definition 2.3. *A strongly possible functional dependency, in notation $X \to_{sp} Y$, holds in table T over schema R if there exists a strongly possible world T' of T such that $T' \models X \to Y$. That is, for any $t'_1, t'_2 \in T'$ $t'_1[X] = t'_2[X]$ implies $t'_1[Y] = t'_2[Y]$. The set of attributes X is a strongly possible key, if there exists a strongly possible world T' of T such that X is a key in T', in notation $sp\langle X \rangle$. That is, for any $t'_1, t'_2 \in T'$ $t'_1[X] = t'_2[X]$ implies $t'_1 = t'_2$.*

Note that this is not equivalent with spFD $X \to_{sp} R$, since we use the bag semantics. For example, {Course Name, Year} is a strongly possible key of Table 2 as the strongly possible world in Table 3 shows it.

Table 2. Incomplete Dataset

Course Name	Year	Lecturer	Credits	Semester
Mathematics	2019	$\perp$	5	1
Datamining	2018	Sarah	7	$\perp$
$\perp$	2019	Sarah	$\perp$	2

If $T = \{t_1, t_2, \ldots, t_p\}$ and $T' = \{t'_1, t'_2, \ldots, t'_p\}$ is an spWorld of it with $t_i \sim_w t'_i$, then t'_i is called an sp-extension or in short an extension of t_i. Let $X \subseteq R$ be a set of attributes and let $t_i \sim_w t'_i$ such that for each $A \in R$: $t'_i[A] \in VD(A)$, then $t'_i[X]$ is an strongly possible extension of t_i on X (sp-extension)

Table 3. Complete Dataset

Course Name	Year	Lecturer	Credits	Semester
Mathematics	2019	Sarah	5	1
Datamining	2018	Sarah	7	2
Datamining	2019	Sarah	7	2

3 Related Work

Giannella et al. [11] measure the approximate degree of functional dependencies. They developed the IFD approximation measure and compared it with the other two measures: g_3 (minimum number of tuples need to be removed so that the dependency holds) and τ (the probability of a correct guess of an FD satisfaction) introduced in [16] and [12] respectively. They developed analytical bounds on the measure differences and compared these measures analysis on five datasets. The authors show that when measures are meant to define the knowledge degree of X determines Y (prediction or classification), then IFD and τ measures are more appropriate than g_3. On the other hand, when measures are meant to define the number of "violating" tuples in an FD, then, g_3 measure is more appropriate than IFD and τ. This paper extends the earlier work of [5] that utilized the g_3 measure for spKeys by calculating the minimum number of tuples to be removed from a table so that an spKey holds if it is not. The same paper proposed the g_4 measure that is derived from g_3 by emphasizing the effect of each connected component in the table's corresponding bipartite graph (where vertices of the first class of the graph represent the table's tuples and the second class represent all the possible combinations of the attributes' active domains). In this paper, we propose a new measure g_5 to approximate FDs by adding new tuples with unique values rather than deleting tuples as in g_3.

Several other researchers worked on approximating FDs in the literature. King et al. [15] provided an algorithmic method to discover functional and approximate functional dependencies in relational databases. The method provided is based upon the mathematical theory of partitions of row identification numbers from the relation, then determining non-trivial minimal dependencies from the partitions. They showed that the operations that need to be done on partitions are both simple and fast.

In [26], Varkonyi et al. introduced a structure called Sequential Indexing Tables (SIT) to detect an FD regarding the last attribute in their sequence. SIT is a fast approach so it can process large data quickly. The structure they used does not scale efficiently with the number of the attributes and the sizes of their domains, however. Other methods, such as TANE and FastFD face the same problem [23]. TANE was introduced by Huhtala [13] to discover functional and approximate dependencies by taking into consideration partitions and deriving valid dependencies from these partitions in a breadth-first or level-wise manner.

Bra, P. De, and Jan Paredaens gave a new decomposition theory for functional dependencies in [8]. They break up a relation into two subrelations whose union is the given relation and a functional dependency that holds in one subrelation is not in the other.

In [25], Tusor et al. presented the Parallelized Sequential Indexing Tables method that is memory-efficient for large datasets to find exact and approximate functional dependencies. Their method uses the same principle of Sequential Indexing Tables in storing data, but their training approach and operation are different.

Pyro is an algorithm to discover all approximate FDs in a dataset presented by Kruse [19]. Pyro verifies samples of agree sets and prunes the search spaces with the discovered FDs. On the other hand, based on the concept of "agree sets", Lopes et al. [22] developed an algorithm to find a minimum cover of a set of FDs for a given table by applying the so-called "Luxenburger basis" to develop a basis of the set of approximate FDs in the table.

Simovici et al. [24] provide an algorithm to find purity dependencies such that, for a fixed right-hand side (Y), the algorithm applies a level-wise search on the left-hand sides (X) so that $X \rightarrow Y$ has a purity measure below a user-defined threshold. Other algorithms were proposed in [14,21] to discover all FDs that hold in a given table by searching through the lattice of subsets of attributes.

In [27], Jef Wijsen summarizes and discusses some theoretical developments and concepts in Consistent query answering CQA (when a user queries a database that is inconsistent with respect to a set of constraints). Database repairing was modeled by an acyclic binary relation $\leq_{db}$ on the set of consistent database instances, where $r_1 \leq_{db} r_2$ means that r_1 is at least as close to db as r_2. One possible distance is the number of tuples to be added and/or removed. In addition to that, Bertossi studied the main concepts of database repairs and CQA in [6], and emphasis on tracing back the origin, motivation, and early developments. J. Biskup and L. Wiese present and analyze an algorithm called preCQE that is able to correctly compute a solution instance, for a given original database instance, that obeys the formal properties of inference-proofness and distortion minimality of a set of appropriately formed constraints in [7].

4 SPKey Approximation

In [5], the authors studied strongly possible keys, and the main motivation is to uniquely identify tuples in incomplete tables, if it is possible, by using the already shown values only to fill up the occurrences of NULLs. Consider the relational schema $R =$ and $K \subseteq R$. Furthermore, let T be an instance over R with NULLs. Let T' be the set of total tuples $T' = \{t' \in \Pi_{i=1}^{b} VD_i^T : \exists t \in T \text{ such that } t[K] \sim_w t'[K]\}$, furthermore let $G = (T, T'; E)$ be the bipartite graph, called the K-extension graph of T, defined by $\{t, t'\} \in E \iff t[K] \sim_w t'[K]$. Finding a matching of G that covers all the tuples in T (if exists) provides the set of tuples in T' to replace the incomplete tuples in T with, to verify that K is an spKey. A polynomial-time algorithm was given in [3] to find such matching. It is a non-trivial application of the well-known matching algorithms, as $|T'|$ is usually an exponential function of the size of the input table T.

The Approximate Strongly Possible Key (ASP Key) was defined in [5] as follows.

Definition 4.1. *Attribute set K is an approximate strongly possible key of ratio a in table T, in notation $asp_a^- \langle K \rangle$, if there exists a subset S of the tuples T such that $T \setminus S$ satisfies $sp \langle K \rangle$, and $|S|/|T| \leq a$. The minimum a such that $asp_a^- \langle K \rangle$ holds is denoted by $g_3(K)$.*

The measure $g_3(K)$ represents the approximation which is the ratio of the number of tuples needed to be removed over the total number of tuples so that $sp\langle K\rangle$ holds. The measure $g_3(K)$ has a value between 0 and 1, and it is exactly 0 when $sp\langle K\rangle$ holds in T, which means we don't need to remove any tuples. For this, we used the g_3 measure introduced in [16], to determine the degree to which ASP key is approximate. For example, the g_3 measure of $sp\langle X\rangle$ on Table 4 is 0.5, as we are required to remove two out of four tuples to satisfy the key constraint as shown in Table 5.

It was shown in [5] that the g_3 approximation measure for strongly possible keys satisfies

$$g_3(K) = \frac{|T| - \nu(G)}{|T|}.$$

where $\nu(G)$ denotes the maximum matching size in the K-extension graph G. The smaller value of $g_3(K)$, the closer K is to being an spKey.

For the bipartite graph G defined above, let $\mathscr{C}$ be the collection of all the connected components in G that satisfy the spKey, i.e. for which there exists a matching that covers all tuples in the set ($\forall_{C\in\mathscr{C}} \nexists X \subseteq C \cap T$ such that $|X| > N(X)$ by Hall's Theorem). Let $D \subseteq G$ be defined as $D = G \setminus \bigcup_{\forall C\in\mathscr{C}} C$, and let $\mathscr{C}'$ be the set of connected components of D. Let V_C denote the set of vertices in a connected component C. The approximation measure of strongly possible keys may be more appropriate by considering the effect of each connected component in the bipartite graph on the matching. We consider the effect of the components of $\mathscr{C}$ to get doubled in the approximation measure, as these components represent that part of the data that do not require tuple removal. So a derived version of the g_3 measure was proposed and named g_4 considering these components' effects,

$$g_4(K) = \frac{|T| - (\sum_{C\in\mathscr{C}}(|V_C|) + \sum_{C'\in\mathscr{C}'}\nu(C'))}{|T| + \sum_{C\in\mathscr{C}}|V_C|}.$$

Furthermore, it was proved that for a set of attributes K in any table, we have either $g_3(K) = g_4(K)$ or $1 < g_3(K)/g_4(K) < 2$. Moreover, there exist tables of an arbitrarily large number of tuples with $g_3(K)/g_4(K) = \frac{p}{q}$ for any rational number $1 \leq \frac{p}{q} < 2$.

In this paper, we extend our investigation on approximating spKeys by considering adding new tuples instead of removing them to satisfy an spKey if possible. Removing a non-total tuple t_1 means that there exist another total and/or non-total tuple(s) that share the same strongly possible extension with t_2. The following proposition shows that we can always remove only non-total tuples if the total part of the table satisfies the key.

Proposition 4.1. *Let T be an instance over schema R and let $K \subseteq R$. If the K-total part of the table T satisfies the key $sp\langle K\rangle$, then there exists a minimum set of tuples U to be removed that are all non-K-total so that $T \setminus U$ satisfies $sp\langle K\rangle$.*

Proof. Observe that a minimum set of tuples to be removed is $T \setminus X$ for a subset X of the set of vertices (tuples) covered by a particular maximum matching of the K-extension graph. Let M be a maximum matching, and assume that t_1 is total and not covered by M. Then, the unique neighbour t'_1 of t_1 in T' is covered by an edge (t_2, t'_1) of $\mathcal{M}$. Then t_2 is non-total since the K-total part satisfies $sp\langle K\rangle$, so we replace the edge (t_2, t') by the edge (t_1, t') to get matching M_1 of size $|M_1| = |M|$, and M_1 covers one more total tuple. Repeat this until all total tuples are covered.

4.1 Measure g_5 for SpKeys

The g_3 approximation measure for spKeys was introduced in [5]. In this section, we introduce a new approximation measure for spKeys. As we consider the active domain to be the source of the values to replace each null with, adding a new tuple to the table may increase the number of the values in the active domain of an attribute. for example, consider Table 4, the active domain of the attribute X_1 is $\{2\}$ and it changed to $\{2, 3\}$ after adding a tuple with new values as shown in Table 6.

Table 4. Incomplete Table to measure $sp\langle X\rangle$

X	
X_1	X_2
$\perp$	1
2	$\perp$
2	$\perp$
2	2

Table 5. The table after removing $(asp^-_a\langle X\rangle)$

X	
X_1	X_2
$\perp$	1
2	2

Table 6. The table after adding $(asp^+_b\langle X\rangle)$

X	
X_1	X_2
$\perp$	1
2	$\perp$
2	$\perp$
2	2
3	3

In the following definition, we define the g_5 measure as the ratio of the minimum number of tuples that need to be added over the total number of tuples to have the spKey satisfied.

Definition 4.2. *Attribute set K is an add-approximate strongly possible key of ratio b in table T, in notation $asp^+_b\langle K\rangle$, if there exists a set of tuples S such that the table TS satisfies $sp\langle K\rangle$, and $|S|/|T| \leq b$. The minimum b such that $asp^+_b\langle K\rangle$ holds is denoted by $g_5(K)$.*

The measure $g_5(K)$ represents the approximation which is the ratio of the number of tuples needed to be added over the total number of tuples so that $sp\langle K\rangle$ holds. The value of the measure $g_3(K)$ ranges between 0 and 1, and it is exactly 0 when $sp\langle K\rangle$ holds in T, which means we do not have to add any tuple. For example, the g_5 measure of $sp\langle X\rangle$ on Table 4 is 0.25, as it is enough to add one tuple to satisfy the key constraint as shown in Table 6.

Let T be a table and $U \subseteq T$ be the set of the tuples that we need to remove so that the spKey holds in T, i.e., we need to remove $|U|$ tuples, while by adding a tuple with new values, we may make more than one of the tuples in U satisfy the spKey using the new added values for their NULLs. In other words, we may need to add a fewer number of tuples than the number of tuples we need to remove to satisfy an spKey in the same given table. For example, Table 4 requires removing two tuples to satisfy $sp\langle X \rangle$, while adding one tuple is enough.

On the other hand, one may think about mixed modification of both adding and deleting tuples for Keys approximation, by finding the minimum number of tuples needs to be either added or removed. If first the additions are performed, then after that by Proposition 4.1, it is always true that we can remove only non-total tuples; then, instead of any tuple removal, we may add a new tuple with distinct values. Therefore, mixed modification in that way would not change the approximation measure, as it is always equivalent to tuples addition only. However, if the order of removals and additions count, then it is a topic of further research whether the removals can be substituted by additions.

The values of the two measures, g_3 and g_5, range between 0 and 1, and they are both equal to 0 if the spKey holds (we do not have to add or remove any tuples). Proposition 4.2 proves that the value of g_3 measure is always larger than or equal to the value of g_5 measure.

Proposition 4.2. *For any $K \subseteq R$ with $|K| \geq 2$, we have $g_3(K) \geq g_5(K)$.*

Proof. Indeed, we can always remove non-total tuples for g_3 by Proposition 4.1. Let the tuples to be removed be $U = \{t_1, t_2, \ldots t_u\}$. Assume that T^* is an spWorld of $T \setminus U$, which certifies that $T \setminus U \models sp\langle K \rangle$ For each tuple $t_i \in U$, we add tuple $t'_i = (z_i, z_i, \ldots, z_i)$ where z_i is a value that does not occur in any other tuple originally of T or added. The purpose of adding t'_i is twofold. First it is intended to introduce a completely new active domain value for each attribute. Second, their special structure ensures that they will never agree with any other tuple in the spWorld constructed below for the extended instance. Let t_i" be a tuple such that exactly one NULL in K of t_i is replaced by z_i, any other NULLs of t_i are imputed by values from the original active domain of the attributes. It is not hard to see that tuples in $T^* \cup \{t'_1, t'_2 \ldots, t'_u\} \cup \{t_1", t_2" \ldots, t_u"\}$ are pairwise distinct on K.

According to Proposition 4.2 we have $0 \leq g_3(K) - g_5(K) < 1$ and the difference is a rational number. What is not immediate is that for any rational number $0 \leq \frac{p}{q} < 1$ there exist a table T and $K \subseteq R$ such that $g_3(K) - g_5(K) = \frac{p}{q}$ in table T.

Proposition 4.3. *Let $0 \leq \frac{p}{q} < 1$ be a rational number. Then there exists a table T with an arbitrarily large number of rows and $K \subseteq R$ such that $g_3(K) - g_5(K) = \frac{p}{q}$ in table T.*

Proof. We may assume without loss of generality that $K = R$, since $T' \models sp\langle K \rangle$ if and only if we can make the tuples pairwise distinct on K by imputing values from the active domains, that is values in $R \setminus K$ are irrelevant. Let T be the following $q \times (p + 2)$ table (with $x = q - p - 1$).

$$T = \begin{matrix} 1\ 1\ 1 \ldots 1 \\ 1\ 1\ 1 \ldots 2 \\ \vdots \\ 1\ 1\ 1 \ldots x \\ \perp\ 1 \ldots 1\ 1 \\ 1\ \perp \ldots 1\ 1 \\ \ddots \\ 1\ 1 \ldots \perp\ 1 \end{matrix} \begin{matrix} \left.\vphantom{\begin{matrix}1\\1\\ \vdots\\1\end{matrix}}\right\} q-p-1 \\ \left.\vphantom{\begin{matrix}1\\1\\ \ddots\\1\end{matrix}}\right\} p+1 \end{matrix} \quad (1)$$

Since the active domain of the first $p+1$ attributes is only $\{1\}$, we have to remove $p+1$ rows so $g_3(K) = \frac{p+1}{q}$. On the other hand it is enough to add one new row $(2, 2, \ldots, 2, q-p)$ so $g_5(K) = \frac{1}{q}$. Since $\frac{p}{q} = \frac{cp}{cq}$ for any positive integer c, the number of rows in the table could be arbitrarily large.

The tables constructed in the proof of Proposition 4.3 have an arbitrarily large number of rows, however, the price for this is that the number of columns is also not bounded. The question arises naturally whether there are tables with a fixed number of attributes but with an arbitrarily large number of rows that satisfy $g_3(K) - g_5(K) = \frac{p}{q}$ for any rational number $0 \leq \frac{p}{q} < 1$? The following theorem answers this problem.

Theorem 4.1. *Let $0 \leq \frac{p}{q} < 1$ be a rational number. Then there exist tables over schema $\{A_1, A_2\}$ with arbitrarily large number of rows, such that $g_3(\{A_1, A_2\}) - g_5(\{A_1, A_2\}) = \frac{p}{q}$.*

Proof. The proof is divided into three cases according to whether $\frac{p}{q} < \frac{1}{2}$, $\frac{p}{q} = \frac{1}{2}$ or $\frac{p}{q} > \frac{1}{2}$. In each case, the number of rows of the table will be an increasing function of q and one just has to note that q can be chosen arbitrarily large without changing the value of the fraction $\frac{p}{q}$.

Case $\frac{p}{q} < \frac{1}{2}$ Let $T_{<.5}$ be defined as

$$T_{<.5} = \begin{matrix} q-p-1 \left\{ \begin{matrix} 1 & 1 \\ 1 & 2 \\ \vdots & \vdots \\ 1 & q-p-1 \end{matrix} \right. \\ p+1 \left\{ \begin{matrix} \perp & \perp \\ \perp & \perp \\ \vdots & \vdots \\ \perp & \perp \end{matrix} \right. \end{matrix}$$

Clearly, $g_3(K) = \frac{p+1}{q}$, as all the tuples with NULLs have to be removed. On the other hand, if tuple $(2, q-p)$ is added, then the total number of active domain combinations is $2 \cdot (q-p)$, out of which $q-p$ is used up in the table, so there are $q-p$ possible pairwise distinct tuples to replace the NULLs. Since $\frac{p}{q}<\frac{1}{2}$, we have that $q-p\geq p+1$ so all the tuples in the $q+1$-rowed table can be made pairwise distinct. Thus, $g_3(K)-g_5(K)=\frac{p+1}{q}-\frac{1}{q}$.

Case $\frac{p}{q}=\frac{1}{2}$ Let $T_{=.5}$ be defined as

$$T_{=.5}=\begin{matrix} q-p-2\left\{\begin{matrix}1 & 1\\ 1 & 2\\ \vdots & \vdots\\ 1 & q-p-2\end{matrix}\right. \\ p+2\left\{\begin{matrix}\bot & \bot\\ \bot & \bot\\ \vdots & \vdots\\ \bot & \bot\end{matrix}\right.\end{matrix}$$

Table $T_{=.5}$ contains all possible combinations of the active domain values, so we have to remove every tuple containing NULLs, so $g_3(K)=\frac{p+2}{q}$. On the other hand, if we add just one new tuple (say $(2,q-p-1)$), then the largest number of active domain combinations is $2\cdot(q-p-1)$ that can be achieved. There are already $q-p-1$ pairwise distinct total tuples in the extended table, so only $q-p-1<p+2$ would be available to replace the NULLs. On the other hand, adding two new tuples, $(2,q-p-1)$ and $(3,q-p)$ creates a pool of $3\cdot(q-p)$ combinations of active domains, which is more than $(q-p-1)+p+2$ that is needed.

Case $\frac{p}{q}>\frac{1}{2}$ Table T is defined similarly to the previous cases, but we need more careful analysis of the numbers.

$$T=\begin{matrix} b\left\{\begin{matrix}1 & 1\\ 1 & 2\\ \vdots & \vdots\\ 1 & b\end{matrix}\right. \\ x\left\{\begin{matrix}\bot & \bot\\ \bot & \bot\\ \vdots & \vdots\\ \bot & \bot\end{matrix}\right.\end{matrix} \tag{2}$$

Clearly, $g_3(K)=\frac{x}{x+b}$. Let us assume that y tuples are needed to be added. The maximum number of active domain combinations is $(y+1)(y+b)$ obtained by adding tuples $(2,b+1),(3,b+2),\ldots,(y+1,y+b)$. This is enough to replace all tuples with NULLs if

$$(y+1)(y+b)\geq x+y+b. \tag{3}$$

On the other hand, $y-1$ added tuples are not enough, so

$$y(y-1+b)<x+y-1+b. \tag{4}$$

Since the total number of active domain combinations must be less than the tuples in the extended table. We have $\frac{p}{q} = g_3(K) - g_5(K) = \frac{x-y}{x+b}$ that is for some positive integer c we must have $cp = x - y$ and $cq = x + b$ if $gcd(p,q) = 1$. This can be rewritten as

$$y = x - cp\,; \quad y + b = c(q-p)\,; \quad b = cq - x\,; \quad x + y + b = y + cq. \tag{5}$$

Using (5) we obtain that (3) is equivalent with

$$y \geq \frac{cp}{c(q-p)-1}. \tag{6}$$

If c is large enough then $\lceil \frac{cp}{c(q-p)-1} \rceil = \lceil \frac{p}{q-p} \rceil$ so if $y = \lceil \frac{p}{q-p} \rceil$ is chosen then (6) and consequently (3) holds. On the other hand, (4) is equivalent to

$$y < \frac{cq-1}{c(q-p)-2}. \tag{7}$$

The right hand side of (7) tends to $\frac{q}{q-p}$ as c tends to infinity. Thus, for large enough c we have $\lfloor \frac{cq-1}{c(q-p)-2} \rfloor = \lfloor \frac{q}{q-p} \rfloor$. Thus, if

$$y = \lceil \frac{p}{q-p} \rceil \leq \lfloor \frac{q}{q-p} \rfloor \tag{8}$$

and $\frac{q}{q-p}$ is not an integer, then both (3) and (4) are satisfied for large enough c. Observe that $\frac{p}{q-p} + 1 = \frac{q}{q-p}$, thus (8) always holds. Also, if $\frac{q}{q-p}$ is indeed an integer, then we have strict inequality in (8) that implies (7) and consequently (4).

5 spFD Approximation

In this section, we measure to which extent a table satisfies a Strongly Possible Functional Dependency (spFD) $X \rightarrow_{sp} Y$ if $T \not\models X \rightarrow_{sp} Y$.

Similarly to Sect. 4, we assume that the X-total part of the table satisfies the FD $X \rightarrow Y$, so we can always consider adding tuples. The measures g_3 and g_5 are defined analogously to the spKey case.

Definition 5.1. *For the attribute sets X and Y, $\sigma : X \rightarrow_{sp} Y$ is a remove-approximate strongly possible functional dependency of ratio a in a table T, in notation.*

$T \models\approx_a^- X \rightarrow_{sp} Y$, if there exists a set of tuples S such that the table $T \setminus S \models X \rightarrow_{sp} Y$, and $|S|/|T| \leq a$. Then, $g_3(\sigma)$ is the smallest a such that $T \models\approx_a^- \sigma$ holds.

The measure $g_3(\sigma)$ represents the approximation which is the ratio of the number of tuples needed to be removed over the total number of tuples so that $T \models X \rightarrow_{sp} Y$ holds.

Definition 5.2. *For the attribute sets X and Y, $\sigma : X \rightarrow_{sp} Y$ is an add-approximate strongly possible functional dependency of ratio b in a table T, in notation $T \models \approx_b^+ X \rightarrow_{sp} Y$, if there exists a set of tuples S such that the table $T \cup S \models X \rightarrow_{sp} Y$, and $|S|/|T| \leq b$. Then, $g_5(\sigma)$ is the smallest b such that $T \models \approx_b^+ \sigma$ holds.*

The measure $g_5(\sigma)$ represents the approximation which is the ratio of the number of tuples needed to be added over the total number of tuples so that $T \models X \rightarrow_{sp} Y$ holds. For example, consider Table 7. We are required to remove at least 2 tuples so that $X \rightarrow_{sp} Y$ holds, as it is easy to check that if we remove only one tuple, then $T \not\models X \rightarrow_{sp} Y$, but on the other hand, the table obtained by removing tuples 4 and 5, shown in Table 8 satisfies $X \rightarrow_{sp} Y$. It is enough to add only one tuple to satisfy the dependency as the table in Table 9 shows.

Table 7. Incomplete Table to measure $(X \rightarrow_{sp} Y))$

X		Y
X_1	X_2	
⊥	1	1
2	⊥	1
2	⊥	1
2	1	2
2	1	2
2	2	2

Table 8. The table after removing $(^-_a X \rightarrow_{sp} Y)$

X		Y
X_1	X_2	
⊥	1	1
2	⊥	1
2	⊥	1
2	2	2

Table 9. The table after adding $(^+_b X \rightarrow_{sp} Y)$

X		Y
X_1	X_2	
⊥	1	1
2	⊥	1
2	⊥	1
2	1	2
2	1	2
2	2	2
3	3	3

5.1 The Difference of G3 and G5 for SpFDs

The same table may get different approximation measure values for g_3 and g_5. For example, the g_3 approximation measure for Table 7 is 0.334 (it requires removing at least 2 tuples out of 6), while its g_5 approximation measure is 0.167 (it requires adding at least one tuple with new values).

The following theorem proves that it is always true that the g_3 measure value of a table is greater than or equal to the g_5 for spFDs.

Theorem 5.1. *Let T be a table over schema R, $\sigma : X \rightarrow_{sp} Y$ for some $X, Y \subseteq R$. Then $g_3(\sigma) \geq g_5(\sigma)$.*

The proof is much more complicated than the one in the case of spKeys, because we cannot assume that there always exists a minimum set of non-total tuples to be removed for g_3, as the table in Table 10 shows. In this table the third tuple alone forms a minimum set of tuples to be removed to satisfy the dependency and it has no NULL.

Table 10. X-total tuple needs to be removed

X		**Y**
X_1	X_2	
1	$\perp$	1
1	$\perp$	1
1	1	2
1	1	$\perp$
1	2	3

From that table, we need to remove the third row to have $X \to_{sp} Y$ satisfied. Let us note that adding row $(3, 3, 3)$ gives the same result, so $g_3(X \to_{sp} Y) = g_5(X \to_{sp} Y) = 1$. However, there exist no spWorlds that realize the g_3 and g_5 measure values and agree on those tuples that are not removed for g_3.

Proof. of Theorem 5.1 Without loss of generality, we may assume that $X \cap Y = \emptyset$, because $T \models X \to_{sp} Y \iff T \models X \backslash Y \to_{sp} Y \backslash X$. Also, it is enough to consider attributes in $X \cup Y$. Let $U = \{t_1, t_2, \ldots, t_p\}$ be a minimum set of tuples to be removed from T. Let T' be the spWorld of $T \setminus U$ that satisfies $X \to Y$. Let us assume that $t_1, \ldots t_a$ are such that $t_i[X]$ is not total for $1 \leq i \leq a$. Furthermore, let $t_{a+1}[X] = \ldots = t_{j_1}[X]$, $t_{j_1+1}[X] = \ldots = t_{j_2}[X]$, $\ldots$, $t_{j_f+1}[X] = \ldots = t_p[X]$ be the maximal sets of tuples that have the same total projection on X. We construct a collection of tuples $\{s_1, \ldots s_{a+f+1}\}$, together with an spWorld T^* of $T \cup \{s_1, \ldots, s_{a+f+1}\}$ that satisfies $X \to Y$ as follows.

Case 1. $1 \leq i \leq a$. Let z_i be a value not occurring in T neither in every tuple s_j constructed so far. Let $s_i[A] = z_i$ for $\forall A \in X$ and $s_i[B] = t_i[B]$ for $B \in R \setminus X$. The corresponding sp-extensions $s_i^*, t_i^* \in T^*$ are given by setting $s_i^*[B] = t_i^*[B] = \beta$ where $\beta \in VD_B$ arbitrarily fixed if $t_i[B] = \perp$ in case $B \in R \setminus X$, furthermore $t_i^*[A] = z_i$ if $A \in X$ and $t_i[A] = \perp$.

Case 2. X-total tuples. For each such set $t_{j_{g-1}+1}[X] = \ldots = t_{j_g}[X]$ ($g \in \{1, 2, \ldots, f+1\}$) we construct a tuple s_{a+g}. Let $v_1^g, v_2^g, \ldots v_{k_g}^g \in T \setminus U$ be the tuples whose sp-extension $v_j^{g\prime}$ in T' satisfies $v_j^{g\prime}[X] = t_{j_g}[X]$ for $1 \leq j \leq k_g$. Let $v_1^g, v_2^g, \ldots v_\ell^g$ be those that are also X-total. Since the X-total part of the table satisfies $X \to_{sp} Y$, $t_{j_{g-1}+1}, \ldots t_{j_g}, v_1^g, v_2^g, \ldots v_\ell^g$ can be sp-extended to be identical on Y. Let us take those extensions in T^*.

Let s_{a+g} be defined as $s_{a+g}[A] = z_{a+g}$ where z_{a+g} is a value not used before for $A \in X$, furthermore $s_{a+g}[B] = v_{\ell+1}^g[B]$ for $B \in R \backslash X$. The sp-extensions are given as $v_q^{g*}[A] = z_{a+g}$ if $v_q^{g*}[A] = \perp$ and $A \in X$, otherwise $v_q^{g*}[A] = v_q^{g\prime}[A]$ for $\ell + 1 \leq q \leq k_g$. Finally, let $s_{a+g}^*[B] = v_1^{g\prime}[B]$ for $B \in R \setminus X$.

For any tuple $t \in T \setminus U$ for which no sp-extension has been defined yet, let us keep its extension in T', that is let $t^* = t'$.

Claim. $T^* \models X \to_{sp} Y$. Indeed, let $t^1, t^2 \in T \cup \{s_1, \ldots, s_{a+f+1}\}$ be two tuples such that their sp-extensions in T^* agree on X, that is $t^{1*}[X] = t^{2*}[X]$. If $t^{1*}[X]$ contains a new value z_j for some $1 \leq j \leq a+f+1$, then by definition of the sp-extensions above, we have $t^{1*}[Y] = t^{2*}[Y]$. Otherwise, either both t^1, t^2 are X-total, so again by definition of the sp-extensions above, we have $t^{1*}[Y] = t^{2*}[Y]$, or at least one of them is not X-total, and then $t^{1*} = t^{1'}$ and $t^{2*} = t^{2'}$. But in this latter case using $T' \models X \to_{sp} Y$ we get $t^{1*}[Y] = t^{2*}[Y]$.

The values g_3 and g_5 are similarly independent of each other for spFDs as in the case of spKeys.

Theorem 5.2. *For any rational number* $0 \leq \frac{p}{q} < 1$ *there exists tables with an arbitrarily large number of rows and bounded number of columns that satisfy* $g_3(\sigma) - g_5(\sigma) = \frac{p}{q}$ *for* $\sigma\colon X \to_{sp} Y$.

Table 11. $g_3 - g_5 = \frac{p}{q}$

$T =$

X		**Y**
X_1	X_2	
1	1	1
1	2	2
⋮	⋮	⋮
1	b	b
$\perp$	$\perp$	$b+1$
$\perp$	$\perp$	$b+2$
⋮	⋮	⋮
$\perp$	$\perp$	$b+x$

Proof. Consider the following table T. We clearly have $g_3(X \to_{sp} Y) = \frac{x}{x+b}$ for T as all tuples with NULLs must be removed. On the other hand, by adding new tuples and so extending the active domains, we need to be able to make at least $x+b$ pairwise distinct combinations of X-values. If y tuples are added, then we can extend the active domains to the sizes $|VD_1| = y+1$ and $|VD_2| = y+b$. Also, if y is the minimum number of tuples to be added, then

$$g_3(X \to_{sp} Y) - g_5(X \to_{sp} Y) = \frac{x-y}{x+b} = \frac{p}{q} \tag{9}$$

if $cp = x - y$ and $cq = x + b$ for some positive integer c. From here $y = x - cp$ and $y + b = c(q-p)$ Thus, what we need is

$$(y+1)(y+b) = (y+1)c(q-p) \geq cq \tag{10}$$

and, to make sure that $y-1$ added tuples are not enough,

$$y(y+b-1) = y(c(q-p)-1) \leq cq-1. \tag{11}$$

Easy calculation shows that (10) is equivalent with $y \geq \frac{p}{q-p}$, so we take $y = \left\lceil \frac{p}{q-p} \right\rceil$. On the other hand, (11) is equivalent with $y \leq \frac{cq-1}{c(q-p)-1}$. Now, similarly to Case 3 of the proof of Theorem 4.1 observe that $\frac{cq-1}{c(q-p)-1} \to \infty$ as $c \to \infty$, so, if c is large enough, then (11) holds.

5.2 Semantic Comparison of g_3 and g_5

In this section, we compare the g_3 and g_5 measures to analyze their applicability and usability for different cases. The goal is to specify when it is semantically better to consider adding or removing rows for approximation for both spFDs and spKeys.

Considering the teaching table in Table 12, we have the two strongly possible constraints $Semester\,TeacherID \to_{sp} CourseID$ and $sp\langle Semester\,TeacherID\rangle$. It requires adding one row so that $asp^+_a\langle Semester\,TeacherID\rangle =^+_a Semester\,TeacherID \to_{sp} CourseID$. But on the other hand, it requires removing 3 out of the 6 rows. Then, it would be more convenient to add a new row rather than removing half of the table, which makes the remaining rows not useful for analysis for some cases.

Adding new tuples to satisfy some violated strongly possible constraints ensures that we make the minimum changes. In addition to that, in the case of deletion, some active domain values may be removed. There are some cases where it may be more appropriate to remove rather than add tuples, however. This is to preserve semantics of the data and to avoid using values that are out of the appropriate domain of the attributes while adding new tuples with new unseen values. For example, Table 13 represents the grade records for some students in a course that imply the key (*Name*, *Group*) and the dependency *Points Assignment* $\to$ *Result*, while both of $sp\langle$NameGroup$\rangle$ and *Points Assignment* $\to_{sp}$ *Result* are violated by the table. Then, adding one tuple with the new values (Dummy, 3, 3, Maybe, Hopeless) is enough to satisfy the two strongly possible constraints, while they can also be satisfied by removing the last two tuples. However, it is not convenient to use these new values for the attributes, since they are probably not contained in the intended domains. Hence, removing two tuples is semantically more acceptable than adding one tuple.

If g_3 is much larger than g_5 for a table, it is better to add rows than remove them. Row removal may leave only a short version of the table which may not give a useful data analysis, as is the case in Table 11. If g_3 and g_5 are close to each other, it is mostly better to add rows, but when the attributes' domains are restricted to a short-range, then it may be better to remove rows rather than adding new rows with "noise" values that are semantically not related to the meaning of the data, as is the case in Table 13.

Table 12. Incomplete teaching table

Semester	TeacherID	CourseID
First	1	1
⊥	1	2
First	2	3
⊥	2	4
First	3	5
⊥	3	6

Table 13. Incomplete course grading table

Name	Group	Points	Assignment	Result
Bob	1	2	Submitted	Pass
Sara	1	1	Not Submitted	Fail
Alex	1	2	Not Submitted	Fail
John	1	1	Submitted	Pass
⊥	1	1	⊥	Retake
Alex	⊥	2	⊥	Retake

6 Conclusion and Future Directions

Two approximation measures for spKeys and spFDs were investigated. The first one, g_3, is the ratio of the minimum number of rows to be removed, and was introduced for functional dependencies in tables without NULL values in [11] and for spKeys in [2]. In the present paper, we extended the definition for spFDs, as well. A new measure g_5 was also introduced here, which measures the ratio of the minimum number of tuples to be added to satisfy a strongly possible constraint. This measure is only meaningful for strongly possible constraints because ordinary functional dependencies or possible functional dependencies cannot be made valid by adding tuples. However, the new tuples may extend the active domains of the attributes and hence may make some strongly possible constraints satisfied. Note that any add-approximate spKey or spFD is a possible key, respectively possible FD. Thus, the g_5 measure measures the minimum number of "extra" attribute values one has to use in a possible world satisfying the constraint.

We proved that the value of g_5 is at most as large as the value of g_3 for both spKeys and spFDs. Otherwise, however, the two measures are independent of each other, as their difference can take any non-negative rational value less than one.

The referees suggested considering tuple removal and addition concurrently, or tuple modification. If first the additions are performed, then after that by Proposition 4.1, it is always true that we can remove only non-total tuples;

then, instead of any tuple removal, we may add a new tuple with distinct values. Therefore, mixed modification in that way would not change the approximation measure, as it is always equivalent to tuples addition only. However, if the order of removals and additions count, then it is a topic of further research whether the removals can be substituted by additions. Also, Proposition 4.1 is only valid for spKeys, so mixed modifications are interesting research problem for spFDs. One tuple modification can easily be replaced by one removal and one addition. The question remains open whether one can gain more with tuple modifications than the above replacement. A future research direction is to tackle algorithmic and complexity questions. It was proven in [3] that checking whether for a given subset $K \subseteq R$ and table T, $T \models sp\langle K\rangle$ holds can be decided in polynomial time. However, the questions whether $g_3(sp\langle K\rangle) \leq q$ and $g_5(sp\langle K\rangle) \leq q$ are not known to be polynomial. The problem is that we would have to check all possible tables $T' \subset T$ with $|T'|/|T| \geq 1 - q$ which could mean exponentially many tables. On the other hand, it is clear that both problems, $g_3(sp\langle K\rangle) \leq q$ and $g_5(sp\langle K\rangle) \leq q$ are in NP.

The analogous question for spFDs, that is whether $T \models X \rightarrow_{sp} Y$ for a table T and subsets $X, Y \subseteq R$, is itself NP-complete [3]. This suggests that the problem of bounding the approximation measures g_3 and g_5 for spFDs is also intractable. However, it is a topic of further study to really prove it.

We studied handling missing values for Multi-valued Dependencies (spMVDs) in [4]. An interesting future research direction can be measuring approximation ratio of spMVDs.

Acknowledgement. The authors are indebted to the unknown referees for their careful reading of the paper. The authors are thankful for the many suggestions of improvements and calling their attention to several related works.

References

1. Al-Atar, M., Sali, A.: Strongly possible functional dependencies for SQL. Acta Cybernetica (2022)
2. Alattar, M., Sali, A.: Keys in relational databases with nulls and bounded domains. In: Welzer, T., Eder, J., Podgorelec, V., Kamišalić Latifić, A. (eds.) ADBIS 2019. LNCS, vol. 11695, pp. 33–50. Springer, Cham (2019). https://doi.org/10.1007/978-3-030-28730-6_3
3. Alattar, M., Sali, A.: Functional dependencies in incomplete databases with limited domains. In: Herzig, A., Kontinen, J. (eds.) FoIKS 2020. LNCS, vol. 12012, pp. 1–21. Springer, Cham (2020). https://doi.org/10.1007/978-3-030-39951-1_1
4. Alattar, M., Sali, A.: Multivalued dependencies in incomplete databases with limited domain: properties and rules. In: 16th International Miklos Ivanyi PhD & DLA Symposium, p. 226 (2020)
5. Alattar, M., Sali, A.: Strongly possible keys for SQL. J. Data Semant. **9**(2), 85–99 (2020)
6. Bertossi, L.: Database repairs and consistent query answering: origins and further developments. In: Proceedings of the 38th ACM SIGMOD-SIGACT-SIGAI Symposium on Principles of Database Systems, pp. 48–58 (2019)

7. Biskup, J., Wiese, L.: A sound and complete model-generation procedure for consistent and confidentiality-preserving databases. Theoret. Comput. Sci. **412**(31), 4044–4072 (2011)
8. De Bra, P., Paredaens, J.: Conditional dependencies for horizontal decompositions. In: Diaz, J. (ed.) ICALP 1983. LNCS, vol. 154, pp. 67–82. Springer, Heidelberg (1983). https://doi.org/10.1007/BFb0036898
9. De Keijzer, A., Van Keulen, M.: A possible world approach to uncertain relational data. In: Proceedings of 15th International Workshop on Database and Expert Systems Applications, pp. 922–926. IEEE (2004)
10. Farhangfar, A., Kurgan, L.A., Pedrycz, W.: A novel framework for imputation of missing values in databases. IEEE Trans. Syst. Man Cybern. Part A Syst. Hum. **37**(5), 692–709 (2007)
11. Giannella, C., Robertson, E.: On approximation measures for functional dependencies. Inf. Syst. **29**(6), 483–507 (2004)
12. Goodman, L.A., Kruskal, W.H.: Measures of association for cross classifications. In: Goodman, L.A., Kruskal, W.H. (eds.) Measures of Association for Cross Classifications, pp. 2–34. Springer, New York (1979). https://doi.org/10.1007/978-1-4612-9995-0_1
13. Huhtala, Y., Kärkkäinen, J., Porkka, P., Toivonen, H.: Tane: an efficient algorithm for discovering functional and approximate dependencies. Comput. J. **42**(2), 100–111 (1999)
14. Kantola, M., Mannila, H., Räihä, K.-J., Siirtola, H.: Discovering functional and inclusion dependencies in relational databases. Int. J. Intell. Syst. **7**(7), 591–607 (1992)
15. King, R.S., Legendre, J.J.: Discovery of functional and approximate functional dependencies in relational databases. J. Appl. Math. Decis. Sci. **7**(1), 49–59 (2003)
16. Kivinen, J., Mannila, H.: Approximate inference of functional dependencies from relations. Theoret. Comput. Sci. **149**(1), 129–149 (1995)
17. Köhler, H., Leck, U., Link, S., Zhou, X.: Possible and certain keys for SQL. VLDB J. **25**(4), 571–596 (2016)
18. Köhler, H., Link, S., Zhou, X.: Possible and certain SQL keys. Proc. VLDB Endow. **8**(11), 1118–1129 (2015)
19. Kruse, S., Naumann, F.: Efficient discovery of approximate dependencies. Proc. VLDB Endow. **11**(7), 759–772 (2018)
20. Lipski Jr, W.: On databases with incomplete information. J. ACM (JACM) **28**(1), 41–70 (1981)
21. Lopes, S., Petit, J.-M., Lakhal, L.: Efficient discovery of functional dependencies and armstrong relations. In: Zaniolo, C., Lockemann, P.C., Scholl, M.H., Grust, T. (eds.) EDBT 2000. LNCS, vol. 1777, pp. 350–364. Springer, Heidelberg (2000). https://doi.org/10.1007/3-540-46439-5_24
22. Lopes, S., Petit, J.-M., Lakhal, L.: Functional and approximate dependency mining: database and FCA points of view. J. Exp. Theor. Artif. Intell. **14**(2–3), 93–114 (2002)
23. Papenbrock, T., et al.: Functional dependency discovery: an experimental evaluation of seven algorithms. Proc. VLDB Endow. **8**(10), 1082–1093 (2015)
24. Simovici, D.A., Cristofor, D., Cristofor, L.: Impurity measures in databases. Acta Informatica **38**(5), 307–324 (2002)
25. Tusor, B., Várkonyi-Kóczy, A.R.: Memory efficient exact and approximate functional dependency extraction with parsit. In: 2020 IEEE 24th International Conference on Intelligent Engineering Systems (INES), pp. 133–138. IEEE (2020)

26. Várkonyi-Kóczy, A.R., Tusor, B., Tóth, J.T.: A multi-attribute classification method to solve the problem of dimensionality. In: Jabłoński, R., Szewczyk, R. (eds.) Recent Global Research and Education: Technological Challenges. AISC, vol. 519, pp. 403–409. Springer, Cham (2017). https://doi.org/10.1007/978-3-319-46490-9_54
27. Wijsen, J.: Foundations of query answering on inconsistent databases. ACM SIGMOD Rec. **48**(3), 6–16 (2019)
28. Zimányi, E., Pirotte, A.: Imperfect information in relational databases. In: Motro, A., Smets, P. (eds.) Uncertainty Management in Information Systems, pp. 35–87. Springer, Boston (1997). https://doi.org/10.1007/978-1-4615-6245-0_3

The Fault-Tolerant Cluster-Sending Problem

Jelle Hellings[1(✉)] and Mohammad Sadoghi[2]

[1] McMaster University, 1280 Main St. W., Hamilton, ON L8S 4L7, Canada
jhellings@mcmaster.ca

[2] Exploratory Systems Lab, Department of Computer Science, University of California, Davis, USA

Abstract. The emergence of blockchains is fueling the development of resilient data management systems that can deal with *Byzantine failures* due to crashes, bugs, or even malicious behavior. As traditional resilient systems lack the scalability required for modern data, several recent systems explored using *sharding*. Enabling these sharded designs requires two basic primitives: a primitive to reliably make decisions within a cluster and a primitive to reliably communicate between clusters. Unfortunately, such communication has not yet been formally studied.

In this work, we improve on this situation by formalizing the *cluster-sending problem*: the problem of sending a message from one resilient system to another in a fault-tolerant manner. We also establish lower bounds on the complexity of cluster-sending under both crashes and Byzantine failures. Finally, we present *worst-case optimal* cluster-sending protocols that meet these lower bounds in practical settings. As such, our work provides a strong foundation for the future development of sharded resilient data management systems.

Keywords: Byzantine Failures · Sharding · Message Sending · Communication Lower Bounds · Worst-Case Optimal Communication

1 Introduction

The emergence of blockchain technology is fueling interest in the development of new data management systems that can manage data between fully-independent parties (*federated data management)* and can provide services continuously, even during *Byzantine failures* (e.g., network failure, hardware failure, software failure, or malicious attacks) [5,13,14,18,20,22]. Recently, this has led to the development of several resilient data management systems based on *permissioned blockchain technology* [6–9].

Unfortunately, systems based on traditional fully-replicated *consensus-based* permissioned blockchain technology lack the scalability required for modern data management. Consequently, several recent systems have proposed to combine sharding with consensus-based designs (e.g., `AHL` [3], `ByShard` [10], and

I. Varzinczak (Ed.): FoIKS 2022, LNCS 13388, pp. 168–186, 2022.
https://doi.org/10.1007/978-3-031-11321-5_10

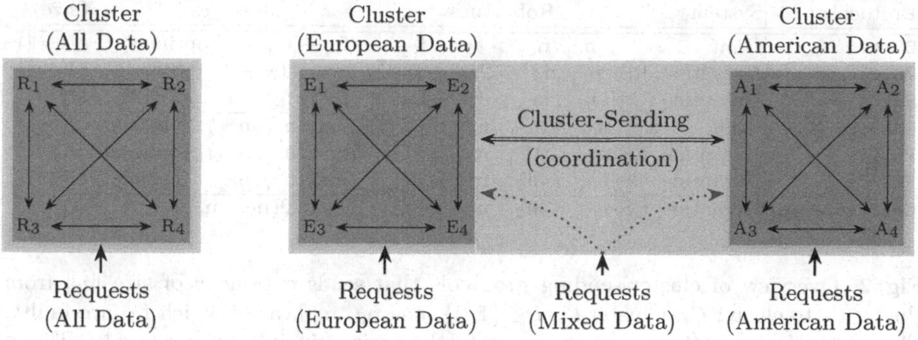

Fig. 1. *Left*, a traditional fully-replicated resilient system in which all four replicas each hold all data. *Right*, a *sharded* design in which each resilient cluster of four replicas holds only a part of the data. (Color figure online)

`Chainspace` [1]). These systems all follow a familiar sharded design: the data is split up into individual pieces called *shards* and each shard is managed by different independent blockchain-driven clusters. To illustrate the benefits of sharding, consider a system with a *sharded design* in which data is kept in *local Byzantine fault-tolerant clusters*, e.g., as sketched in Fig. 1 by storing data relevant to American customers on systems located in the United States, whereas systems located in Europe contain data relevant to European customers. Compared to the traditional fully-replicated design of blockchain systems, this sharded design will improve *storage scalability* by distributing data storage and improve *performance scalability* by enabling concurrent transaction processing, e.g., transactions on American and European data can be performed independently of each other.

At the core of any sharded data processing system are two crucial primitives [17]. First, individual shards need primitives to independently make *decisions*, e.g., to execute transactions that only affect data held within that shard. In the setting where each shard is a fault-tolerant cluster, such per-shard decision making is formalized by the well-known *consensus problem*, which can be solved by practical consensus protocols such as `Pbft` [2]. Second, shards need primitives to *communicate* between each other, e.g., to coordinate the execution of transactions that affect data held by multiple shards. Unfortunately, even though inter-shard communication is a fundamental basic primitive, it has not yet been studied in much detail. Indeed, existing sharded blockchain-inspired data processing systems typically use expensive ad-hoc techniques to enable coordination between shards (e.g., `Chainspace` [1] uses expensive all-to-all broadcasts).

In this work, we improve on this situation by formalizing the problem of inter-shard communication in permissioned fault-tolerant systems: the *cluster-sending problem*. In specific, we fully formalize the cluster-sending problem in Sect. 2. Then, in Sect. 3, we prove strict *lower bounds on the complexity* of the cluster-sending problem that are *linear* in terms of the number of messages (when faulty replicas only crash) and in terms of the number of signatures (when faulty replicas can be malicious and messages are signed via public-key cryptography).

Protocol	System	Robustness	Messages	(*size*)
BS-cs	Omit	$\mathbf{n}_{C_1}, \mathbf{n}_{C_2} > \mathbf{f}_{C_1} + \mathbf{f}_{C_2}$	$\mathbf{f}_{C_1} + \mathbf{f}_{C_2} + 1$ (optimal)	$\mathcal{O}(\|v\|)$
BS-rs	Byzantine, RS	$\mathbf{n}_{C_1}, \mathbf{n}_{C_2} > 2\mathbf{f}_{C_1} + \mathbf{f}_{C_2}$	$2\mathbf{f}_{C_1} + \mathbf{f}_{C_2} + 1$ (optimal)	$\mathcal{O}(\|v\|)$
BS-cs	Byzantine, CS	$\mathbf{n}_{C_1}, \mathbf{n}_{C_2} > \mathbf{f}_{C_1} + \mathbf{f}_{C_2}$	$\mathbf{f}_{C_1} + \mathbf{f}_{C_2} + 1$ (optimal)	$\mathcal{O}(\|v\|)$
PBS-cs	Omit	$\mathbf{n}_{C_1} > 3\mathbf{f}_{C_1}$, $\mathbf{n}_{C_2} > 3\mathbf{f}_{C_2}$	$\mathcal{O}(\max(\mathbf{n}_{C_1}, \mathbf{n}_{C_2}))$ (optimal)	$\mathcal{O}(\|v\|)$
PBS-rs	Byzantine, RS	$\mathbf{n}_{C_1} > 4\mathbf{f}_{C_1}$, $\mathbf{n}_{C_2} > 4\mathbf{f}_{C_2}$	$\mathcal{O}(\max(\mathbf{n}_{C_1}, \mathbf{n}_{C_2}))$ (optimal)	$\mathcal{O}(\|v\|)$
PBS-cs	Byzantine, CS	$\mathbf{n}_{C_1} > 3\mathbf{f}_{C_1}$, $\mathbf{n}_{C_2} > 3\mathbf{f}_{C_2}$	$\mathcal{O}(\max(\mathbf{n}_{C_1}, \mathbf{n}_{C_2}))$ (optimal)	$\mathcal{O}(\|v\|)$
Chainspace [1]	Byzantine, CS	$\mathbf{n}_{C_1} > 3\mathbf{f}_{C_1}$, $\mathbf{n}_{C_2} > 3\mathbf{f}_{C_2}$	$\mathcal{O}(\mathbf{n}_{C_1} \cdot \mathbf{n}_{C_2})$	$\mathcal{O}(\|v\|)$

Fig. 2. Overview of cluster-sending protocols that sends a value v of size $\|v\|$ from cluster $\mathcal{C}_1$ to cluster $\mathcal{C}_2$. Cluster $\mathcal{C}_i$, $i \in \{1, 2\}$, has $\mathbf{n}_{\mathcal{C}_i}$ replicas of which $\mathbf{f}_{\mathcal{C}_i}$ are faulty. The protocol names (first column) indicate the main principle the protocol relies on (BS for *bijective sending*, and PBS for *partitioned bijective sending*), and the specific variant the protocol is designed for (variant -cs is designed to use cluster signing, and variant -rs is designed to use replica signing). The system column describe the type of Byzantine behavior the protocol must deal with ("Omit" for systems in which Byzantine replicas can drop messages, and "Byzantine" for systems in which Byzantine replicas have arbitrary behavior) and the signature scheme present in the system ("RS" is shorthand for *replica signing*, and "CS" is shorthand for *cluster signing*).

Next, in Sects. 4 and 5, we introduce *bijective sending* and *partitioned bijective sending*, powerful techniques to provide *worst-case optimal* cluster-sending between clusters of roughly the same size (bijective sending) and of arbitrary sizes (partitioned bijective sending). Finally, in Sect. 6, we evaluate the behavior of the proposed cluster-sending protocols via an in-depth evaluation. In this evaluation, we show that our worst-case optimal cluster-sending protocols have exceptionally low communication costs in comparison with existing ad-hoc approaches from the literature. A full overview of all environmental conditions in which we study the cluster-sending problem and the corresponding worst-case optimal cluster-sending protocols we propose can be found in Fig. 2.

Our cluster-sending problem is closely related to cross-chain coordination in *permissionless blockchains* such as Bitcoin [16] and Ethereum [21], e.g., as provided via atomic swaps [11], atomic commitment [24], and cross-chain deals [12]. Unfortunately, such permissionless solutions are not fit for a permissioned environment. Although *cluster-sending* can be solved using well-known permissioned techniques such as consensus, interactive consistency, Byzantine broadcasts, and message broadcasting [2,4], the best-case costs for these primitives are much higher than the worst-case costs of our cluster-sending protocols, making them unsuitable for cluster-sending. As such, the cluster-sending problem is an independent problem and our initial results on this problem provide novel directions for the design and implementation of high-performance resilient data management systems.

2 Formalizing the Cluster-Sending Problem

A *cluster* $\mathcal{C}$ is a set of replicas. We write $\mathsf{f}(\mathcal{C}) \subseteq \mathcal{C}$ to denote the set of *faulty replicas* in $\mathcal{C}$ and $\mathsf{nf}(\mathcal{C}) = \mathcal{C} \setminus \mathsf{f}(\mathcal{C})$ to denote the set of *non-faulty replicas* in $\mathcal{C}$. We write $\mathbf{n}_{\mathcal{C}} = |\mathcal{C}|$, $\mathbf{f}_{\mathcal{C}} = |\mathsf{f}(\mathcal{C})|$, and $\mathbf{nf}_{\mathcal{C}} = |\mathsf{nf}(\mathcal{C})|$ to denote the number of

	Ping round-trip times (ms)						*Bandwidth* (Mbit/s)					
	O	*I*	*M*	*B*	*T*	*S*	*O*	*I*	*M*	*B*	*T*	*S*
Oregon (*O*)	≤1	38	65	136	118	161	7998	669	371	194	188	136
Iowa (*I*)		≤1	33	98	153	172		10004	752	243	144	120
Montreal (*M*)			≤1	82	186	202			7977	283	111	102
Belgium (*B*)				≤1	252	270				9728	79	66
Taiwan (*T*)					≤1	137					7998	160
Sydney (*S*)						≤1						7977

Fig. 3. Real-world communication costs in Google Cloud, using clusters of `n1` machines deployed in six different regions, in terms of the ping round-trip times (which determines *latency*) and bandwidth (which determines *throughput*). These measurements are reproduced from Gupta et al. [8]. (Color figure online)

replicas, faulty replicas, and non-faulty replicas in the cluster, respectively. We extend the notations $\mathsf{f}(\cdot)$, $\mathsf{nf}(\cdot)$, $\mathbf{n}_{(\cdot)}$, $\mathbf{f}_{(\cdot)}$, and $\mathbf{nf}_{(\cdot)}$ to arbitrary sets of replicas. We assume that all replicas in each cluster have a predetermined order (e.g., on identifier or on public address), which allows us to deterministically select any number of replicas in a unique order from each cluster. In this work, we consider faulty replicas that can *crash*, *omit* messages, or behave *Byzantine*. A *crashing replica* executes steps correctly up till some point, after which it does not execute anything. An *omitting replica* executes steps correctly, but can decide to not send a message when it should or decide to ignore messages it receives. A *Byzantine replica* can behave in arbitrary, possibly coordinated and malicious, manners.

A *cluster system* $\mathfrak{S}$ is a finite set of clusters such that communication between replicas in a cluster is *local* and communication between clusters is *non-local*. We assume that there is no practical bound on local communication (e.g., within a single data center), while global communication is limited, costly, and to be avoided (e.g., between data centers in different continents). If $\mathcal{C}_1, \mathcal{C}_2 \in \mathfrak{S}$ are distinct clusters, then we assume that $\mathcal{C}_1 \cap \mathcal{C}_2 = \emptyset$: no replica is part of two distinct clusters. Our abstract model of a cluster system—in which we distinguish between unbounded local communication and costly global communication—is supported by practice. E.g., the ping round-trip time and bandwidth measurements of Fig. 3 imply that message latencies between clusters are at least 33–270 times higher than within clusters, while the maximum throughput is 10–151 times lower, both implying that communication between clusters is *up-to-two orders of magnitude* more costly than communication within clusters.

Definition 1. *Let $\mathfrak{S}$ be a system and $\mathcal{C}_1, \mathcal{C}_2 \in \mathfrak{S}$ be two clusters with non-faulty replicas ($\mathsf{nf}(\mathcal{C}_1) \neq \emptyset$ and $\mathsf{nf}(\mathcal{C}_2) \neq \emptyset$). The* cluster-sending problem *is the problem of sending a value v from $\mathcal{C}_1$ to $\mathcal{C}_2$ such that:* ***(1.)*** *all non-faulty replicas in $\mathcal{C}_2$* receive *the value v;* ***(2.)*** *all non-faulty replicas in $\mathcal{C}_1$* confirm *that the value v was received by all non-faulty replicas in $\mathcal{C}_2$; and* ***(3.)*** *non-faulty replicas in $\mathcal{C}_2$ can only receive a value v if all non-faulty replicas in $\mathcal{C}_1$* agree *upon sending v.*

In the following, we assume *asynchronous reliable communication*: all messages sent by non-faulty replicas eventually arrive at their destination. None of the protocols we propose rely on message delivery timings for their correctness.

We assume that communication is *authenticated*: on receipt of a message m from replica $\mathrm{R} \in \mathcal{C}$, one can determine that R did send m if $\mathrm{R} \in \mathsf{nf}(\mathcal{C})$ and if $\mathrm{R} \in \mathsf{nf}(\mathcal{C})$, then one can only determine that m was sent by R if R did send m. Hence, faulty replicas are only able to impersonate each other. We study the *cluster-sending problem* for Byzantine systems in two types of environments:

1. A system provides *replica signing* if every replica R can *sign* arbitrary messages m, resulting in a certificate $\langle m \rangle_{\mathrm{R}}$. These certificates are non-forgeable and can be constructed only if R cooperates in constructing them. Based on only the certificate $\langle m \rangle_{\mathrm{R}}$, anyone can verify that m was supported by R.
2. A system provides *cluster signing* if it is equipped with a *signature scheme* that can be used to *cluster-sign* arbitrary messages m, resulting in a certificate $\langle m \rangle_{\mathcal{C}}$. These certificates are non-forgeable and can be constructed whenever all non-faulty replicas in $\mathsf{nf}(\mathcal{C})$ cooperate in constructing them. Based on only the certificate $\langle m \rangle_{\mathcal{C}}$, anyone can verify that m was originally supported by all non-faulty replicas in $\mathcal{C}$.

In practice, *replica signing* can be implemented using *digital signatures*, which rely on a public-key cryptography infrastructure [15], and *cluster signing* can be implemented using threshold signatures, which are available for some public-key cryptography infrastructures [19]. Let m be a message, $\mathcal{C} \in \mathfrak{S}$ a cluster, and $\mathrm{R} \in \mathcal{C}$ a replica. We write $\|v\|$ to denote the size of any arbitrary value v. We assume that the size of certificates $\langle m \rangle_{\mathrm{R}}$, obtained via replica signing, and certificates $\langle m \rangle_{\mathcal{C}}$, obtained via cluster signing, are both linearly upper-bounded by $\|m\|$. More specifically, $\|(m, \langle m \rangle_{\mathrm{R}})\| = \mathcal{O}(\|m\|)$ and $\|(m, \langle m \rangle_{\mathcal{C}})\| = \mathcal{O}(\|m\|)$.

When necessary, we assume that replicas in each cluster $\mathcal{C} \in \mathfrak{S}$ can reach agreement on a value using an off-the-shelf *consensus protocol* [2,23]. In general, these protocols require $\mathbf{n}_{\mathcal{C}} > 2\mathbf{f}_{\mathcal{C}}$ (crash failures) or $\mathbf{n}_{\mathcal{C}} > 3\mathbf{f}_{\mathcal{C}}$ (Byzantine failures), which we assume to be the case for all *sending clusters*. Finally, in this paper we use the notation $i \operatorname{sgn} j$, with $i, j \geq 0$ and sgn the sign function, to denote i if $j > 0$ and 0 otherwise.

3 Lower Bounds for Cluster-Sending

In the previous section, we formalized the cluster-sending problem. The cluster-sending problem can be solved intuitively using *message broadcasts* (e.g., as used by `Chainspace` [1]), a principle technique used in the implementation of Byzantine primitives such as consensus and interactive consistency to assure that all non-faulty replicas reach the same conclusions. Unfortunately, broadcast-based protocols have a high communication cost that is quadratic in the size of the clusters involved. To determine whether we can do better than broadcasting, we will study the *lower bound* on the communication cost for any protocol solving the cluster-sending problem.

First, we consider systems with only crash failures, in which case we can lower bound the number of messages exchanged. As systems with omit failures or Byzantine failures can behave as-if they have only crash failures, these lower

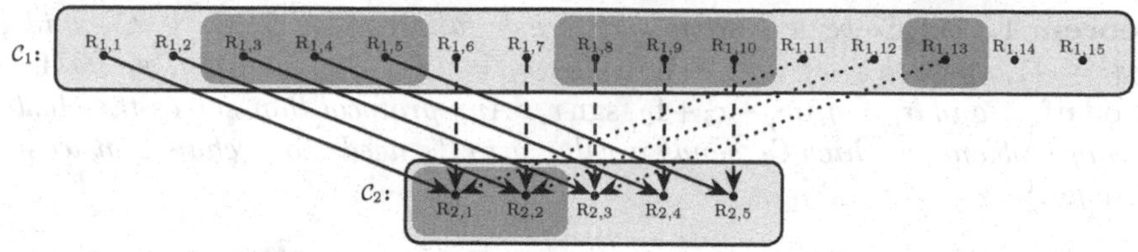

Fig. 4. A run of a protocol that sends messages from $\mathcal{C}_1$ and $\mathcal{C}_2$. The protocol P sends 13 messages, which is one message short of guaranteeing successful cluster-sending. Hence, to thwart cluster-sending in this particular run we can crash (highlighted using a red background) $\mathbf{f}_{\mathcal{C}_1} = 7$ and $\mathbf{f}_{\mathcal{C}_2} = 2$ replicas in $\mathcal{C}_1$ and $\mathcal{C}_2$, respectively. (Color figure online)

bounds apply to all environments. Any lower bound on the number of messages exchanged is determined by the maximum number of messages that can get *lost* due to crashed replicas that do not send or receive messages. If some replicas need to send or receive *multiple* messages, the capabilities of crashed replicas to lose messages is likewise multiplied, as the following example illustrates.

Example 1. Consider a system $\mathfrak{S}$ with clusters $\mathcal{C}_1, \mathcal{C}_2 \in \mathfrak{S}$ such that $\mathbf{n}_{\mathcal{C}_1} = 15$, $\mathbf{f}_{\mathcal{C}_1} = 7$, $\mathbf{n}_{\mathcal{C}_2} = 5$, and $\mathbf{f}_{\mathcal{C}_2} = 2$. We assume that $\mathfrak{S}$ only has crash failures and that the cluster $\mathcal{C}_1$ wants to send value v to $\mathcal{C}_2$. We will argue that any correct cluster-sending protocol P needs to send at least 14 messages in the worst case, as we can always assure that up-to-13 messages will get lost by crashing $\mathbf{f}_{\mathcal{C}_1}$ replicas in $\mathcal{C}_1$ and $\mathbf{f}_{\mathcal{C}_2}$ replicas in $\mathcal{C}_2$.

Consider the messages of a protocol P that wants to send only 13 messages from $\mathcal{C}_1$ to $\mathcal{C}_2$, e.g., the run in Fig. 4. Notice that $13 > \mathbf{n}_{\mathcal{C}_2}$. Hence, the run of P can only send 13 messages to replicas in $\mathcal{C}_2$ if some replicas in $\mathcal{C}_2$ will receive several messages. Neither P nor the replicas in $\mathcal{C}_1$ know which replicas in $\mathcal{C}_2$ have crashed. Hence, in the worst case, the $\mathbf{f}_{\mathcal{C}_2} = 2$ replicas in $\mathcal{C}_2$ that received the most messages have crashed. As we are sending 13 messages and $\mathbf{n}_{\mathcal{C}_2} = 5$, the two replicas that received the most messages must have received at least 6 messages in total. Hence, out of the 13 messages sent, at least 6 can be considered lost. In the run of Fig. 4, this loss would happen if $\mathrm{R}_{2,1}$ and $\mathrm{R}_{2,2}$ crash. Consequently, at most $13 - 6 = 7$ messages will arrive at non-faulty replicas. These messages are sent by at most 7 distinct replicas. As $\mathbf{f}_{\mathcal{C}_1} = 7$, all these sending replicas could have crashed. In the run of Fig. 4, this loss would happen if $\mathrm{R}_{1,3}$, $\mathrm{R}_{1,4}$, $\mathrm{R}_{1,5}$, $\mathrm{R}_{1,8}$, $\mathrm{R}_{1,9}$, $\mathrm{R}_{1,10}$, and $\mathrm{R}_{1,13}$ crash. Hence, we can thwart any run of P that intends to send 13 messages by crashing $\mathbf{f}_{\mathcal{C}_1}$ replicas in $\mathcal{C}_1$ and $\mathbf{f}_{\mathcal{C}_2}$ replicas in $\mathcal{C}_2$. Consequently, none of the messages of the run will be sent and received by non-faulty replicas, assuring that cluster-sending does not happen.

At least $\mathbf{f}_{\mathcal{C}_1} + 1$ replicas in $\mathcal{C}_1$ need to send messages to non-faulty replicas in $\mathcal{C}_2$ to assure that at least a *single* such message is sent by a non-faulty replica in $\mathsf{nf}(\mathcal{C}_1)$ and, hence, is guaranteed to arrive. We combine this with a thorough analysis along the lines of Example 1 to arrive at the following lower bounds:

Theorem 1. *Let $\mathfrak{S}$ be a system with crash failures, let $\mathcal{C}_1, \mathcal{C}_2 \in \mathfrak{S}$, and let $\{i, j\} = \{1, 2\}$ such that $\mathbf{n}_{\mathcal{C}_i} \geq \mathbf{n}_{\mathcal{C}_j}$. Let $q_i = (\mathbf{f}_{\mathcal{C}_i} + 1) \operatorname{div} \mathbf{nf}_{\mathcal{C}_j}$, $r_i = (\mathbf{f}_{\mathcal{C}_i} + 1) \bmod \mathbf{nf}_{\mathcal{C}_j}$, and $\sigma_i = q_i \mathbf{n}_{\mathcal{C}_j} + r_i + \mathbf{f}_{\mathcal{C}_j} \operatorname{sgn} r_i$. Any protocol that solves the cluster-sending problem in which $\mathcal{C}_1$ sends a value v to $\mathcal{C}_2$ needs to exchange at least σ_i messages.*[1]

Proof. The proof uses the same reasoning as Example 1: if a protocol sends at most $\sigma_i - 1$ messages, then we can choose $\mathbf{f}_{\mathcal{C}_1}$ replicas in $\mathcal{C}_1$ and $\mathbf{f}_{\mathcal{C}_2}$ replicas in $\mathcal{C}_2$ that will crash and thus assure that each of the $\sigma_i - 1$ messages is either sent by a crashed replica in $\mathcal{C}_1$ or received by a crashed replica in $\mathcal{C}_2$.

We assume $i = 1$, $j = 2$, and $\mathbf{n}_{\mathcal{C}_1} \geq \mathbf{n}_{\mathcal{C}_2}$. The proof is by contradiction. Hence, assume that a protocol $\mathtt{P}$ can solve the cluster-sending problem using at most $\sigma_1 - 1$ messages. Consider a run of $\mathtt{P}$ that sends messages M. Without loss of generality, we can assume that $|M| = \sigma_1 - 1$. Let R be the top $\mathbf{f}_{\mathcal{C}_2}$ receivers of messages in M, let $S = \mathcal{C}_2 \setminus R$, let $M_R \subset M$ be the messages received by replicas in R, and let $N = M \setminus M_R$. We notice that $\mathbf{n}_R = \mathbf{f}_{\mathcal{C}_2}$ and $\mathbf{n}_S = \mathbf{nf}_{\mathcal{C}_2}$.

First, we prove that $|M_R| \geq q_1 \mathbf{f}_{\mathcal{C}_2} + \mathbf{f}_{\mathcal{C}_2} \operatorname{sgn} r_1$, this by contradiction. Assume $|M_R| = q_1 \mathbf{f}_{\mathcal{C}_2} + \mathbf{f}_{\mathcal{C}_2} \operatorname{sgn} r_1 - v$, $v \geq 1$. Hence, we must have $|N| = q_1 \mathbf{nf}_{\mathcal{C}_2} + r_1 + v - 1$. Based on the value r_1, we distinguish two cases. The first case is $r_1 = 0$. In this case, $|M_R| = q_1 \mathbf{f}_{\mathcal{C}_2} - v < q_1 \mathbf{f}_{\mathcal{C}_2}$ and $|N| = q_1 \mathbf{nf}_{\mathcal{C}_2} + v - 1 \geq q_1 \mathbf{nf}_{\mathcal{C}_2}$. As $q_1 \mathbf{f}_{\mathcal{C}_2} > |M_R|$, there must be a replica in R that received at most $q_1 - 1$ messages. As $|N| \geq q_1 \mathbf{nf}_{\mathcal{C}_2}$, there must be a replica in S that received at least q_1 messages. The other case is $r_1 > 0$. In this case, $|M_R| = q_1 \mathbf{f}_{\mathcal{C}_2} + \mathbf{f}_{\mathcal{C}_2} - v < (q_1 + 1)\mathbf{f}_{\mathcal{C}_2}$ and $|N| = q_1 \mathbf{nf}_{\mathcal{C}_2} + r_1 + v - 1 > q_1 \mathbf{nf}_{\mathcal{C}_2}$. As $(q_1 + 1)\mathbf{f}_{\mathcal{C}_2} > |M_R|$, there must be a replica in R that received at most q_1 messages. As $|N| > q_1 \mathbf{nf}_{\mathcal{C}_2}$, there must be a replica in S that received at least $q_1 + 1$ messages. In both cases, we identified a replica in S that received more messages than a replica in R, a contradiction. Hence, we must conclude that $|M_R| \geq q_1 \mathbf{f}_{\mathcal{C}_2} + \mathbf{f}_{\mathcal{C}_2} \operatorname{sgn} r_1$ and, consequently, $|N| \leq q_1 \mathbf{nf}_{\mathcal{C}_2} + r_1 - 1 \leq \mathbf{f}_{\mathcal{C}_1}$. As $\mathbf{n}_R = \mathbf{f}_{\mathcal{C}_2}$, all replicas in R could have crashed, in which case only the messages in N are actually received. As $|N| \leq \mathbf{f}_{\mathcal{C}_1}$, all messages in N could be sent by replicas that have crashed. Hence, in the worst case, no message in M is successfully sent by a non-faulty replica in $\mathcal{C}_1$ and received by a non-faulty replica in $\mathcal{C}_2$, implying that $\mathtt{P}$ fails. □

The above lower bounds guarantee that at least one message can be delivered. Next, we look at systems with Byzantine failures and replica signing. In this case, at least $2\mathbf{f}_{\mathcal{C}_1} + 1$ replicas in $\mathcal{C}_1$ need to send a replica certificate to non-faulty replicas in $\mathcal{C}_2$ to assure that at least $\mathbf{f}_{\mathcal{C}_1} + 1$ such certificates are sent by non-faulty replicas and, hence, are guaranteed to arrive. Via a similar analysis to the one of Theorem 1, we arrive at:

---

[1] Example 1 showed that the impact of faulty replicas is minimal if we minimize the number of messages each replica exchanges. Let $\mathbf{n}_{\mathcal{C}_1} > \mathbf{n}_{\mathcal{C}_2}$. If the number of messages sent to $\mathbf{n}_{\mathcal{C}_2}$ is not a multiple of $\mathbf{n}_{\mathcal{C}_2}$, then minimizing the number of messages received by each replica in $\mathbf{n}_{\mathcal{C}_2}$ means that some replicas in $\mathbf{n}_{\mathcal{C}_2}$ will receive *one* more message than others: each replica in $\mathbf{n}_{\mathcal{C}_2}$ will receive at least q_1 messages, while the term $r_1 + \mathbf{f}_{\mathcal{C}_2} \operatorname{sgn} r_1$ specifies the number of replicas in $\mathbf{n}_{\mathcal{C}_2}$ that will receive $q_1 + 1$ messages.

Theorem 2. *Let $\mathfrak{S}$ be a system with Byzantine failures and replica signing, let $\mathcal{C}_1, \mathcal{C}_2 \in \mathfrak{S}$, and let $\{i, j\} = \{1, 2\}$ such that $\mathbf{n}_{\mathcal{C}_i} \geq \mathbf{n}_{\mathcal{C}_j}$. Let $q_1 = (2\mathbf{f}_{\mathcal{C}_1} + 1) \operatorname{div} \mathbf{nf}_{\mathcal{C}_2}$, $r_1 = (2\mathbf{f}_{\mathcal{C}_1} + 1) \bmod \mathbf{nf}_{\mathcal{C}_2}$, and $\tau_1 = q_1\mathbf{n}_{\mathcal{C}_2} + r_1 + \mathbf{f}_{\mathcal{C}_2} \operatorname{sgn} r_1$; and let $q_2 = (\mathbf{f}_{\mathcal{C}_2} + 1) \operatorname{div} (\mathbf{nf}_{\mathcal{C}_1} - \mathbf{f}_{\mathcal{C}_1})$, $r_2 = (\mathbf{f}_{\mathcal{C}_2} + 1) \bmod (\mathbf{nf}_{\mathcal{C}_1} - \mathbf{f}_{\mathcal{C}_1})$, and $\tau_2 = q_2\mathbf{n}_{\mathcal{C}_1} + r_2 + 2\mathbf{f}_{\mathcal{C}_1} \operatorname{sgn} r_2$. Any protocol that solves the cluster-sending problem in which $\mathcal{C}_1$ sends a value v to $\mathcal{C}_2$ needs to exchange at least τ_i messages.*[2]

Proof. For simplicity, we assume that each certificate is sent to $\mathcal{C}_2$ in an individual message independent of the other certificates. Hence, each certificate has a sender and a signer (both replicas in $\mathcal{C}_1$) and a receiver (a replica in $\mathcal{C}_2$).

First, we prove the case for $\mathbf{n}_{\mathcal{C}_1} \geq \mathbf{n}_{\mathcal{C}_2}$ using contradiction. Assume that a protocol P can solve the cluster-sending problem using at most $\tau_1 - 1$ certificates. Consider a run of P that sends messages C, each message representing a single certificate, with $|C| = \tau_1 - 1$. Following the proof of Theorem 1, one can show that, in the worst case, at most $\mathbf{f}_{\mathcal{C}_1}$ messages are sent by non-faulty replicas in $\mathcal{C}_1$ and received by non-faulty replicas in $\mathcal{C}_2$. Now consider the situation in which the faulty replicas in $\mathcal{C}_1$ mimic the behavior in C by sending certificates for another value v' to the same receivers. For the replicas in $\mathcal{C}_2$, the two runs behave the same, as in both cases at most $\mathbf{f}_{\mathcal{C}_1}$ certificates for a value, possibly signed by distinct replicas, are received. Hence, either both runs successfully send values, in which case v' is received by $\mathcal{C}_2$ without agreement, or both runs fail to send values. In both cases, P fails to solve the cluster-sending problem.

Next, we prove the case for $\mathbf{n}_{\mathcal{C}_2} \geq \mathbf{n}_{\mathcal{C}_1}$ using contradiction. Assume that a protocol P can solve the cluster-sending problem using at most $\tau_2 - 1$ certificates. Consider a run of P that sends messages C, each message representing a single certificate, with $|C| = \tau_2 - 1$. Let R be the top $2\mathbf{f}_{\mathcal{C}_1}$ signers of certificates in C, let $C_R \subset C$ be the certificates signed by replicas in R, and let $D = C \setminus C_R$. Via a contradiction argument similar to the one used in the proof of Theorem 1, one can show that $|C_R| \geq 2q_2\mathbf{f}_{\mathcal{C}_1} + 2\mathbf{f}_{\mathcal{C}_1} \operatorname{sgn} r$ and $|D| \leq q_2(\mathbf{nf}_{\mathcal{C}_1} - \mathbf{f}_{\mathcal{C}_1}) + r - 1 = \mathbf{f}_{\mathcal{C}_2}$. As $|D| \leq \mathbf{f}_{\mathcal{C}_2}$, all replicas receiving these certificates could have crashed. Hence, the only certificates that are received by $\mathcal{C}_2$ are in C_R. Partition C_R into two sets of certificates $C_{R,1}$ and $C_{R,2}$ such that both sets contain certificates signed by at most $\mathbf{f}_{\mathcal{C}_1}$ distinct replicas. As the certificates in $C_{R,1}$ and $C_{R,2}$ are signed by $\mathbf{f}_{\mathcal{C}_1}$ distinct replicas, one of these sets can contain only certificates signed by Byzantine replicas. Hence, either $C_{R,1}$ or $C_{R,2}$ could certify a non-agreed upon value v', while only the other set certifies v. Consequently, the replicas in $\mathcal{C}_2$ cannot distinguish between receiving an agreed-upon value v or a non-agreed-upon-value v'. We conclude that P fails to solve the cluster-sending problem. □

[2] Tolerating Byzantine failures in an environment with replica signatures leads to an asymmetry between the sending cluster $\mathcal{C}_1$, in which $2\mathbf{f}_{\mathcal{C}_1} + 1$ replicas need to send, and the receiving cluster $\mathcal{C}_2$, in which only $\mathbf{f}_{\mathcal{C}_2} + 1$ replicas need to receive. This asymmetry results in two distinct cases based on the relative cluster sizes.

4 Cluster-Sending via Bijective Sending

In the previous section, we established lower bounds for the cluster-sending problem. Next, we develop *bijective sending*, a powerful technique that allows the design of efficient cluster-sending protocols that match these lower bounds.

Protocol for the sending cluster $\mathcal{C}_1$:

All replicas in $\mathsf{nf}(\mathcal{C}_1)$ agree on v and construct $\langle v \rangle_{\mathcal{C}_1}$.
Choose replicas $S_1 \subseteq \mathcal{C}_1$ and $S_2 \subseteq \mathcal{C}_2$ with $\mathbf{n}_{S_2} = \mathbf{n}_{S_1} = \mathbf{f}_{\mathcal{C}_1} + \mathbf{f}_{\mathcal{C}_2} + 1$.
Choose a bijection $b : S_1 \rightarrow S_2$.
for $\text{R}_1 \in S_1$ **do**
    R_1 sends $(v, \langle v \rangle_{\mathcal{C}_1})$ to $b(\text{R}_1)$.

Protocol for the receiving cluster $\mathcal{C}_2$:

event $\text{R}_2 \in \mathsf{nf}(\mathcal{C}_2)$ receives $(w, \langle w \rangle_{\mathcal{C}_1})$ from $\text{R}_1 \in \mathcal{C}_1$ **do**
    Broadcast $(w, \langle w \rangle_{\mathcal{C}_1})$ to all replicas in $\mathcal{C}_2$.
event $\text{R}_2' \in \mathsf{nf}(\mathcal{C}_2)$ receives $(w, \langle w \rangle_{\mathcal{C}_1})$ from $\text{R}_2 \in \mathcal{C}_2$ **do**
    R_2' considers w *received.*

Fig. 5. BS-cs, the bijective sending cluster-sending protocol that sends a value v from $\mathcal{C}_1$ to $\mathcal{C}_2$. We assume Byzantine failures and a system that provides cluster signing.

First, we present a bijective sending protocol for systems with Byzantine failures and cluster signing. Let $\mathcal{C}_1$ be a cluster in which the non-faulty replicas have reached *agreement* on sending a value v to a cluster $\mathcal{C}_2$ and have access to a cluster certificate $\langle v \rangle_{\mathcal{C}_1}$. Let $\mathcal{C}_i$, $i \in \{1, 2\}$, be the cluster with the most replicas. To assure that at least a single non-faulty replica in $\mathcal{C}_1$ sends a message to a non-faulty replica in $\mathcal{C}_2$, we use the lower bound of Theorem 1: we choose σ_i distinct replicas $S_1 \subseteq \mathcal{C}_1$ and replicas $S_2 \subseteq \mathcal{C}_2$ and instruct each replica in $S_1 \subseteq \mathcal{C}_1$ to send v to a distinct replica in $\mathcal{C}_2$. By doing so, we guarantee that at least a single message is sent and received by non-faulty replicas and, hence, guarantee successful cluster-sending. To be able to choose S_1 and S_2 with $\mathbf{n}_{S_1} = \mathbf{n}_{S_2} = \sigma_i$, we need $\sigma_i \leq \min(\mathbf{n}_{\mathcal{C}_1}, \mathbf{n}_{\mathcal{C}_2})$, in which case we have $\sigma_i = \mathbf{f}_{\mathcal{C}_1} + \mathbf{f}_{\mathcal{C}_2} + 1$. The pseudo-code for this *bijective sending* protocol for systems that provide *cluster signing* (BS-cs), can be found in Fig. 5. Next, we illustrate bijective sending:

Example 2. Consider system $\mathfrak{S} = \{\mathcal{C}_1, \mathcal{C}_2\}$ of Fig. 6 with $\mathcal{C}_1 = \{\text{R}_{1,1}, \ldots, \text{R}_{1,8}\}$, $\mathsf{f}(\mathcal{C}_1) = \{\text{R}_{1,1}, \text{R}_{1,3}, \text{R}_{1,4}\}$, $\mathcal{C}_2 = \{\text{R}_{2,1}, \ldots, \text{R}_{2,7}\}$, and $\mathsf{f}(\mathcal{C}_2) = \{\text{R}_{2,1}, \text{R}_{2,3}\}$. We have $\mathbf{f}_{\mathcal{C}_1} + \mathbf{f}_{\mathcal{C}_2} + 1 = 6$ and we choose $S_1 = \{\text{R}_{1,2}, \ldots, \text{R}_{1,7}\}$, $S_2 = \{\text{R}_{2,1}, \ldots, \text{R}_{2,6}\}$, and $b = \{\text{R}_{1,i} \mapsto \text{R}_{2,i-1} \mid 2 \leq i \leq 7\}$. Replica $\text{R}_{1,2}$ sends a valid message to $\text{R}_{2,1}$. As $\text{R}_{2,1}$ is faulty, it might ignore this message. Replicas $\text{R}_{1,3}$ and $\text{R}_{1,4}$ are faulty and might not send a valid message. Additionally, $\text{R}_{2,3}$ is faulty and might ignore any message it receives. The messages sent from $\text{R}_{1,5}$ to $\text{R}_{2,4}$, from $\text{R}_{1,6}$ to $\text{R}_{2,5}$, and from $\text{R}_{1,7}$ to $\text{R}_{2,6}$ are all sent by non-faulty replicas to non-faulty replicas. Hence, these messages all arrive correctly.

Having illustrated the concept of bijective sending, as employed by BS-cs, we are now ready to prove correctness of BS-cs:

Proposition 1. *Let $\mathfrak{S}$ be a system with Byzantine failures and cluster signing and let $\mathcal{C}_1, \mathcal{C}_2 \in \mathfrak{S}$. If $\mathbf{n}_{\mathcal{C}_1} > 2\mathbf{f}_{\mathcal{C}_1}$, $\mathbf{n}_{\mathcal{C}_1} > \mathbf{f}_{\mathcal{C}_1} + \mathbf{f}_{\mathcal{C}_2}$, and $\mathbf{n}_{\mathcal{C}_2} > \mathbf{f}_{\mathcal{C}_1} + \mathbf{f}_{\mathcal{C}_2}$, then* BS-cs *satisfies Definition 1 and sends $\mathbf{f}_{\mathcal{C}_1} + \mathbf{f}_{\mathcal{C}_2} + 1$ messages, of size $\mathcal{O}(\|v\|)$ each, between $\mathcal{C}_1$ and $\mathcal{C}_2$.*

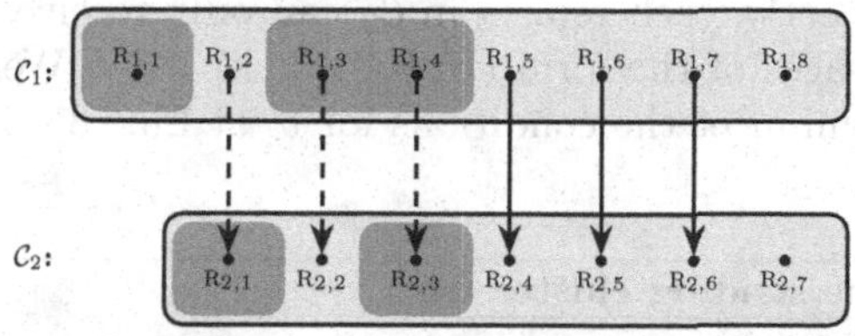

Fig. 6. Bijective sending from $\mathcal{C}_1$ to $\mathcal{C}_2$. The faulty replicas are highlighted using a red background. The edges connect replicas $\mathtt{R} \in \mathcal{C}_1$ with $b(\mathtt{R}) \in \mathcal{C}_2$. Each solid edge indicates a message sent and received by non-faulty replicas. Each dashed edge indicates a message sent or received by a faulty replica. (Color figure online)

Proof. Choose $S_1 \subseteq \mathcal{C}_1$, $S_2 \subseteq \mathcal{C}_2$, and $b : S_1 \to S_2$ in accordance with BS-cs (Fig. 5). We have $\mathbf{n}_{S_1} = \mathbf{n}_{S_2} = \mathbf{f}_{\mathcal{C}_1} + \mathbf{f}_{\mathcal{C}_2} + 1$. Let $T = \{b(\mathtt{R}) \mid \mathtt{R} \in \mathsf{nf}(S_1)\}$. By construction, we have $\mathbf{nf}_{S_1} = \mathbf{n}_T \geq \mathbf{f}_{\mathcal{C}_2} + 1$. Hence, we have $\mathbf{nf}_T \geq 1$. Due to Line 5, each replica in $\mathsf{nf}(T)$ will receive the message $(v, \langle v \rangle_{\mathcal{C}_1})$ from a distinct replica in $\mathsf{nf}(S_1)$ and broadcast $(v, \langle v \rangle_{\mathcal{C}_1})$ to all replicas in $\mathcal{C}_2$. As $\mathbf{nf}_T \geq 1$, each replica $\mathtt{R}_2' \in \mathsf{nf}(\mathcal{C}_2)$ will receive $(v, \langle v \rangle_{\mathcal{C}_1})$ from a replica in $\mathcal{C}_2$ and meet the condition at Line 8, proving *receipt* and *confirmation*. Finally, we have *agreement*, as $\langle v \rangle_{\mathcal{C}_1}$ is non-forgeable. □

To provide cluster-sending in environments with only replica signing, we combine the principle idea of bijective sending with the lower bound on the number of replica certificates exchanged, as provided by Theorem 2. Let $\mathcal{C}_i$, $i \in \{1, 2\}$, be the cluster with the most replicas. To assure that at least $\mathbf{f}_{\mathcal{C}_1} + 1$ non-faulty replicas in $\mathcal{C}_1$ send replica certificates to non-faulty replicas in $\mathcal{C}_2$, we choose sets of replicas $S_1 \subseteq \mathcal{C}_1$ and $S_2 \subseteq \mathcal{C}_2$ with $\mathbf{n}_{S_1} = \mathbf{n}_{S_2} = \tau_i$. To be able to choose S_1 and S_2 with $\mathbf{n}_{S_1} = \mathbf{n}_{S_2} = \tau_i$, we need $\tau_i \leq \min(\mathbf{n}_{\mathcal{C}_1}, \mathbf{n}_{\mathcal{C}_2})$, in which case we have $\tau_i = 2\mathbf{f}_{\mathcal{C}_1} + \mathbf{f}_{\mathcal{C}_2} + 1$. The pseudo-code for this *bijective sending* protocol for systems that provide *replica signing* (BS-rs), can be found in Fig. 7. Next, we prove the correctness of BS-rs:

Proposition 2. *Let $\mathfrak{S}$ be a system with Byzantine failures and replica signing and let $\mathcal{C}_1, \mathcal{C}_2 \in \mathfrak{S}$. If $\mathbf{n}_{\mathcal{C}_1} > 2\mathbf{f}_{\mathcal{C}_1} + \mathbf{f}_{\mathcal{C}_2}$ and $\mathbf{n}_{\mathcal{C}_2} > 2\mathbf{f}_{\mathcal{C}_1} + \mathbf{f}_{\mathcal{C}_2}$, then* BS-rs *satisfies Definition 1 and sends $2\mathbf{f}_{\mathcal{C}_1} + \mathbf{f}_{\mathcal{C}_2} + 1$ messages, of size $\mathcal{O}(\|v\|)$ each, between $\mathcal{C}_1$ and $\mathcal{C}_2$.*

Proof. Choose $S_1 \subseteq \mathcal{C}_1$, $S_2 \subseteq \mathcal{C}_2$, and $b : S_1 \rightarrow S_2$ in accordance with BS-rs (Fig. 7). We have $\mathbf{n}_{S_1} = \mathbf{n}_{S_2} = 2\mathbf{f}_{\mathcal{C}_1} + \mathbf{f}_{\mathcal{C}_2} + 1$. Let $T = \{b(\mathrm{R}) \mid \mathrm{R} \in \mathsf{nf}(S_1)\}$. By construction, we have $\mathbf{nf}_{S_1} = \mathbf{n}_T \geq \mathbf{f}_{\mathcal{C}_1} + \mathbf{f}_{\mathcal{C}_2} + 1$. Hence, we have $\mathbf{nf}_T \geq \mathbf{f}_{\mathcal{C}_1} + 1$. Due to Line 5, each replica in $\mathsf{nf}(T)$ will receive the message $(v, \langle v \rangle_{\mathrm{R}_1})$ from a distinct replica $\mathrm{R}_1 \in \mathsf{nf}(S_1)$ and meet the condition at Line 8, proving *receipt* and *confirmation*.

Next, we prove *agreement*. Consider a value v' not agreed upon by $\mathcal{C}_1$. Hence, no non-faulty replicas $\mathsf{nf}(\mathcal{C}_1)$ will sign v'. Due to non-forgeability of replica certificates, the only certificates that can be constructed for v' are of the form $\langle v' \rangle_{\mathrm{R}_1}$, $\mathrm{R}_1 \in \mathsf{f}(\mathcal{C}_1)$. Consequently, each replica in $\mathcal{C}_2$ can only receive and broadcast up to $\mathbf{f}_{\mathcal{C}_1}$ distinct messages of the form $(v', \langle v' \rangle_{\mathrm{R}_1'})$, $\mathrm{R}_1' \in \mathcal{C}_1$. We conclude that no non-faulty replica will meet the conditions for v' at Line 8. □

Protocol for the sending cluster $\mathcal{C}_1$:

1: All replicas in $\mathsf{nf}(\mathcal{C}_1)$ agree on v.
2: Choose replicas $S_1 \subseteq \mathcal{C}_1$ and $S_2 \subseteq \mathcal{C}_2$ with $\mathbf{n}_{S_2} = \mathbf{n}_{S_1} = 2\mathbf{f}_{\mathcal{C}_1} + \mathbf{f}_{\mathcal{C}_2} + 1$.
3: Choose bijection $b : S_1 \rightarrow S_2$.
4: **for** $\mathrm{R}_1 \in S_1$ **do**
5: $\quad$ R_1 sends $(v, \langle v \rangle_{\mathrm{R}_1})$ to $b(\mathrm{R}_1)$.

Protocol for the receiving cluster $\mathcal{C}_2$:

6: **event** $\mathrm{R}_2 \in \mathsf{nf}(\mathcal{C}_2)$ receives $(w, \langle w \rangle_{\mathrm{R}_1'})$ from $\mathrm{R}_1' \in \mathcal{C}_1$ **do**
7: $\quad$ Broadcast $(w, \langle w \rangle_{\mathrm{R}_1'})$ to all replicas in $\mathcal{C}_2$.
8: **event** $\mathrm{R}_2' \in \mathsf{nf}(\mathcal{C}_2)$ receives $\mathbf{f}_{\mathcal{C}_1} + 1$ messages $(w, \langle w \rangle_{\mathrm{R}_1'})$:
(i) each message is sent by a replica in $\mathcal{C}_2$;
(ii) each message carries the same value w; and
(iii) each message has a distinct signature $\langle w \rangle_{\mathrm{R}_1'}$, $\mathrm{R}_1' \in \mathcal{C}_1$
do
9: $\quad$ R_2' considers w *received.*

Fig. 7. BS-rs, the bijective sending cluster-sending protocol that sends a value v from $\mathcal{C}_1$ to $\mathcal{C}_2$. We assume Byzantine failures and a system that provides replica signing.

5 Cluster-Sending via Partitioning

Unfortunately, the worst-case optimal bijective sending techniques introduced in the previous section are limited to similar-sized clusters:

Example 3. Consider a system $\mathfrak{S}$ with Byzantine failures and cluster certificates. The cluster $\mathcal{C}_1 \in \mathfrak{S}$ wants to send value v to $\mathcal{C}_2 \in \mathfrak{S}$ with $\mathbf{n}_{\mathcal{C}_1} \geq \mathbf{n}_{\mathcal{C}_2}$. To do so, BS-cs requires $\sigma_1 = \mathbf{f}_{\mathcal{C}_1} + \mathbf{f}_{\mathcal{C}_2} \leq \mathbf{n}_{\mathcal{C}_2}$. Hence, BS-cs requires that $\mathbf{f}_{\mathcal{C}_1}$ is upper-bounded by $\mathbf{nf}_{\mathcal{C}_2} \leq \mathbf{n}_{\mathcal{C}_2}$, which is independent of the size of cluster $\mathcal{C}_1$.

Next, we show how to generalize bijective sending to arbitrary-sized clusters. We do so by *partitioning* the larger-sized cluster into a set of smaller clusters, and then letting sufficient of these smaller clusters participate independently in bijective sending. First, we introduce the relevant partitioning notation.

Definition 2. *Let $\mathfrak{S}$ be a system, let $\mathcal{P}$ be a subset of the replicas in $\mathfrak{S}$, let $c > 0$ be a constant, let $q = \mathbf{n}_{\mathcal{P}} \operatorname{div} c$, and let $r = \mathbf{n}_{\mathcal{P}} \bmod c$. A c-partition $\mathsf{partition}(\mathcal{P}) = \{P_1, \dots, P_q, P'\}$ of $\mathcal{P}$ is a partition of the set of replicas $\mathcal{P}$ into sets $P_1, \dots, P_q, P'$ such that $\mathbf{n}_{P_i} = c$, $1 \leq i \leq q$, and $\mathbf{n}_{P'} = r$.*

Example 4. Consider system $\mathfrak{S} = \{\mathcal{C}\}$ of Fig. 8 with $\mathcal{C} = \{\mathrm{R}_1, \dots, \mathrm{R}_{11}\}$ and $\mathsf{f}(\mathcal{C}) = \{\mathrm{R}_1, \dots, \mathrm{R}_5\}$. The set $\mathsf{partition}(\mathcal{C}) = \{P_1, P_2, P'\}$ with $P_1 = \{\mathrm{R}_1, \dots, \mathrm{R}_4\}$, $P_2 = \{\mathrm{R}_5, \dots, \mathrm{R}_8\}$, and $P' = \{\mathrm{R}_9, \mathrm{R}_{10}, \mathrm{R}_{11}\}$ is a 4-partition of $\mathcal{C}$. We have $\mathsf{f}(P_1) = P_1$, $\mathsf{nf}(P_1) = \emptyset$, and $\mathbf{n}_{P_1} = \mathbf{f}_{P_1} = 4$. Likewise, we have $\mathsf{f}(P_2) = \{\mathrm{R}_5\}$, $\mathsf{nf}(P_2) = \{\mathrm{R}_6, \mathrm{R}_7, \mathrm{R}_8\}$, $\mathbf{n}_{P_2} = 4$, and $\mathbf{f}_{P_2} = 1$.

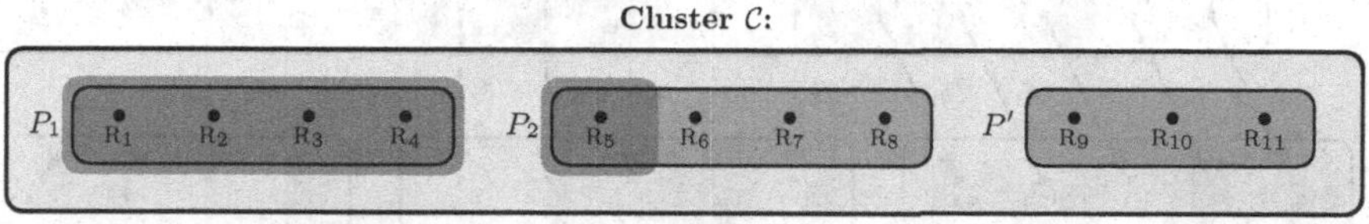

Fig. 8. An example of a 4-partition of a cluster $\mathcal{C}$ with 11 replicas, of which the first five are faulty. The three partitions are grouped in blue boxes, the faulty replicas are highlighted using a red background. (Color figure online)

Next, we apply partitioning to BS-cs. Let $\mathcal{C}_1$ be a cluster in which the non-faulty replicas have reached agreement on sending a value v to a cluster $\mathcal{C}_2$ and constructed $\langle v \rangle_{\mathcal{C}_1}$. First, we consider the case $\mathbf{n}_{\mathcal{C}_1} \geq \mathbf{n}_{\mathcal{C}_2}$. In this case, we choose a set $P \subseteq \mathcal{C}_1$ of σ_1 replicas in $\mathcal{C}_1$ to sent v to replicas in $\mathcal{C}_2$. To minimize the number of values v received by faulty replicas in $\mathcal{C}_2$, we minimize the number of values v sent to each replica in $\mathcal{C}_2$. Conceptually, we do so by constructing an $\mathbf{n}_{\mathcal{C}_2}$-partition of the σ_1 replicas in P and instruct each resultant set in the partition to perform bijective sending. The pseudo-code for the resultant *sender-partitioned bijective sending* protocol for systems that provide cluster signing, named SPBS-(σ_1,cs), can be found in Fig. 10. In a similar fashion, we can apply partitioning to BS-rs, in which case we instruct τ_1 replicas in $\mathcal{C}_1$ to send v to replicas in $\mathcal{C}_2$, which yields the *sender-partitioned bijective sending* protocol SPBS-(τ_1,rs) for systems that provide replica signing. Next, we illustrate sender-partitioned bijective sending:

Example 5. We continue from Example 1. Hence, we have $\mathcal{C}_1 = \{\mathrm{R}_{1,1}, \dots, \mathrm{R}_{1,15}\}$ and $\mathcal{C}_2 = \{\mathrm{R}_{2,1}, \dots, \mathrm{R}_{2,5}\}$ with $\mathsf{f}(\mathcal{C}_1) = \{\mathrm{R}_{1,3}, \mathrm{R}_{1,4}, \mathrm{R}_{1,5}, \mathrm{R}_{1,8}, \mathrm{R}_{1,9}, \mathrm{R}_{1,10}, \mathrm{R}_{1,13}\}$ and $\mathsf{f}(\mathcal{C}_2) = \{\mathrm{R}_{2,1}, \mathrm{R}_{2,2}\}$. We assume that $\mathfrak{S}$ provides cluster signing and we apply sender-partitioned bijective sending. We have $\mathbf{n}_{\mathcal{C}_1} > \mathbf{n}_{\mathcal{C}_2}$, $q_1 = 8 \operatorname{div} 3 = 2$, $r_1 = 8 \bmod 3 = 2$, and $\sigma_1 = 2 \cdot 5 + 2 + 2 = 14$. We choose the replicas $\mathcal{P} = \{\mathrm{R}_{1,1}, \dots, \mathrm{R}_{1,14}\} \subseteq \mathcal{C}_1$ and the $\mathbf{n}_{\mathcal{C}_2}$-partition $\mathsf{partition}(\mathcal{P}) = \{P_1, P_2, P'\}$ with $P_1 = \{\mathrm{R}_{1,1}, \mathrm{R}_{1,2}, \mathrm{R}_{1,3}, \mathrm{R}_{1,4}, \mathrm{R}_{1,5}\}$, $P_2 = \{\mathrm{R}_{1,6}, \mathrm{R}_{1,7}, \mathrm{R}_{1,8}, \mathrm{R}_{1,9}, \mathrm{R}_{1,10}\}$, and $P' = \{\mathrm{R}_{1,11}, \mathrm{R}_{1,12}, \mathrm{R}_{1,13}, \mathrm{R}_{1,14}\}$. Hence, SPBS-($\sigma_1$,cs) will perform three rounds of bijective sending. In the first two rounds, SPBS-(σ_1,cs) will send to all replicas in $\mathcal{C}_2$. In the last round, SPBS-(σ_1,cs) will send to the replicas $Q = \{\mathrm{R}_{2,1}, \mathrm{R}_{2,2}, \mathrm{R}_{2,3}, \mathrm{R}_{2,4}\}$. We choose bijections $b_1 = \{\mathrm{R}_{1,1} \mapsto \mathrm{R}_{2,1}, \dots, \mathrm{R}_{1,5} \mapsto \mathrm{R}_{2,5}\}$,

$b_2 = \{R_{1,6} \mapsto R_{2,1}, \ldots, R_{1,10} \mapsto R_{2,5}\}$, and $b' = \{R_{1,11} \mapsto R_{2,1}, \ldots, R_{1,14} \mapsto R_{2,4}\}$. In the first two rounds, we have $\mathbf{f}_{P_1} + \mathbf{f}_{C_2} = \mathbf{f}_{P_2} + \mathbf{f}_{C_2} = 3+2 = 5 = \mathbf{n}_{C_2}$. Due to the particular choice of bijections b_1 and b_2, these rounds will fail cluster-sending. In the last round, we have $\mathbf{f}_{P'} + \mathbf{f}_Q = 1 + 2 = 3 < \mathbf{n}_{P'} = \mathbf{n}_Q$. Hence, these two sets of replicas satisfy the conditions of BS-cs, can successfully apply bijective sending, and we will have successful cluster-sending (as the non-faulty replica $R_{1,14} \in C_1$ will send v to the non-faulty replica $R_{2,4} \in C_2$). We have illustrated the described working of SPBS-(σ_1,cs) in Fig. 9.

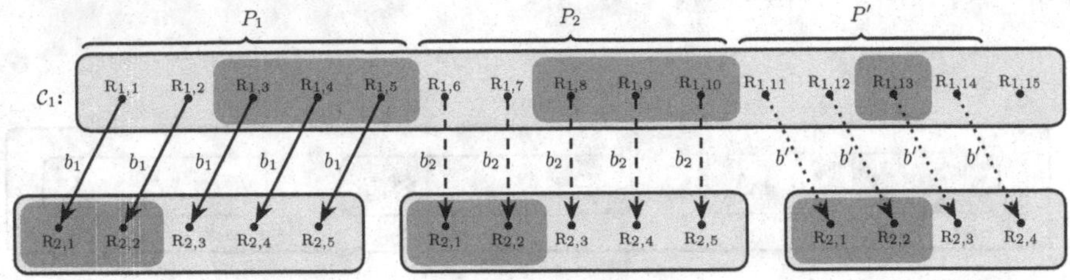

Fig. 9. An example of SPBS-(σ_1,cs) with $\sigma_1 = 14$ and $\text{partition}(\mathcal{P}) = \{P_1, P_2, P'\}$. Notice that only the instance of bijective sending with the replicas in P' and bijection b' will succeed in cluster-sending. (Color figure online)

Protocol for the sending cluster C_1:

The agreement step of BS-ζ for value v.
Choose replicas $\mathcal{P} \subseteq C_1$ with $\mathbf{n}_{\mathcal{P}} = \alpha$ and choose $\mathbf{n}_{C_2}$-partition $\text{partition}(\mathcal{P})$ of $\mathcal{P}$.
for $P \in \text{partition}(\mathcal{P})$ **do**
  Choose replicas $Q \subseteq C_2$ with $\mathbf{n}_Q = \mathbf{n}_P$ and choose bijection $b : P \to Q$.
  for $R_1 \in P$ **do**
    Send v from R_1 to $b(R_1)$ via the send step of BS-ζ.

Protocol for the receiving cluster C_2:

See the protocol for the receiving cluster in BS-ζ.

Fig. 10. SPBS-(α,ζ), $\zeta \in \{\text{cs}, \text{rs}\}$, the sender-partitioned bijective sending cluster-sending protocol that sends a value v from C_1 to C_2. We assume the same system properties as BS-ζ.

Next, we prove the correctness of sender-partitioned bijective sending:

Proposition 3. *Let $\mathfrak{S}$ be a system with Byzantine failures, let $C_1, C_2 \in \mathfrak{S}$, let σ_1 be as defined in Theorem 1, and let τ_1 be as defined in Theorem 2.*

1. *If $\mathfrak{S}$ provides cluster signing and $\sigma_1 \leq \mathbf{n}_{C_1}$, then* SPBS-($\sigma_1$,cs) *satisfies Definition 1 and sends σ_1 messages, of size $\mathcal{O}(\|v\|)$ each, between C_1 and C_2.*
2. *If $\mathfrak{S}$ provides replica signing and $\tau_1 \leq \mathbf{n}_{C_1}$, then* SPBS-($\tau_1$,rs) *satisfies Definition 1 and sends τ_1 messages, of size $\mathcal{O}(\|v\|)$ each, between C_1 and C_2.*

Proof. Let $\beta = (\mathbf{f}_{\mathcal{C}_1} + 1)$ in the case of cluster signing and let $\beta = (2\mathbf{f}_{\mathcal{C}_1} + 1)$ in the case of replica signing. Let $q = \beta \operatorname{div} \mathbf{nf}_{\mathcal{C}_2}$ and $r = \beta \bmod \mathbf{nf}_{\mathcal{C}_2}$. We have $\alpha = q\mathbf{n}_{\mathcal{C}_2} + r + \mathbf{f}_{\mathcal{C}_2} \operatorname{sgn} r$. Choose $\mathcal{P}$ and choose $\mathsf{partition}(\mathcal{P}) = \{P_1, \dots, P_q, P'\}$ in accordance with $\texttt{SPBS-}(\alpha,\zeta)$ (Fig. 10). For each $P \in \mathcal{P}$, choose a Q and b in accordance with $\texttt{SPBS-}(\alpha,\zeta)$, and let $z(P) = \{\mathrm{R} \in P \mid b(\mathrm{R}) \in \mathsf{f}(Q)\}$. As each such b has a distinct domain, the union of them is a surjection $f : \mathcal{P} \rightarrow \mathcal{C}_2$. By construction, we have $\mathbf{n}_{P'} = r + \mathbf{f}_{\mathcal{C}_2} \operatorname{sgn} r$, $\mathbf{n}_{z(P')} \leq \mathbf{f}_{\mathcal{C}_2} \operatorname{sgn} r$, and, for every i, $1 \leq i \leq q$, $\mathbf{n}_{P_i} = \mathbf{n}_{\mathcal{C}_2}$ and $\mathbf{n}_{z(P_i)} = \mathbf{f}_{\mathcal{C}_2}$. Let $V = \mathcal{P} \setminus \left(\bigcup_{P \in \mathsf{partition}(\mathcal{P})} z(P)\right)$. We have

Protocol for the sending cluster $\mathcal{C}_1$:

The agreement step of $\texttt{BS-}\zeta$ for value v.
Choose replicas $\mathcal{P} \subseteq \mathcal{C}_2$ with $\mathbf{n}_{\mathcal{P}} = \alpha$ and choose $\mathbf{n}_{\mathcal{C}_1}$-partition $\mathsf{partition}(\mathcal{P})$ of $\mathcal{P}$.
for $P \in \mathsf{partition}(\mathcal{P})$ **do**
    Choose replicas $Q \subseteq \mathcal{C}_1$ with $\mathbf{n}_Q = \mathbf{n}_P$ and choose bijection $b : Q \rightarrow P$.
    for $\mathrm{R}_1 \in Q$ **do**
        Send v from R_1 to $b(\mathrm{R}_1)$ via the send step of $\texttt{BS-}\zeta$.

Protocol for the receiving cluster $\mathcal{C}_2$:

See the protocol for the receiving cluster in $\texttt{BS-}\zeta$.

Fig. 11. $\texttt{RPBS-}(\alpha,\zeta)$, $\zeta \in \{\texttt{cs}, \texttt{rs}\}$, the receiver-partitioned bijective sending cluster-sending protocol that sends a value v from $\mathcal{C}_1$ to $\mathcal{C}_2$. We assume the same system properties as $\texttt{BS-}\zeta$.

$$\mathbf{n}_V \geq \mathbf{n}_{\mathcal{P}} - (q\mathbf{f}_{\mathcal{C}_2} + \mathbf{f}_{\mathcal{C}_2} \operatorname{sgn} r) = (q\mathbf{n}_{\mathcal{C}_2} + r + \mathbf{f}_{\mathcal{C}_2} \operatorname{sgn} r) - (q\mathbf{f}_{\mathcal{C}_2} + \mathbf{f}_{\mathcal{C}_2} \operatorname{sgn} r) = q\mathbf{nf}_{\mathcal{C}_2} + r = \beta.$$

Let $T = \{f(\mathrm{R}) \mid \mathrm{R} \in \mathsf{nf}(V)\}$. By construction, we have $\mathbf{nf}_T = \mathbf{n}_T$. To complete the proof, we consider cluster signing and replica signing separately. First, the case for cluster signing. As $\mathbf{n}_V \geq \beta = \mathbf{f}_{\mathcal{C}_1} + 1$, we have $\mathbf{nf}_V \geq 1$. By construction, the replicas in $\mathsf{nf}(T)$ will receive the messages $(v, \langle v \rangle_{\mathcal{C}_1})$ from the replicas $\mathrm{R}_1 \in \mathsf{nf}(V)$. Hence, analogous to the proof of Proposition 1, we can prove *receipt*, *confirmation*, and *agreement*. Finally, the case for replica signing. As $\mathbf{n}_V \geq \beta = 2\mathbf{f}_{\mathcal{C}_1} + 1$, we have $\mathbf{nf}_V \geq \mathbf{f}_{\mathcal{C}_1} + 1$. By construction, the replicas in $\mathsf{nf}(T)$ will receive the messages $(v, \langle v \rangle_{\mathrm{R}_1})$ from each replica $\mathrm{R}_1 \in \mathsf{nf}(V)$. Hence, analogous to the proof of Proposition 2, we can prove *receipt*, *confirmation*, and *agreement*. □

Finally, we consider the case $\mathbf{n}_{\mathcal{C}_1} \leq \mathbf{n}_{\mathcal{C}_2}$. In this case, we apply partitioning to $\texttt{BS-cs}$ by choosing a set P of σ_2 replicas in $\mathcal{C}_2$, constructing an $\mathbf{n}_{\mathcal{C}_1}$-partition of P, and instruct $\mathcal{C}_1$ to perform bijective sending with each set in the partition. The pseudo-code for the resultant *receiver-partitioned bijective sending* protocol for systems that provide cluster signing, named $\texttt{RPBS-}(\sigma_2\texttt{,cs})$, can be found in Fig. 11. In a similar fashion, we can apply partitioning to $\texttt{BS-rs}$, which yields the *receiver-partitioned bijective sending* protocol $\texttt{RPBS-(}\tau_2\texttt{,rs)}$ for systems that provide replica signing. Next, we prove the correctness of these instances of receiver-partitioned bijective sending:

Proposition 4. *Let $\mathfrak{S}$ be a system with Byzantine failures, let $\mathcal{C}_1, \mathcal{C}_2 \in \mathfrak{S}$, let σ_2 be as defined in Theorem 1, and let τ_2 be as defined in Theorem 2.*

1. *If $\mathfrak{S}$ provides cluster signing and $\sigma_2 \leq \mathbf{n}_{\mathcal{C}_2}$, then $\texttt{RPBS-(}\sigma_2\texttt{,cs)}$ satisfies Definition 1 and sends σ_2 messages, of size $\mathcal{O}(\|v\|)$ each, between $\mathcal{C}_1$ and $\mathcal{C}_2$.*
2. *If $\mathfrak{S}$ provides replica signing and $\tau_2 \leq \mathbf{n}_{\mathcal{C}_2}$, then $\texttt{RPBS-(}\tau_2\texttt{,rs)}$ satisfies Definition 1 and sends τ_2 messages, of size $\mathcal{O}(\|v\|)$ each, between $\mathcal{C}_1$ and $\mathcal{C}_2$.*

Proof. Let $\beta = \mathbf{nf}_{\mathcal{C}_1}$ and $\gamma = 1$ in the case of cluster signing and let $\beta = (\mathbf{nf}_{\mathcal{C}_1} - \mathbf{f}_{\mathcal{C}_1})$ and $\gamma = 2$ in the case of replica signing. Let $q = (\mathbf{f}_{\mathcal{C}_2} + 1) \operatorname{div} \beta$ and $r = (\mathbf{f}_{\mathcal{C}_2} + 1) \bmod \beta$. We have $\alpha = q\mathbf{n}_{\mathcal{C}_1} + r + \gamma\mathbf{f}_{\mathcal{C}_1} \operatorname{sgn} r$. Choose $\mathcal{P}$ and choose $\mathsf{partition}(\mathcal{P}) = \{P_1, \dots, P_q, P'\}$ in accordance with $\texttt{RPBS-(}\alpha\texttt{,}\zeta\texttt{)}$ (Fig. 11). For each $P \in \mathcal{P}$, choose a Q and b in accordance with $\texttt{RPBS-(}\alpha\texttt{,}\zeta\texttt{)}$, and let $z(P) = \{\mathrm{R} \in P \mid b^{-1}(\mathrm{R}) \in \mathsf{f}(Q)\}$. As each such b^{-1} has a distinct domain, the union of them is a surjection $f^{-1} : \mathcal{P} \to \mathcal{C}_1$. By construction, we have $\mathbf{n}_{P'} = r + \gamma\mathbf{f}_{\mathcal{C}_1} \operatorname{sgn} r$, $\mathbf{n}_{z(P')} \leq \mathbf{f}_{\mathcal{C}_1} \operatorname{sgn} r$, and, for every i, $1 \leq i \leq q$, $\mathbf{n}_{P_i} = \mathbf{n}_{\mathcal{C}_1}$ and $\mathbf{n}_{z(P_i)} = \mathbf{f}_{\mathcal{C}_1}$. Let $T = \mathcal{P} \setminus \left(\bigcup_{P \in \mathsf{partition}(\mathcal{P})} z(P)\right)$. We have

$$\begin{aligned}\mathbf{n}_T \geq \mathbf{n}_{\mathcal{P}} - (q\mathbf{f}_{\mathcal{C}_1} + \mathbf{f}_{\mathcal{C}_1} \operatorname{sgn} r) &= (q\mathbf{n}_{\mathcal{C}_1} + r + \gamma\mathbf{f}_{\mathcal{C}_1} \operatorname{sgn} r) - (q\mathbf{f}_{\mathcal{C}_1} + \mathbf{f}_{\mathcal{C}_1} \operatorname{sgn} r)\\ &= q\mathbf{nf}_{\mathcal{C}_1} + r + (\gamma - 1)\mathbf{f}_{\mathcal{C}_1} \operatorname{sgn} r.\end{aligned}$$

To complete the proof, we consider cluster signing and replica signing separately.

First, the case for cluster signing. We have $\beta = \mathbf{nf}_{\mathcal{C}_1}$ and $\gamma = 1$. Hence, $\mathbf{n}_T \geq q\mathbf{nf}_{\mathcal{C}_1} + r + (\gamma - 1)\mathbf{f}_{\mathcal{C}_1} \operatorname{sgn} r = q\beta + r = \mathbf{f}_{\mathcal{C}_2} + 1$. We have $\mathbf{nf}_T \geq \mathbf{n}_T - \mathbf{f}_{\mathcal{C}_2} \geq 1$. Let $V = \{f^{-1}(\mathrm{R}) \mid \mathrm{R} \in \mathsf{nf}(T)\}$. By construction, we have $\mathbf{nf}_V = \mathbf{n}_V$ and we have $\mathbf{nf}_V \geq 1$. Consequently, the replicas in $\mathsf{nf}(T)$ will receive the messages $(v, \langle v \rangle_{\mathcal{C}_1})$ from the replicas $\mathrm{R}_1 \in \mathsf{nf}(V)$. Analogous to the proof of Proposition 1, we can prove *receipt*, *confirmation*, and *agreement*.

Finally, the case for replica signing. We have $\beta = \mathbf{nf}_{\mathcal{C}_1} - \mathbf{f}_{\mathcal{C}_1}$ and $\gamma = 2$. Hence, $\mathbf{n}_T \geq q\mathbf{nf}_{\mathcal{C}_1} + r + (\gamma - 1)\mathbf{f}_{\mathcal{C}_1} \operatorname{sgn} r = q(\beta + \mathbf{f}_{\mathcal{C}_1}) + r + \mathbf{f}_{\mathcal{C}_1} \operatorname{sgn} r = (q\beta + r) + q\mathbf{f}_{\mathcal{C}_1} + \mathbf{f}_{\mathcal{C}_1} \operatorname{sgn} r = (\mathbf{f}_{\mathcal{C}_2} + 1) + q\mathbf{f}_{\mathcal{C}_1} + \mathbf{f}_{\mathcal{C}_1} \operatorname{sgn} r$. We have $\mathbf{nf}_T \geq q\mathbf{f}_{\mathcal{C}_1} + \mathbf{f}_{\mathcal{C}_1} \operatorname{sgn} r + 1 = (q + \operatorname{sgn} r)\mathbf{f}_{\mathcal{C}_1} + 1$. As there are $(q + \operatorname{sgn} r)$ non-empty sets in $\mathsf{partition}(\mathcal{P})$, there must be a set $P \in \mathcal{P}$ with $\mathbf{n}_{P \cap \mathbf{nf}_T} \geq \mathbf{f}_{\mathcal{C}_1} + 1$. Let b be the bijection chosen earlier for P and let $V = \{b^{-1}(\mathrm{R}) \mid \mathrm{R} \in (P \cap \mathbf{nf}_T)\}$. By construction, we have $\mathbf{nf}_V = \mathbf{n}_V$ and we have $\mathbf{nf}_V \geq \mathbf{f}_{\mathcal{C}_1} + 1$. Consequently, the replicas in $\mathsf{nf}(T)$ will receive the messages $(v, \langle v \rangle_{\mathrm{R}_1})$ from each replica $\mathrm{R}_1 \in \mathsf{nf}(V)$. Hence, analogous to the proof of Proposition 2, we can prove *receipt*, *confirmation*, and *agreement*. □

The bijective sending cluster-sending protocols, the sender-partitioned bijective cluster-sending protocols, and the receiver-partitioned bijective cluster-sending protocols each deal with differently-sized clusters. By choosing the applicable protocols, we have the following:

Corollary 1. *Let $\mathfrak{S}$ be a system, let $\mathcal{C}_1, \mathcal{C}_2 \in \mathfrak{S}$, let σ_1 and σ_2 be as defined in Theorem 1, and let τ_1 and τ_2 be as defined in Theorem 2. Consider the cluster-sending problem in which $\mathcal{C}_1$ sends a value v to $\mathcal{C}_2$.*

1. *If $\mathbf{n}_{\mathcal{C}} > 3\mathbf{f}_{\mathcal{C}}$, $\mathcal{C} \in \mathfrak{S}$, and $\mathfrak{S}$ has crash failures, omit failures, or Byzantine failures and cluster signing, then* `BS-cs`, `SPBS-(`σ_1`,cs)`, *and* `RPBS-(`σ_2`,cs)` *are a solution to the cluster-sending problem with optimal message complexity. These protocols solve the cluster-sending problem using $\mathcal{O}(\max(\mathbf{n}_{\mathcal{C}_1}, \mathbf{n}_{\mathcal{C}_2}))$ messages, of size $\mathcal{O}(\|v\|)$ each.*
2. *If $\mathbf{n}_{\mathcal{C}} > 4\mathbf{f}_{\mathcal{C}}$, $\mathcal{C} \in \mathfrak{S}$, and $\mathfrak{S}$ has Byzantine failures and replica signing, then* `BS-rs`, `SPBS-(`τ_1`,rs)`, *and* `RPBS-(`τ_2`,rs)` *are a solution to the cluster-sending problem with optimal replica certificate usage. These protocols solve the cluster-sending problem using $\mathcal{O}(\max(\mathbf{n}_{\mathcal{C}_1}, \mathbf{n}_{\mathcal{C}_2}))$ messages, of size $\mathcal{O}(\|v\|)$ each.*

6 Performance Evaluation

In the previous sections, we introduced *worst-case optimal* cluster-sending protocols. To gain further insight in the performance attainable by these protocols, we implemented these protocols in a simulated sharded resilient system environment that allows us to control the faulty replicas in each cluster. In the experiments, we used equal-sized clusters, which corresponds to the setup used by recent sharded consensus-based system proposals [1,3,10]. Hence, we used only the bijective cluster-sending protocols `BS-cs` and `BS-rs`. As a baseline of comparison, we also evaluated the *broadcast-based cluster-sending protocol* of `Chainspace` [1] that can perform cluster-sending using $\mathbf{n}_{\mathcal{C}_1} \cdot \mathbf{n}_{\mathcal{C}_2}$ messages. We refer to Fig. 2 for a theoretical comparison between our cluster-sending protocols and the protocol utilized by `Chainspace`. Furthermore, we have implemented `MC-cs` and `MC-rs`, two multicast-based cluster-sending protocols, one using cluster signing and the other using replica signing), that work similar to the protocol of `Chainspace`, but minimize the number of messages to provide cluster-sending.

In the experiment, we measured the number of messages exchanged as a function of the number of faulty replicas. In specific, we measured the number of messages exchanged in 10 000 runs of the cluster-sending protocols under consideration. In each run we measure the number of messages exchanged when sending a value v from a cluster $\mathcal{C}_1$ to a cluster $\mathcal{C}_2$ with $\mathbf{n}_{\mathcal{C}_1} = \mathbf{n}_{\mathcal{C}_2} = 3\mathbf{f}_{\mathcal{C}_1} + 1 = 3\mathbf{f}_{\mathcal{C}_2} + 1$, and we aggregate this data over 10 000 runs. Furthermore, we measured in each run the number of messages exchanged between non-faulty replicas, as these messages are necessary to guarantee cluster-sending. The results of the experiment can be found in Fig. 12.

As is clear from the results, our worst-case optimal cluster-sending protocols are able to out-perform existing cluster-sending protocols by a wide margin, which is a direct consequence of the difference between quadratic message complexity (`Chainspace`, `MC-cs`, and `MC-rs`) and a worst-case optimal linear message complexity (`BS-cs` and `BS-rs`). As can be seen in Fig. 12, *right*, our protocols do so by massively cutting back on sending messages between faulty replicas, while still ensuring that in all cases sufficient messages are exchanged between non-faulty replicas (thereby assuring cluster-sending).

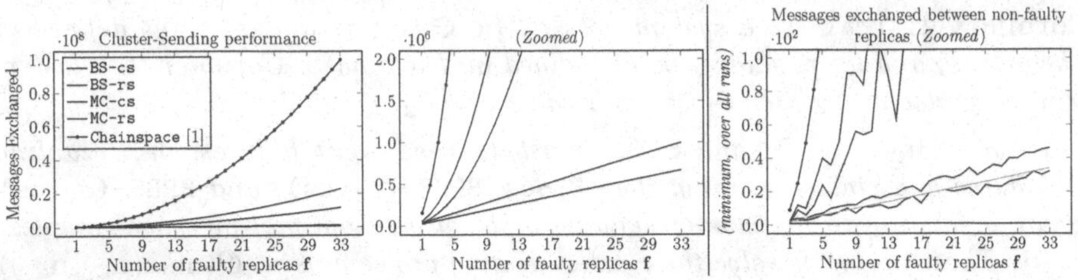

Fig. 12. A comparison of the number of message exchange steps as a function of the number of faulty replicas in both clusters by our *worst-case optimal* cluster-sending protocols BS-cs and BS-rs, and by three protocols based on the literature. For each protocol, we measured the number of message exchange steps to send 10 000 values between two equally-sized clusters, each cluster having $n = 3\mathbf{f} + 1$ replicas. The dashed lines in the plot on the *right* indicate the minimum number of messages that need to be exchanged between non-faulty replicas for the protocols BS-cs and BS-rs, respectively, to guarantee cluster-sending (no protocol can do better). (Color figure online)

7 Conclusion

In this paper, we identified and formalized the *cluster-sending problem*, a fundamental primitive in the design and implementation of blockchain-inspired sharded fault-tolerant data processing systems. Not only did we formalize the cluster-sending problem, we also proved lower bounds on the complexity of this problem. Furthermore, we developed bijective sending and partitioned bijective sending, two powerful techniques that can be used in the construction of practical cluster-sending protocols with optimal complexity that matches the lower bounds established. We believe that our work provides a strong foundation for future blockchain-inspired sharded fault-tolerant data processing systems that can deal with Byzantine failures and the challenges of large-scale data processing.

Our fundamental results open a number of key research avenues to further high-performance fault-tolerant data processing. First, we are interested in further improving our understanding of cluster-sending, e.g., by establishing lower bounds on cluster-sending in the absence of public-key cryptography and in the absence of reliable networks. Second, we are interested in improved cluster-sending protocols that can perform cluster-sending with less-than a linear number of messages, e.g., by using randomization or by optimizing for cases without failures. Finally, we are interested in putting cluster-sending protocols to practice by incorporating them in the design of practical sharded fault-tolerant systems, thereby moving even closer to general-purpose high-performance fault-tolerant data processing.

References

1. Al-Bassam, M., Sonnino, A., Bano, S., Hrycyszyn, D., Danezis, G.: Chainspace: a sharded smart contracts platform (2017). http://arxiv.org/abs/1708.03778
2. Castro, M., Liskov, B.: Practical byzantine fault tolerance and proactive recovery. ACM Trans. Comput. Syst. **20**(4), 398–461 (2002). https://doi.org/10.1145/571637.571640
3. Dang, H., Dinh, T.T.A., Loghin, D., Chang, E.C., Lin, Q., Ooi, B.C.: Towards scaling blockchain systems via sharding. In: Proceedings of the 2019 International Conference on Management of Data. pp. 123–140. ACM (2019). https://doi.org/10.1145/3299869.3319889
4. Dolev, D., Strong, H.R.: Authenticated algorithms for byzantine agreement. SIAM J. Comput. **12**(4), 656–666 (1983). https://doi.org/10.1137/0212045
5. Gordon, W.J., Catalini, C.: Blockchain technology for healthcare: Facilitating the transition to patient-driven interoperability. Comput. Struct. Biotechnol. J. **16**, 224–230 (2018). https://doi.org/10.1016/j.csbj.2018.06.003
6. Gupta, S., Hellings, J., Sadoghi, M.: Fault-Tolerant distributed transactions on Blockchain. Synthesis Lectures on Data Management, Morgan & Claypool (2021). https://doi.org/10.1007/978-3-031-01877-0
7. Gupta, S., Hellings, J., Sadoghi, M.: RCC: resilient concurrent consensus for high-throughput secure transaction processing. In: 2021 IEEE 37th International Conference on Data Engineering (ICDE), pp. 1392–1403. IEEE (2021). https://doi.org/10.1109/ICDE51399.2021.00124
8. Gupta, S., Rahnama, S., Hellings, J., Sadoghi, M.: ResilientDB: global scale resilient blockchain fabric. Proc. VLDB Endow. **13**(6), 868–883 (2020). https://doi.org/10.14778/3380750.3380757
9. Gupta, S., Rahnama, S., Hellings, J., Sadoghi, M.: Proof-of-execution: reaching consensus through fault-tolerant speculation. In: Proceedings of the 24th International Conference on Extending Database Technology (EDBT), pp. 301–312. OpenProceedings.org (2021). https://doi.org/10.5441/002/edbt.2021.27
10. Hellings, J., Sadoghi, M.: ByShard: sharding in a byzantine environment. Proc. VLDB Endow. **14**(11), 2230–2243 (2021). https://doi.org/10.14778/3476249.3476275
11. Herlihy, M.: Atomic cross-chain swaps. In: Proceedings of the 2018 ACM Symposium on Principles of Distributed Computing. pp. 245–254. ACM (2018). https://doi.org/10.1145/3212734.3212736
12. Herlihy, M., Liskov, B., Shrira, L.: Cross-chain deals and adversarial commerce. Proc. VLDB Endow. **13**(2), 100–113 (2019). https://doi.org/10.14778/3364324.3364326
13. Kamel Boulos, M.N., Wilson, J.T., Clauson, K.A.: Geospatial blockchain: promises, challenges, and scenarios in health and healthcare. Int. J. Health Geogr. **17**(1), 1211–1220 (2018). https://doi.org/10.1186/s12942-018-0144-x
14. Lao, L., Li, Z., Hou, S., Xiao, B., Guo, S., Yang, Y.: A survey of IoT applications in blockchain systems: architecture, consensus, and traffic modeling. ACM Comput. Surv. **53**(1) (2020). https://doi.org/10.1145/3372136
15. Menezes, A.J., Vanstone, S.A., Oorschot, P.C.V.: Handbook of Applied Cryptography, 1st edn. CRC Press Inc., Boca Raton (1996)
16. Nakamoto, S.: Bitcoin: a peer-to-peer electronic cash system. https://bitcoin.org/en/bitcoin-paper

17. Özsu, M.T., Valduriez, P.: Principles of Distributed Database Systems. (2020). https://doi.org/10.1007/978-3-030-26253-2
18. Rejeb, A., Keogh, J.G., Zailani, S., Treiblmaier, H., Rejeb, K.: Blockchain technology in the food industry: a review of potentials, challenges and future research directions. Logistics **4**(4) (2020). https://doi.org/10.3390/logistics4040027
19. Shoup, V.: Practical threshold signatures. In: Preneel, B. (ed.) EUROCRYPT 2000. LNCS, vol. 1807, pp. 207–220. Springer, Heidelberg (2000). https://doi.org/10.1007/3-540-45539-6_15
20. Treiblmaier, H., Beck, R. (eds.): Business Transformation Through Blockchain. Springer, Cham (2019). https://doi.org/10.1007/978-3-319-98911-2
21. Wood, G.: Ethereum: a secure decentralised generalised transaction ledger. https://gavwood.com/paper.pdf, EIP-150 revision
22. Wu, M., Wang, K., Cai, X., Guo, S., Guo, M., Rong, C.: A comprehensive survey of blockchain: from theory to IoT applications and beyond. IEEE Internet Things J. **6**(5), 8114–8154 (2019). https://doi.org/10.1109/JIOT.2019.2922538
23. Yin, M., Malkhi, D., Reiter, M.K., Gueta, G.G., Abraham, I.: HotStuff: BFT consensus with linearity and responsiveness. In: Proceedings of the 2019 ACM Symposium on Principles of Distributed Computing, pp. 347–356. ACM (2019). https://doi.org/10.1145/3293611.3331591
24. Zakhary, V., Agrawal, D., El Abbadi, A.: Atomic commitment across blockchains. Proc. VLDB Endow. **13**(9), 1319–1331 (2020). https://doi.org/10.14778/3397230.3397231

Optimizing Multiset Relational Algebra Queries Using Weak-Equivalent Rewrite Rules

Jelle Hellings[1(✉)], Yuqing Wu[2], Dirk Van Gucht[3], and Marc Gyssens[4]

[1] McMaster University, 1280 Main St., W., Hamilton, ON L8S 4L7, Canada
jhellings@mcmaster.ca
[2] Pomona College, 185 E 6th St., Claremont, CA 91711, USA
[3] Indiana University, 919 E 10th St., Bloomington, IN 47408, USA
[4] Hasselt University, Martelarenlaan 42, 3500 Hasselt, Belgium

Abstract. Relational query languages rely heavily on costly join operations to combine tuples from multiple tables into a single resulting tuple. In many cases, the cost of query evaluation can be reduced by *manually optimizing* (parts of) queries to use cheaper semi-joins instead of joins. Unfortunately, existing database products can only apply such optimizations *automatically* in rather limited cases.

To improve on this situation, we propose a framework for automatic query optimization via *weak-equivalent rewrite rules* for a multiset relational algebra (that serves as a faithful formalization of core SQL). The weak-equivalent rewrite rules we propose aim at replacing joins by semi-joins. To further maximize their usability, these rewrite rules do so by only providing "weak guarantees" on the evaluation results of rewritten queries. We show that, in the context of certain operators, these weak-equivalent rewrite rules still provide strong guarantees on the final evaluation results of the rewritten queries.

Keywords: Query Optimization · Relational Algebra · Multiset Semantics · Semi-Joins

1 Introduction

To combine tables, SQL relies on join operations that are costly to evaluate. To reduce the high costs of joins, a significant portion of query optimization and query planning is aimed at evaluating joins as efficient as possible. Still, it is well-known that some complex join-based SQL queries can be further optimized manually by rewriting these queries into queries that involve semi-joins implicitly. As an illustration, we consider the following example involving a graph represented as a binary relation. The SQL query

```
SELECT DISTINCT S.nfrom
FROM edges S, edges R, edges T, edges U
WHERE S.nto = R.nfrom AND R.nto = T.nfrom AND T.nto = U.nfrom;
```


I. Varzinczak (Ed.): FoIKS 2022, LNCS 13388, pp. 187–205, 2022.
https://doi.org/10.1007/978-3-031-11321-5_11

computes the sources of paths of length four. A straightforward way to evaluate this query is to compute the joins, then project the join result onto *S.nfrom*, and, finally, remove duplicates. The cost of this straightforward evaluation is very high: in the worst-case, the join result is quartic in size with respect to the size of the number of edges (the size of the *Edges* relation). The typical way to manually optimize this query is to rewrite it as follows:

```
SELECT DISTINCT nfrom FROM edges
WHERE nto IN (SELECT nfrom FROM edges
              WHERE nto IN (SELECT nfrom FROM edges
                            WHERE nto IN (SELECT nfrom
                                          FROM edges)));
```

In most relational database systems, the `WHERE ... IN ...` clauses in the rewritten query are evaluated using a semi-join algorithm. In doing so, this query can be evaluated in linear time with respect to the number of edges, which is a significant improvement. E.g., when evaluated on a randomly generated database with 75 000 rows (each row a single edge) managed by PostgreSQL 14.1, the original query could not finish in a reasonable amount of time, whereas the manually rewritten semi-join style query finished in 90 ms.

We believe that query optimizers should not require users to manually rewrite queries to enforce particular evaluation strategies: manual rewriting goes against the advantages of using high-level declarative languages such as SQL. Instead, we want query optimizers to be able to recognize situations in which semi-join rewriting is appropriate, and apply such optimizing rewritings automatically.

Traditional approaches towards query optimization for SQL and the relational algebra usually employ two basic steps [12]. First, the query involved is rewritten. The rewrite rules used in these rewrites guarantee *strong-equivalence*: the original subquery and the rewritten subquery always evaluate to the same result. Examples of such rules are the well-known push-down rules for selection and projection, which can reduce the size of intermediate query results significantly. Second, the order of execution of the operations, appropriate algorithms to perform each operation, and data access methods are chosen to evaluate the rewritten query.

Unfortunately, requiring strong-equivalence imposes a severe restriction on the rewrite rules that can be considered. As a consequence, there are significant limitations to query optimization using traditional query rewriting: often, these rewrite rules only manipulate the order of operations. More lucrative rewritings, such as replacing expensive join operations by semi-joins, are not considered because such rewrites cannot guarantee strong-equivalence.

To improve on this situation, we propose the concept of *weak-equivalence*, which is a relaxation of strong-equivalence. Weak-equivalent rewrite rules only guarantee that the original subquery and the rewritten subquery evaluate to the same result up to duplicate elimination (with respect to the attributes of interest). The rewrite rules we propose are aimed at eliminating joins in favor of semi-joins and eliminating the need for deduplication altogether. To illustrate the benefits of weak-equivalent rewrites, we present two examples.

As a first example, consider a university database containing the relation *Course*, with attributes *id* and *name*, and the relation *Enroll*, with, among its attributes, *cid* (the course id). Other attributes of *Enroll* refer to students. Now consider the task of rewriting the relational algebra query

$$\pi_{C.name}(q) \text{ with } q = \rho_C(Course) \bowtie_{C.id=E.cid} \rho_E(Enroll)$$

that computes the set of names of courses in which students are effectively enrolled. As the end result of this query is a projection on *only* the *name* attribute of *Course* (which we assume to be a key), any rewrite of the subquery q can forgo any of the other attributes. E.g., although subquery q yields a completely different result than subquery $q' = \rho_C(Course) \ltimes_{C.id=E.cid} \rho_E(Enroll)$, their projection onto $C.name$ is identical.

As a second example, consider a sales database containing the relation *Customer*, with among its attributes *cname*, the relation *Product*, with among its attributes *pname* and *type*, and the relation *Bought*, with among its attributes *cname* and *pname*. We refer to Fig. 1 for an instance of this sales database. Now consider the following query:

```
SELECT DISTINCT C.cname, P.type
FROM customer C, bought B, product P
WHERE C.cname = B.cname AND B.pname = P.pname AND
      P.type = 'food';
```

which our rules can rewrite into the following query:

```
SELECT cname, 'food' AS type FROM customer WHERE cname IN (
    SELECT cname FROM bought WHERE pname IN (
        SELECT pname FROM product WHERE type = 'food'));
```

Observe that the rewritten query does not perform any joins, even though information from several relations is combined and returned. Moreover, the need for deduplication is eliminated, as the available key on *cname* guarantees that no duplicates are possible in the resultant query. In this particular example, the original query joins table *Bought* with two tables on their primary keys. Hence, all intermediate results remain small if query evaluation chooses a proper join order. Still, even in this case, the rewritten query evaluates 15%–20% faster on a randomly generated database with 500 customers, 24 077 products, and 100 000 sale records in *Bought*.

The latter example also illustrates that the applicability of weak-equivalent rewrite rules may depend on structural properties, such as keys, of the input relations and of the intermediate query results. Therefore, we also propose techniques to derive these properties from the available schema information.

The automatic use of semi-join algorithms in query evaluation has already been studied and applied before. In these approaches, however, semi-joins are typically only employed as a preprocessing step for joins. E.g., in distributed databases, semi-joins are used as an intermediate step to reduce the size of

Customer		*Product*		*Bought*		
cname	*age*	*pname*	*type*	*cname*	*pname*	*price*
Alice	19	apple	food	Alice	apple	0.35
Bob	20	apple	fruit	Alice	apple	0.35
Eve	21	banana	fruit	Bob	apple	0.45
		car	non-food	Bob	banana	0.50
				Eve	car	10000

Fig. 1. A database instance for a sales database that has customers, a categorization of products according to their types, and transaction information for each sale.

relations before joining them and, as a consequence, reducing the communication overhead of distributing (intermediate) relations to other computational nodes [3]. The semi-join has a similar role in the well-known acyclic multi-join algorithm of Yannakakis [13]. We take a different approach: using semi-joins, we aim at eliminating join operations altogether instead of merely reducing their cost.

The usage of weak-equivalent rewrite rules for query optimization is inspired by the projection-equivalent rewriting techniques for graph query languages on binary relations proposed by Hellings et al. [6,7]. In the current paper, we not only adapt these projection-equivalent rewriting techniques to the setting of relational algebra, but we also extend them to effectively deal with multiset semantics [2,4,5,9–11]. The latter is essential, because we want to use our weak-equivalent rewriting rules to optimize SQL queries, and SQL has multiset semantics. Furthermore, we integrate rewrite techniques based on derived structural knowledge on the schema and knowledge derived from selection conditions, which are both not applicable in the setting of binary relations.

Finally, we note that there have been previous SQL-inspired studies for rewriting and optimizing relational algebra with multiset semantics, e.g., [5,11]. These studies do not cover the main optimizations explored in this work, however.

2 Preliminaries

We consider disjoint infinitely enumerable sets $\mathfrak{U}$ and $\mathfrak{N}$ of *constants* and *names*, respectively. A *relation schema* is a finite set of names, which are used as attributes. Let $\mathrm{A} \subseteq \mathfrak{N}$ be a relation schema. A *tuple over* A is a function $\mathsf{t} : \mathrm{A} \to \mathfrak{U}$, a *relation over* $\mathrm{A} \subseteq \mathfrak{N}$ is a set of tuples over A, and a *multiset relation over* A is a pair $\langle \mathscr{R}; \tau \rangle$, in which $\mathscr{R}$ is a relation over A and $\tau : \mathscr{R} \to \mathbb{N}^+$ is a function mapping tuples $\mathsf{t} \in \mathscr{R}$ to the number of copies of t in the multiset relation. We say that $\langle \mathscr{R}; \tau \rangle$ is a *set relation* if, for every $\mathsf{t} \in \mathscr{R}$, we have $\tau(\mathsf{t}) = 1$. We write $(\mathsf{t} : n) \in \langle \mathscr{R}; \tau \rangle$ for *tuple-count pair* $(\mathsf{t} : n)$ to indicate $\mathsf{t} \in \mathscr{R}$ with $\tau(\mathsf{t}) = n$. We write $\mathsf{t} \in \langle \mathscr{R}; \tau \rangle$ to indicate $\mathsf{t} \in \mathscr{R}$ and we write $\mathsf{t} \notin \langle \mathscr{R}; \tau \rangle$ to indicate $\mathsf{t} \notin \mathscr{R}$.

Let t, t_1, and t_2 be tuples over the relation schema A and let $\mathrm{B} \subseteq \mathrm{A}$. The *restriction of* t *to* B, denoted by $\mathsf{t}|_{\mathrm{B}}$, is defined by $\mathsf{t}|_{\mathrm{B}} = \{a \mapsto \mathsf{t}(a) \mid a \in \mathrm{B}\}$. Tuples t_1 and t_2 *agree on* B, denoted by $\mathsf{t}_1 \equiv_{\mathrm{B}} \mathsf{t}_2$, if $\mathsf{t}_1|_{\mathrm{B}} = \mathsf{t}_2|_{\mathrm{B}}$. Let $\langle \mathscr{R}; \tau \rangle$ be a multiset relation over relation schema A and let $\mathrm{K} \subseteq \mathrm{A}$. We say that K is a *key* of $\langle \mathscr{R}; \tau \rangle$ if, for every pair of tuples $\mathrm{t}_1, \mathrm{t}_2 \in \langle \mathscr{R}; \tau \rangle$ with $\mathrm{t}_1 \equiv_{\mathrm{K}} \mathrm{t}_2$, we have $\mathrm{t}_1 = \mathrm{t}_2$.

A *database schema* is a 4-tuple $\mathfrak{D} = (\mathbf{N}, \mathbf{A}, \mathbf{K}, \mathbf{S})$, in which $\mathbf{N} \subseteq \mathfrak{N}$ is a set of *relation names*, $\mathbf{A}$ is a function mapping each relation name to a relation schema, $\mathbf{K}$ is a function mapping each relation name to a set of sets of attributes of the corresponding relation schema, and $\mathbf{S}$ is a function mapping relation names to booleans. A *database instance* over schema $\mathfrak{D} = (\mathbf{N}, \mathbf{A}, \mathbf{K}, \mathbf{S})$ is a function $\mathfrak{I}$ mapping each name $R \in \mathbf{N}$ into a multiset relation. This multiset relation $\mathfrak{I}(R)$ has schema $\mathbf{A}(R)$, keys $\mathbf{K}(R)$, and must be a set relation if $\mathbf{S}(R) = \mathtt{true}$.

Let t_1 and t_2 be tuples over relation schemas A_1 and A_2, respectively, such that $\mathrm{t}_1 \equiv_{\mathrm{A}_1 \cap \mathrm{A}_2} \mathrm{t}_2$. The *concatenation of* t_1 *and* t_2, denoted by $\mathrm{t}_1 \cdot \mathrm{t}_2$, is defined by $\mathrm{t}_1 \cdot \mathrm{t}_2 = \mathrm{t}_1 \cup \mathrm{t}_2$.[1] Notice that $\mathrm{t}_1 \cdot \mathrm{t}_2$ is a tuple over $\mathrm{A}_1 \cup \mathrm{A}_2$. Let A be a relation schema. A *condition on* A is either an expression of the form $a_1 = a_2$ or of the form $a = \mathrm{u}$, $a, a_1, a_2 \in \mathrm{A}$ and $\mathrm{u} \in \mathfrak{U}$. A tuple t over A satisfies $a_1 = a_2$ if $\mathrm{t}(a_1) = \mathrm{t}(a_2)$ and satisfies $a = \mathrm{u}$ if $\mathrm{t}(a) = \mathrm{u}$. If E is a set of conditions over A, then tuple t *satisfies* E if t satisfies each condition in E. By $\mathrm{attrs}(E)$, we denote the set of attributes used in conditions in E.

Example 1. The database schema for the sales database of Fig. 1 consist of the named multiset relations *Customer* with schema $\{cname, age\}$, *Product* with schema $\{pname, type\}$, and *Bought* with schema $\{cname, pname, price\}$. *Customer* and *Product* are set relations with key $\{cname\}$ and trivial key $\{pname, type\}$, respectively. *Bought* is not a set relation and only has the trivial key $\{cname, pname, price\}$.

3 Multiset Relational Algebra

In this work, we study the multiset relational algebra. To distinguish a traditional set-based relational algebra operator $\oplus$ from its multiset relational algebra counterpart introduced below, we annotate the latter as $\dot{\oplus}$.

Let $\mathfrak{D} = (\mathbf{N}, \mathbf{A}, \mathbf{K}, \mathbf{S})$ be a database schema and let $\mathfrak{I}$ be a database instance over $\mathfrak{D}$. If e is a multiset relational algebra expression over $\mathfrak{D}$, which we will formally define next, then we write $\mathcal{S}(e; \mathfrak{D})$ to denote the schema of the multiset relation obtained by evaluating e on an instance over $\mathfrak{D}$, and we write $[\![e]\!]_{\mathfrak{I}}$ to denote the evaluation of e on instance $\mathfrak{I}$. The *standard* relational algebra expressions over $\mathfrak{D}$ are built from the following operators:

- *Multiset relation.* If $R \in \mathbf{N}$ is a relation name, then R is an expression with $\mathcal{S}(R; \mathfrak{D}) = \mathbf{A}(R)$ and $[\![R]\!]_{\mathfrak{I}} = \mathfrak{I}(R)$.
- *Selection.*[2] If e is an expression and E is a set of conditions on $\mathcal{S}(e; \mathfrak{D})$, then $\dot{\sigma}_E(e)$ is an expression with $\mathcal{S}(\dot{\sigma}_E(e); \mathfrak{D}) = \mathcal{S}(e; \mathfrak{D})$ and $[\![\dot{\sigma}_E(e)]\!]_{\mathfrak{I}} = \{(\mathrm{t} : n) \in [\![e]\!]_{\mathfrak{I}} \mid (\mathrm{t} \text{ satisfies } E)\}$.

[1] Every occurrence of the operators $\cup$, $\cap$, and $-$ in this paper is to be interpreted using standard set semantics.

[2] We only consider conjunctions of conditions in the selection operator because more general boolean combinations of conditions do not provide additional opportunities for rewriting in our framework.

- *Projection.* If e is an expression and $\mathrm{B} \subseteq \mathcal{S}(e;\mathfrak{D})$, then $\dot{\pi}_{\mathrm{B}}(e)$ is an expression with $\mathcal{S}(\dot{\pi}_{\mathrm{B}}(e);\mathfrak{D}) = \mathrm{B}$ and $[\![\dot{\pi}_{\mathrm{B}}(e)]\!]_{\mathfrak{I}} = \{(\mathrm{t}|_{\mathrm{B}} : \mathrm{count}(\mathrm{t}|_{\mathrm{B}}, e)) \mid \mathrm{t} \in [\![e]\!]_{\mathfrak{I}}\}$, in which $\mathrm{count}(\mathrm{t}|_{\mathrm{B}}, e) = \sum_{((\mathrm{s}:m)\in[\![e]\!]_{\mathfrak{I}})\wedge(\mathrm{s}\equiv_{\mathrm{B}}\mathrm{t})} m$.
- *Renaming.* If e is an expression, $\mathrm{B} \subseteq \mathfrak{N}$, and $f : \mathcal{S}(e;\mathfrak{D}) \to \mathrm{B}$ is a bijection, then $\dot{\rho}_f(e)$ is an expression with $\mathcal{S}(\dot{\rho}_f(e);\mathfrak{D}) = \mathrm{B}$ and $[\![\dot{\rho}_f(e)]\!]_{\mathfrak{I}} = \{(\mathrm{rename}(\mathrm{t}, f) : m) \mid (\mathrm{t} : m) \in [\![e]\!]_{\mathfrak{I}}\}$, in which $\mathrm{rename}(\mathrm{t}, f) = \{f(a) \mapsto \mathrm{t}(a) \mid a \in \mathcal{S}(e;\mathfrak{D})\}$.
- *Deduplication.* If e is an expression, then $\dot{\delta}(e)$ is an expression with $\mathcal{S}(\dot{\delta}(e);\mathfrak{D}) = \mathbf{A}(e)$ and $[\![\dot{\delta}(e)]\!]_{\mathfrak{I}} = \{(\mathrm{t} : 1) \mid (\mathrm{t} : n) \in [\![e]\!]_{\mathfrak{I}}\}$.
- *Union, intersection, and difference.*[3] If e_1 and e_2 are expressions such that $\mathrm{A} = \mathcal{S}(e_1;\mathfrak{D}) = \mathcal{S}(e_2;\mathfrak{D})$, then, for $\oplus \in \{\cup, \cap, -\}$, $e_1 \dot{\oplus} e_2$ is an expression with $\mathcal{S}(e_1 \dot{\oplus} e_2;\mathfrak{D}) = \mathrm{A}$ and

$$\begin{aligned}
[\![e_1 \dot{\cup} e_2]\!]_{\mathfrak{I}} &= \{(\mathrm{t} : n_1 + n_2) \mid (\mathrm{t} : n_1) \in [\![e_1]\!]_{\mathfrak{I}} \wedge (\mathrm{t} : n_2) \in [\![e_2]\!]_{\mathfrak{I}}\} \cup \\
&\quad \{(\mathrm{t} : n) \in [\![e_1]\!]_{\mathfrak{I}} \mid \mathrm{t} \notin [\![e_2]\!]_{\mathfrak{I}}\} \cup \{(\mathrm{t} : n) \in [\![e_2]\!]_{\mathfrak{I}} \mid \mathrm{t} \notin [\![e_1]\!]_{\mathfrak{I}}\}; \\
[\![e_1 \dot{\cap} e_2]\!]_{\mathfrak{I}} &= \{(\mathrm{t} : \min(n_1, n_2)) \mid (\mathrm{t} : n_1) \in [\![e_1]\!]_{\mathfrak{I}} \wedge (\mathrm{t} : n_2) \in [\![e_2]\!]_{\mathfrak{I}}\}; \\
[\![e_1 \dot{-} e_2]\!]_{\mathfrak{I}} &= \{(\mathrm{t} : n) \in [\![e_1]\!]_{\mathfrak{I}} \mid \mathrm{t} \notin [\![e_2]\!]_{\mathfrak{I}}\} \cup \\
&\quad \{(\mathrm{t} : n_1 - n_2) \mid (n_1 > n_2) \wedge (\mathrm{t} : n_1) \in [\![e_1]\!]_{\mathfrak{I}} \wedge (\mathrm{t} : n_2) \in [\![e_2]\!]_{\mathfrak{I}}\}.
\end{aligned}$$

- *θ-join and natural join.*[4] If e_1 and e_2 are expressions and E is a set of conditions on $\mathrm{A} = \mathcal{S}(e_1;\mathfrak{D}) \cup \mathcal{S}(e_2;\mathfrak{D})$, then $e_1 \dot{\bowtie}_E e_2$ is an expression with $\mathcal{S}(e_1 \dot{\bowtie}_E e_2;\mathfrak{D}) = \mathrm{A}$ and $[\![e_1 \dot{\bowtie}_E e_2]\!]_{\mathfrak{I}}$ is defined by

$$\{(\mathrm{t}_1 \cdot \mathrm{t}_2 : n_1 \cdot n_2) \mid (\mathrm{t}_1 : n_1) \in [\![e_1]\!]_{\mathfrak{I}} \wedge (\mathrm{t}_2 : n_2) \in [\![e_2]\!]_{\mathfrak{I}} \wedge \\ \mathrm{t}_1 \equiv_{\mathcal{S}(e_1;\mathfrak{D})\cap\mathcal{S}(e_2;\mathfrak{D})} \mathrm{t}_2 \wedge (\mathrm{t}_1 \cdot \mathrm{t}_2 \text{ satisfies } E)\}.$$

If $E = \emptyset$, then we simply write $e_1 \dot{\bowtie} e_2$ (the *natural join*).

Example 2. Consider the database instance $\mathfrak{I}$ of Example 1, which is visualized in Fig. 1. The expression $e = \dot{\pi}_{age}(\dot{\sigma}_{type=\texttt{non-food}}(Customer \dot{\bowtie} Bought \dot{\bowtie} Product))$ returns the ages of people that bought non-food products: as Eve bought a car, we have $[\![e]\!]_{\mathfrak{I}} = \{(age \mapsto 21{:}1)\}$. If we change the condition $type = \texttt{non-food}$ into $type = \texttt{food}$, resulting in the expression e', then $[\![e']\!]_{\mathfrak{I}} = \{(age \mapsto 19 : 2), (age \mapsto 20 : 1)\}$, which includes Alice's age twice as she bought two apples. Observe that we have $[\![\dot{\delta}(e')]\!]_{\mathfrak{I}} = \{(age \mapsto 19 : 1), (age \mapsto 20 : 1)\}$.

The extended multiset relational algebra. We aim at reducing the complexity of query evaluation by rewriting joins into semi-joins, which are not part of the standard relational algebra described above. *Extended* relational algebra expressions over $\mathfrak{D}$ are built from the standard relational algebra operators and the following additional ones:

---

[3] The set operators we define have the same semantics as the `UNION ALL`, `INTERSECT ALL`, and `EXCEPT ALL` operators of standard SQL [8]. The semantics of `UNION`, `INTERSECT`, and `EXCEPT` can be obtained using deduplication.

[4] To simplify presentation, the θ-join and the θ-semi-join also perform equi-join on all attributes common to the multiset relations involved.

- *θ-semi-join and semi-join.* If e_1 and e_2 are expressions and E is a set of conditions on $\mathcal{S}(e_1;\mathfrak{D}) \cup \mathcal{S}(e_2;\mathfrak{D})$, then $e_1 \ltimes_E e_2$ is an expression with $\mathcal{S}(e_1 \ltimes_E e_2;\mathfrak{D}) = \mathcal{S}(e_1;\mathfrak{D})$ and $[\![e_1 \ltimes_E e_2]\!]_{\mathfrak{I}}$ is defined by

$$\{(\mathsf{t}_1 : n_1) \in [\![e_1]\!]_{\mathfrak{I}} \mid \exists \mathsf{t}_2 \ (\mathsf{t}_2 \in [\![e_2]\!]_{\mathfrak{I}} \wedge \\ \mathsf{t}_1 \equiv_{\mathcal{S}(e_1;\mathfrak{D}) \cap \mathcal{S}(e_2;\mathfrak{D})} \mathsf{t}_2 \wedge (\mathsf{t}_1 \cdot \mathsf{t}_2 \text{ satisfies } E))\}.$$

  If $E = \emptyset$, then we simply write $e_1 \ltimes e_2$ (the *semi-join*).
- *Max-union.*[5] If e_1 and e_2 are expressions such that $\mathrm{A} = \mathcal{S}(e_1;\mathfrak{D}) = \mathcal{S}(e_2;\mathfrak{D})$, then $e_1 \dot{\sqcup} e_2$ is an expression with $\mathcal{S}(e_1 \dot{\sqcup} e_2;\mathfrak{D}) = \mathrm{A}$ and

$$[\![e_1 \dot{\sqcup} e_2]\!]_{\mathfrak{I}} = \{(\mathsf{t} : \max(n_1, n_2)) \mid (\mathsf{t} : n_1) \in [\![e_1]\!]_{\mathfrak{I}} \wedge (\mathsf{t} : n_2) \in [\![e_2]\!]_{\mathfrak{I}}\} \cup \\ \{(\mathsf{t} : n) \in [\![e_1]\!]_{\mathfrak{I}} \mid \mathsf{t} \notin [\![e_2]\!]_{\mathfrak{I}}\} \cup \{(\mathsf{t} : n) \in [\![e_2]\!]_{\mathfrak{I}} \mid \mathsf{t} \notin [\![e_1]\!]_{\mathfrak{I}}\};$$

- *Attribute introduction.*[6] If e is an expression, B is a set of attributes with $\mathrm{B} \cap \mathcal{S}(e;\mathfrak{D}) = \emptyset$, and $f = \{b := x \mid b \in \mathrm{B} \wedge x \in (\mathcal{S}(e;\mathfrak{D}) \cup \mathfrak{U})\}$ is a set of assignment-pairs, then $i_f(e)$ is an expression with $\mathcal{S}(i_f(e);\mathfrak{D}) = \mathcal{S}(e;\mathfrak{D}) \cup \mathrm{B}$ and $[\![i_f(e)]\!]_{\mathfrak{I}} = \{(\mathsf{t} \cdot \{\mathrm{B} \mapsto \mathrm{value}(\mathsf{t}, x) \mid (b := x) \in f\} : m) \mid (\mathsf{t} : m) \in [\![e]\!]_{\mathfrak{I}}\}$, in which $\mathrm{value}(\mathsf{t}, x) = \mathsf{t}(x)$ if $x \in \mathcal{S}(e;\mathfrak{D})$ and $\mathrm{value}(\mathsf{t}, x) = x$ otherwise.

The extended relational algebra provides all the operators used in our framework and the following example illustrates the use of these operators:

Example 3. Consider the database instance $\mathfrak{I}$ of Examples 1 and 2. Let $e = \dot{\sigma}_{type=\texttt{food}}(\mathit{Customer} \bowtie \mathit{Bought} \bowtie \mathit{product})$. The query $\dot{\delta}(\dot{\pi}_{age}(e))$ is equivalent to $\dot{\delta}(\dot{\pi}_{age}(\mathit{Customer} \ltimes (\mathit{Bought} \ltimes_{type=\texttt{food}} \mathit{Product})))$. The query $\dot{\delta}(\dot{\pi}_{cname,type}(e))$ is equivalent to $i_{type:=\texttt{food}}(\dot{\pi}_{cname}(\mathit{Customer} \ltimes (\mathit{Bought} \ltimes_{type=\texttt{food}} \mathit{Product})))$. Notice that we were able to eliminate joins altogether in these rewritings. In the latter rewriting, we were also able to eliminate deduplication, even though attributes from several relations are involved. This is because *cname* is a key of the set relation *Customer*.

4 Rewriting Queries

Traditional rewrite rules for optimizing queries, such as the selection and projection push-down rewrite rules, guarantee *strong-equivalence*: the original subquery and the rewritten subquery always evaluate to the same result. Requiring this strong form of equivalence severely limits the optimizations one can perform in queries that involve projection and/or deduplication steps, as is illustrated in the following example:

[5] The max-union operators is inspired by the max-based multiset relation union [2].
[6] Attribute introduction is a restricted form of the operator commonly known as *generalized projection* or *extended projection* [1,5].

Example 4. Consider the database instance $\mathfrak{I}$ of Examples 1–3. The query $e = \dot{\delta}(\dot{\pi}_{cname}(Customer \bowtie Bought))$ returns the names of customers that bought a product. This query is equivalent to $e' = \dot{\delta}(\dot{\pi}_{cname}(Customer) \ltimes Bought)$. Observe that the subqueries $Customer \bowtie Bought$ and $\dot{\pi}_{cname}(Customer \bowtie Bought)$ of e are not equivalent to any subqueries of e', however. Hence, this rewriting cannot be achieved using strong-equivalent rewriting only. Finally, as $\{cname\}$ is a key of *Customer*, e and e' are also equivalent to $e'' = \dot{\pi}_{cname}(Customer) \ltimes Bought$.

To be able to discuss the full range of possible optimizations within the scope of projection and deduplication operations, we differentiate between the following notions of query equivalence:

Definition 1. *Let e_1 and e_2 be multiset relational algebra expressions over $\mathfrak{D}$ with $\text{A} = \mathcal{S}(e_1;\mathfrak{D}) \cap \mathcal{S}(e_2;\mathfrak{D})$, and let $\text{B} \subseteq \text{A}$. We say that e_1 and e_2 are* strong-equivalent, *denoted by $e_1 \doteq e_2$, if, for every database instance $\mathfrak{I}$ over $\mathfrak{D}$, we have $[\![e_1]\!]_{\mathfrak{I}} = [\![e_2]\!]_{\mathfrak{I}}$. We say that e_1 and e_2 are* weak-equivalent, *denoted by $e_1 \doteqdot e_2$, if, for every database instance $\mathfrak{I}$ over $\mathfrak{D}$, we have $[\![\dot{\delta}(e_1)]\!]_{\mathfrak{I}} = [\![\dot{\delta}(e_2)]\!]_{\mathfrak{I}}$. We say that e_1 and e_2 are* strong-B-equivalent, *denoted by $e_1 \doteq_{\text{B}} e_2$, if, for every database instance $\mathfrak{I}$ over $\mathfrak{D}$, we have $[\![\dot{\pi}_{\text{B}}(e_1)]\!]_{\mathfrak{I}} = [\![\dot{\pi}_{\text{B}}(e_2)]\!]_{\mathfrak{I}}$. Finally, we say that e_1 and e_2 are* weak-B-equivalent, *denoted by $e_1 \doteqdot_{\text{B}} e_2$, if, for every database instance $\mathfrak{I}$ over $\mathfrak{D}$, we have $[\![\dot{\delta}(\dot{\pi}_{\text{B}}(e_1))]\!]_{\mathfrak{I}} = [\![\dot{\delta}(\dot{\pi}_{\text{B}}(e_2))]\!]_{\mathfrak{I}}$.*

While strong-equivalent expressions always yield the same result, weak-equivalent expressions yield the same result only up to duplicate elimination. For query optimization, the latter is often sufficient at the level of subqueries: structural properties, such as the presence of a key in one of the relations involved, may have as a side effect that any duplicates in subqueries are eliminated in the end result anyway.

Example 5. Consider the queries of Example 4. We have $e \doteq e' \doteq e''$, $e \doteq_{\{cname\}} Customer \ltimes Bought$, $Customer \bowtie Bought \doteqdot_{\{cname,age\}} Customer \ltimes Bought$, and $\dot{\pi}_{cname}(Customer \bowtie Bought) \doteqdot_{\{cname\}} \dot{\pi}_{cname}(Customer) \ltimes Bought$.

Examples 4 and 5 not only show the relevance of non-strong-equivalent rewriting rules, they also show that further optimizations are possible if the expressions evaluate to set relations or satisfy certain keys. Therefore, to facilitate the discussion, we extend the definition of set relations and keys to expressions. Let e be an expression over $\mathfrak{D}$ with $\text{A} = \mathcal{S}(e;\mathfrak{D})$. We say that e is a *set relation* if, for every database instance $\mathfrak{I}$ over $\mathfrak{D}$, $[\![e]\!]_{\mathfrak{I}}$ is a set relation and we say that $\text{B} \subseteq \text{A}$ is a *key* of e if, for every database instance $\mathfrak{I}$ over $\mathfrak{D}$, B is a key of $[\![e]\!]_{\mathfrak{I}}$.

The following simple rules can be derived by straightforwardly applying Definition 1:

Proposition 1. *Let e, e_1, and e_2 be expressions over $\mathfrak{D}$ with $\text{A} = \mathcal{S}(e;\mathfrak{D}) = \mathcal{S}(e_1;\mathfrak{D}) \cap \mathcal{S}(e_2;\mathfrak{D})$ and $\text{B} \subseteq \text{A}$. Let $\hat{=}$ be either $\doteq$ or $\doteqdot$. We have:*

(i) *if $e_1 \hat{=} e_2$, then $e_1 \hat{=}_{\text{B}} e_2$;*

(ii) *if* $e_1 \mathrel{\hat{=}}_{\mathrm{A}} e_2$, *then* $e_1 \mathrel{\hat{=}} e_2$;
(iii) *if* $\mathrm{C} \subseteq \mathrm{B}$ *and* $e_1 \mathrel{\hat{=}}_{\mathrm{B}} e_2$, *then* $e_1 \mathrel{\hat{=}}_{\mathrm{C}} e_2$;
(iv) *if* $e_1 \mathrel{\hat{=}}_{\mathrm{B}} e_2$, *then* $\dot{\pi}_{\mathrm{B}}(e_1) \mathrel{\hat{=}} \dot{\pi}_{\mathrm{B}}(e_2)$;
(v) *if* $e_1 \mathrel{\dot{=}}_{\mathrm{B}} e_2$, *then* $e_1 \mathrel{\tilde{=}}_{\mathrm{B}} e_2$;
(vi) *if* $e_1 \mathrel{\tilde{=}} e_2$, *then* $\dot{\delta}(e_1) \mathrel{\dot{=}} \dot{\delta}(e_2)$;
(vii) *if* $e_1 \mathrel{\tilde{=}}_{\mathrm{B}} e_2$, e_1 *and* e_2 *are set relations, and* B *is a key of* e_1 *and* e_2, *then* $e_1 \mathrel{\dot{=}}_{\mathrm{B}} e_2$; *and*
(viii) $e \mathrel{\dot{=}}_{\mathrm{B}} \dot{\pi}_{\mathrm{B}}(e)$ *and* $e \mathrel{\tilde{=}} \dot{\delta}(e)$.

Proposition 1.*(iii)* allows us to restrict the scope of strong-B-equivalences and weak-B-equivalences to subsets of B.

In the presence of conditions, as enforced by selections, θ-joins, or θ-semi-joins, we can also extend the scope of strong-B-equivalences and weak-B-equivalences. E.g., if $\mathrm{u} \in \mathfrak{U}$ is a constant and e_1 and e_2 are expressions over $\mathfrak{D}$ with $a, a', b \in \mathcal{S}(e_1;\mathfrak{D}) \cap \mathcal{S}(e_2;\mathfrak{D})$ and $e_1 \mathrel{\tilde{=}}_{\{a\}} e_2$, then $\dot{\sigma}_{a=a',b=\mathrm{u}}(e_1) \mathrel{\tilde{=}}_{\{a,a',b\}} \dot{\sigma}_{a=a',b=\mathrm{u}}(e_2)$. Next, we develop the framework for these scope extensions, for which we first introduce the closure of a set of attributes under a set of conditions:

Definition 2. *Let* A *be a relation schema, let* $\mathrm{B} \subseteq \mathrm{A}$, *and let* E *be a set of conditions over* A. *The* closure of B under E, *denoted by* $\mathcal{C}(\mathrm{B}; E)$, *is the smallest superset of* B *such that, for every condition* $(v = w) \in E$ *or* $(w = v) \in E$, $v \in \mathcal{C}(\mathrm{B}; E)$ *if and only if* $w \in \mathcal{C}(\mathrm{B}; E)$.

If $a \in \mathrm{A}$, then we write $\mathcal{C}(a; E)$ for $\mathcal{C}(\{a\}; E)$.

Notice that, besides attributes, $\mathcal{C}(\mathrm{B}; E)$ may also contain constants. E.g., if $(b = \mathrm{u}) \in E$ for $b \in \mathrm{B}$ and $\mathrm{u} \in \mathfrak{U}$, then $\mathrm{u} \in \mathcal{C}(\mathrm{B}; E)$. We denote $\mathcal{C}(\mathrm{B}; E) \cap \mathrm{A}$, the set of attributes in $\mathcal{C}(\mathrm{B}; E)$, by $\mathrm{attrs}(\mathrm{B}; E)$; and we denote $\mathcal{C}(\mathrm{B}; E) - \mathrm{A}$, the set of constants in $\mathcal{C}(\mathrm{B}; E)$, by $\mathrm{consts}(\mathrm{B}; E)$.

Intuitively, the values of a tuple for the attributes in $\mathrm{attrs}(\mathrm{B}; E)$ are uniquely determined by the values of that tuple for the attributes in B. There may be other attributes $c \notin \mathrm{attrs}(\mathrm{B}; E)$ that can only have a single value, however. E.g., if attribute c is constraint by a constant condition of the form $c = \mathrm{u}$, $\mathrm{u} \in \mathfrak{U}$. The values of a tuple for such attributes c are therefore trivially determined by the values of that tuple for the attributes in B. This observation leads to the following definition:

Definition 3. *Let* A *be a relation schema, let* $\mathrm{B} \subseteq \mathrm{A}$, *and let* E *be a set of conditions over* A. *The set of attributes determined by* B, *denoted by* $\mathrm{det}(\mathrm{B}; E)$, *is defined by* $\mathrm{attrs}(\mathrm{B}; E) \cup \{a \in \mathrm{A} \mid \mathrm{consts}(a; E) \neq \emptyset\}$.

The intuition given above behind the notions in Definitions 2 and 3 can be formalized as follows:

Lemma 1. *Let* t *be a tuple over a relation schema* A *and let* E *be a set of conditions over* A. *If* t *satisfies* E, *then, for every* $a \in \mathrm{A}$, *we have*

(i) $t(a) = t(b)$ *for every* $b \in \mathrm{attrs}(a; E)$;

(ii) $t(a) = u$ *for every* $u \in \text{consts}(a; E)$;
(iii) $|\text{consts}(a; E)| \leq 1$; *and*
(iv) $t \equiv_{\text{det}(\text{B};E)} t'$ *if* $\text{B} \subseteq \text{A}$, t' *is a tuple over* A, t' *satisfies* E, *and* $t \equiv_{\text{B}} t'$.

Using Lemma 1, we prove the following rewrite rules for selection:

Theorem 1. *Let* g *and* h *be expressions over* $\mathfrak{D}$ *with* $\text{G} = \mathcal{S}(g; \mathfrak{D})$ *and* $\text{H} = \mathcal{S}(h; \mathfrak{D})$, *let* E *be a set of conditions over* G, *let* $\text{A} \subseteq (\text{H} \cap \text{attrs}(E))$, *let* $\text{B} \subseteq (\text{det}(\text{A}; E) - \text{H})$, *and let* f *be a set of assignment-pairs such that, for each* $b \in \text{B}$, *there exists* $(b := x) \in f$ *with* $x \in (\mathcal{C}(b; E) \cap (\text{A} \cup \mathfrak{U}))$. *We have*

(i) *if* $\dot{\sigma}_E(g) \tilde{\dot{=}}_{\text{A}} h$, *then* $\dot{\sigma}_E(g) \tilde{\dot{=}}_{\text{A}\cup\text{B}} \iota_f(h)$; *and*
(ii) *if* $\dot{\sigma}_E(g) \dot{=}_{\text{A}} h$, *then* $\dot{\sigma}_E(g) \dot{=}_{\text{A}\cup\text{B}} \iota_f(h)$.

Proof. We first prove *(i)*. Assume $\dot{\sigma}_E(g) \tilde{\dot{=}}_{\text{A}} h$ and let $\mathfrak{I}$ be a database instance over $\mathfrak{D}$. We prove $[\![\dot{\delta}(\dot{\pi}_{\text{A}\cup\text{B}}(\dot{\sigma}_E(g)))]\!]_{\mathfrak{I}} = [\![\dot{\delta}(\dot{\pi}_{\text{A}\cup\text{B}}(\iota_f(h)))]\!]_{\mathfrak{I}}$ by showing that there exist n and n' such that

$$(\mathsf{t} : n) \in [\![\dot{\pi}_{\text{A}\cup\text{B}}(\dot{\sigma}_E(g))]\!]_{\mathfrak{I}} \iff (\mathsf{t} : n') \in [\![\dot{\pi}_{\text{A}\cup\text{B}}(\iota_f(h))]\!]_{\mathfrak{I}}.$$

Assume $(\mathsf{t} : n) \in [\![\dot{\pi}_{\text{A}\cup\text{B}}(\dot{\sigma}_E(g))]\!]_{\mathfrak{I}}$, and let $(\mathsf{t}_1 : p_1), \dots, (\mathsf{t}_j : p_j) \in [\![\dot{\sigma}_E(g)]\!]_{\mathfrak{I}}$ be all tuple-count pairs such that, for every i, $1 \leq i \leq j$, $\mathsf{t}_i \equiv_{\text{A}\cup\text{B}} \mathsf{t}$. By construction, we have $n = p_1 + \dots + p_j$. By Lemma 1.*(iv)* and $\text{B} \subseteq \text{det}(\text{A}; E)$, no other t' exists with $\mathsf{t}' \notin \{\mathsf{t}_1, \dots, \mathsf{t}_j\}$, $(\mathsf{t}' : p') \in [\![\dot{\sigma}_E(g)]\!]_{\mathfrak{I}}$ and $\mathsf{t}' \equiv_{\text{A}} \mathsf{t}$ as this would imply $\mathsf{t}' \equiv_{\text{A}\cup\text{B}} \mathsf{t}$. By $\dot{\sigma}_E(g) \tilde{\dot{=}}_{\text{A}} h$, we have $(\mathsf{t}|_{\text{A}} : q) \in [\![\dot{\pi}_{\text{A}}(h)]\!]_{\mathfrak{I}}$. Let $(\mathsf{s}_1 : q_1), \dots, (\mathsf{s}_k : q_k) \in [\![h]\!]_{\mathfrak{I}}$ with, for every i, $1 \leq i \leq k$, $\mathsf{s}_i \equiv_{\text{A}} \mathsf{t}$. By construction, we have $q = q_1 + \dots + q_k$. We prove that, for all $1 \leq i \leq k$, $\mathsf{t} \equiv_{\text{B}} \mathsf{s}_i$. Consider $b \in \text{B}$ with $(b := x) \in f$:

1. If $x \in \text{attrs}(b; E)$, then $x \in \text{A}$ and, by Lemma 1.*(i)*, we have $\mathsf{t}(b) = \mathsf{t}(x)$. By $x \in \text{A}$ and $\mathsf{t} \equiv_{\text{A}} \mathsf{s}_i$, we have $\mathsf{t}(b) = \mathsf{t}(x) = \mathsf{s}_i(x)$. From the semantics of $\iota_{b:=x}$, it follows that $\mathsf{t}(b) = \mathsf{t}(x) = \mathsf{s}_i(x) = \mathsf{s}_i(b)$.
2. If $x \in \text{consts}(b; E)$, then, by Lemma 1.*(iii)*, there exists a constant $\mathsf{u} \in \mathfrak{U}$ such that $\text{consts}(b; E) = \{\mathsf{u}\}$. By Lemma 1.*(ii)*, we have $\mathsf{t}(x) = \mathsf{u}$. From the semantics of $\iota_{b:=\mathsf{u}}$, it follows that $\mathsf{t}(b) = \mathsf{u} = \mathsf{s}_i(b)$.

We conclude that $(\mathsf{t} : q) \in [\![\dot{\pi}_{\text{A}\cup\text{B}}(\iota_f(h))]\!]_{\mathfrak{I}}$ and $n' = q$. To prove *(ii)*, we bootstrap the proof of *(i)*. By $\dot{\sigma}_E(g) \dot{=}_{\text{A}} h$, we have $n = q$. Hence, we have $n = n'$ and $(\mathsf{t} : n) \in [\![\dot{\pi}_{\text{A}\cup\text{B}}(\dot{\sigma}_E(g))]\!]_{\mathfrak{I}}$ if and only if $(\mathsf{t} : n) \in [\![\dot{\pi}_{\text{A}\cup\text{B}}(\iota_f(h))]\!]_{\mathfrak{I}}$. □

In the presence of constant conditions, Theorem 1 can be applied to joins to rewrite them into semi-joins combined with attribute introduction. An example of such rewrites is exhibited in Example 3.

Applications of Theorem 1 involve *attribute introduction*, which makes query results larger. Hence, it is best to delay this operation by pulling the operator up. Proposition 2 presents rewrite rules to do so.

Proposition 2. *Let* e, e_1, *and* e_2 *be expressions over* $\mathfrak{D}$. *We have*

(i) $\dot{\sigma}_E(\iota_f(e)) \dot{=} \iota_f(\dot{\sigma}_E(e))$ *if, for all* $(b := x) \in f$, $b \notin \text{attrs}(E)$;

(ii) $\dot{\pi}_{\mathrm{B}}(\iota_f(e)) \doteq \iota_{f'}(\dot{\pi}_{\mathrm{B}'}(e))$ *if* $f' \subseteq f$ *and* $\mathrm{B} = \mathrm{B}' \cup \{b \mid (b := x) \in f'\}$.
(iii) $\dot{\rho}_g(\iota_f(e)) \doteq \iota_{f'}(\dot{\rho}_{g'}(e))$ *if* $g' = \{a \mapsto g(a) \mid a \in \mathcal{S}(e;\mathfrak{D})\}$ *and*

$$f' = \{g(b) := g(x) \mid (b := x) \in f \wedge x \in \mathcal{S}(e;\mathfrak{D})\} \cup \{g(b) := x \mid (b := x) \in f \wedge x \in \mathfrak{U}\}.$$

(iv) $\dot{\delta}(\iota_f(e)) \doteq \iota_f(\dot{\delta}(e))$;
(v) $\iota_f(e_1) \oplus \iota_f(e_2) \doteq \iota_f(e_1 \oplus e_2)$ *if* $\oplus \in \{\dot{\cup}, \dot{\cap}, \dot{-}\}$;
(vi) $\iota_f(e_1) \dot{\bowtie}_E e_2 \doteq \iota_{f'}(e_1 \dot{\bowtie}_{E'} e_2)$ *if* $f' = \{(b := x) \in f \mid b \notin \mathcal{S}(e_2;\mathfrak{D})\}$ *and* $E' = E \cup \{b = x \mid (b := x) \in f \wedge b \in \mathcal{S}(e_2;\mathfrak{D})\}$;
(vii) $\iota_f(e_1) \dot{\ltimes}_E e_2 \doteq \iota_f(e_1 \dot{\ltimes}_{E'} e_2)$ *if*

$$E' = E \cup \{b = x \mid (b := x) \in f \wedge b \in \mathcal{S}(e_2;\mathfrak{D})\};$$

(viii) $e_1 \dot{\ltimes}_E \iota_f(e_2) \doteq e_1 \dot{\ltimes}_{E'} e_2$ *if*

$$E' = E \cup \{b = x \mid (b := x) \in f \wedge b \in \mathcal{S}(e_1;\mathfrak{D})\};$$

(ix) $\iota_f(e_1) \dot{\sqcup} \iota_f(e_2) \doteq \iota_f(e_1 \dot{\sqcup} e_2)$; *and*
(x) $\iota_f(\iota_g(e)) \doteq \iota_{f \cup g}(e)$ *if, for all* $(b := x) \in f$, *we have* $x \in (\mathcal{S}(e;\mathfrak{D}) \cup \mathfrak{U})$.

Attribute introduction and renaming play complementary roles in the context of projection, as is illustrated by the following example:

Example 6. Consider the expression e with $\mathcal{S}(e;\mathfrak{D}) = \{a, b\}$ and consider the query $\dot{\pi}_{a,c,d}(\iota_{c:=a,d:=b}(e))$. As we do not need b after the projection, we can also rename b to d instead of introducing d. This alternative approach results in $\dot{\pi}_{a,c,d}(\iota_{c:=a}(\dot{\rho}_{a\mapsto a,b\mapsto d}(e))$. In this expression, we can easily push the attribute introduction through the projection, resulting in the expression $\iota_{c:=a}(\dot{\pi}_{a,d}(\dot{\rho}_{a\mapsto a,b\mapsto d}(e)))$.

The following rewrite rules can be used to apply the rewriting of Example 6:

Proposition 3. *Let* e *be an expression over* $\mathfrak{D}$ *and let* ι_f *be an attribute introduction operator applicable to* e. *We have*
(i) *if* $(b := x) \in f$, $x \in \mathcal{S}(e;\mathfrak{D})$, *and* $c = b$ *for all* $(c := x) \in f$, *then*

$$\iota_f(e) \doteq_{(\mathcal{S}(e;\mathfrak{D})\cup\{b\})-\{x\}} \iota_{f\setminus\{b:=x\}}(\dot{\rho}_{\{x\mapsto b\}\cup\{a\mapsto a \mid a\in(\mathcal{S}(e;\mathfrak{D})-\{x\})\}}(e));$$

(ii) *if* $(b_1 := x), (b_2 := x) \in f$, $b_1 \neq b_2$, *then* $\iota_f(e) \doteq \iota_{\{b_2:=b_1\}}(\iota_{f-\{b_2:=x\}}(e))$.

The rewrite rules of Proposition 2 that involve selections and attribute introductions put heavy restrictions on the sets of conditions involved. To alleviate these restrictions, we can use Proposition 3 and the well-known push-down rules for selection [2,5,10,12], to push some attribute introductions through selections.

What we have done up to now is examining how selection conditions interact with the notions of strong-equivalence and weak-equivalence. Next, we will put these results to use. As explained in the Introduction, our focus is twofold:

1. eliminating joins in favor of semi-joins; and
2. eliminating deduplication.

These foci are covered in Sects. 4.1 and 4.2, respectively. In Sect. 4.3, finally, we investigate how the other operators interact with strong-equivalence and weak-equivalence.

4.1 θ-Joins and θ-Semi-Joins

Above, we explored how selection conditions interact with strong-equivalence and weak-equivalence. Since θ-joins and θ-semi-joins implicitly use selection conditions, the techniques developed also apply to θ-joins and θ-semi-joins, which are the focus of this subsection. First, we notice that we can use the following rules to change the conditions involved in selections, θ-joins, and θ-semi-joins.

Proposition 4. *Let E and E' be sets of conditions over the same set of attributes* A. *If we have $\mathcal{C}(a; E) = \mathcal{C}(a; E')$ for every $a \in \mathrm{A}$, then we have*

(i) $\dot{\sigma}_E(e) \doteq \dot{\sigma}_{E'}(e)$;
(ii) $e_1 \mathbin{\dot{\bowtie}}_E e_2 \doteq e_1 \mathbin{\dot{\bowtie}}_{E'} e_2$; *and*
(iii) $e_1 \mathbin{\dot{\ltimes}}_E e_2 \doteq e_1 \mathbin{\dot{\ltimes}}_{E'} e_2$.

The equivalence at the basis of semi-join-based query rewriting in the set-based relational algebra is $e_1 \ltimes e_2 = \pi_{\mathrm{A}_1}(e_1 \bowtie e_2)$ with A_1 the relation schema of the relation to which e_1 evaluates. Notice, however, that the equivalence $e_1 \mathbin{\dot{\ltimes}} e_2 = \dot{\pi}_{\mathrm{A}_1}(e_1 \mathbin{\dot{\bowtie}} e_2)$ does *not* hold, because the multiset semi-join does not take into account the number of occurrences of tuples in $[\![e_2]\!]_{\mathfrak{I}}$:[7]

Example 7. Consider the database instance $\mathfrak{I}$ of Example 1, which is visualized in Fig. 1. We apply the queries $e_1 = \dot{\pi}_{\{cname, age\}}(Customer \mathbin{\dot{\bowtie}} Bought)$ and $e_2 = Customer \mathbin{\dot{\ltimes}} Bought$. We have

$$[\![e_1]\!]_{\mathfrak{I}} = \{(cname \mapsto \texttt{Alice}, age \mapsto 19 : 2), \dots\};$$
$$[\![e_2]\!]_{\mathfrak{I}} = \{(cname \mapsto \texttt{Alice}, age \mapsto 19 : 1), \dots\}.$$

Even though the above rewriting of the projection of a join into the corresponding semi-join is not a strong-equivalent rewriting, we observe that it is still a weak-equivalent rewriting.

We now formalize rewrite rules involving joins and semi-joins:

Theorem 2. *Let g_1, g_2, h_1, and h_2 be expressions over $\mathfrak{D}$ with $\mathrm{G}_1 = \mathcal{S}(g_1; \mathfrak{D})$, $\mathrm{G}_2 = \mathcal{S}(g_2; \mathfrak{D})$, $\mathrm{H}_1 = \mathcal{S}(h_1; \mathfrak{D})$, $\mathrm{H}_2 = \mathcal{S}(h_2; \mathfrak{D})$, $\mathrm{A}_1 \subseteq (\mathrm{G}_1 \cap \mathrm{H}_1)$, $\mathrm{A}_2 \subseteq (\mathrm{G}_2 \cap \mathrm{H}_2)$, and $\mathrm{A}_1 \cap \mathrm{A}_2 = \mathrm{G}_1 \cap \mathrm{G}_2 = \mathrm{H}_1 \cap \mathrm{H}_2$, let E be a set of conditions over $\mathrm{A}_1 \cup \mathrm{A}_2$, let $\mathrm{B} \subseteq (\mathrm{A}_2 - \mathrm{A}_1)$, and let f be a set of assignment-pairs such that, for each $b \in \mathrm{B}$, there exists $(b := x) \in f$ with $x \in (\mathcal{C}(b; E) \cap (\mathrm{A}_1 \cup \mathfrak{U}))$. We have*

(i) $g_1 \mathbin{\dot{\bowtie}}_E g_2 \mathrel{\tilde{\doteq}}_{\mathrm{A}_1 \cup \mathrm{A}_2} h_1 \mathbin{\dot{\bowtie}}_E h_2$ *if* $g_1 \mathrel{\tilde{\doteq}}_{\mathrm{A}_1} h_1$ *and* $g_2 \mathrel{\tilde{\doteq}}_{\mathrm{A}_2} h_2$;
(ii) $g_1 \mathbin{\dot{\bowtie}}_E g_2 \doteq_{\mathrm{A}_1 \cup \mathrm{A}_2} h_1 \mathbin{\dot{\bowtie}}_E h_2$ *if* $g_1 \doteq_{\mathrm{A}_1} h_1$ *and* $g_2 \doteq_{\mathrm{A}_2} h_2$;
(iii) $g_1 \mathbin{\dot{\bowtie}}_E g_2 \mathrel{\tilde{\doteq}}_{\mathrm{A}_1} h_1 \mathbin{\dot{\ltimes}}_E h_2$ *if* $g_1 \mathrel{\tilde{\doteq}}_{\mathrm{A}_1} h_1$ *and* $g_2 \mathrel{\tilde{\doteq}}_{\mathrm{A}_2} h_2$;
(iv) $g_1 \mathbin{\dot{\bowtie}}_E g_2 \doteq_{\mathrm{A}_1} h_1 \mathbin{\dot{\ltimes}}_E h_2$ *if* $g_1 \doteq_{\mathrm{A}_1} h_1$, $g_2 \mathrel{\tilde{\doteq}}_{\mathrm{A}_2} h_2$, g_2 *is a set relation, and* g_2 *has a key* C *with* $\mathrm{C} \subseteq \det(\mathrm{A}_1 \cap \mathrm{A}_2; E)$;

[7] We could have defined a multiset semi-join operator that *does take* into account the number of occurrences of tuples in $[\![e_2]\!]_{\mathfrak{I}}$. With such a semi-join operator, we would no longer be able to sharply reduce the size of intermediate query results, however, and lose some potential to optimize query evaluation.

(v) $g_1 \dot{\bowtie}_E g_2 \stackrel{\sim}{=}_{A_1 \cup B} \iota_f(h_1 \dot{\ltimes}_E h_2)$ *if* $g_1 \stackrel{\sim}{=}_{A_1} h_1$ *and* $g_2 \stackrel{\sim}{=}_{A_2} h_2$*; and*
(vi) $g_1 \dot{\bowtie}_E g_2 \doteq_{A_1 \cup B} \iota_f(h_1 \dot{\ltimes}_E h_2)$ *if* $g_1 \doteq_{A_1} h_1$, $g_2 \stackrel{\sim}{=}_{A_2} h_2$, g_2 *is a set relation, and* g_2 *has a key* C *with* $C \subseteq \det(A_1 \cap A_2; E)$.

Proof (Sketch). We prove *(i)*. Let $\mathfrak{I}$ be a database instance over $\mathfrak{D}$. We prove $[\![\dot{\delta}(\dot{\pi}_{A_1 \cup A_2}(g_1 \dot{\bowtie}_E g_2))]\!]_{\mathfrak{I}} = [\![\dot{\delta}(\dot{\pi}_{A_1 \cup A_2}(h_1 \dot{\bowtie}_E h_2))]\!]_{\mathfrak{I}}$ by showing that there exist n and n' such that

$$(\mathsf{t} : n) \in [\![\dot{\pi}_{A_1 \cup A_2}(g_1 \dot{\bowtie}_E g_2)]\!]_{\mathfrak{I}} \iff (\mathsf{t} : n') \in [\![\dot{\pi}_{A_1 \cup A_2}(h_1 \dot{\bowtie}_E h_2)]\!]_{\mathfrak{I}}.$$

Assume $(\mathsf{t} : n) \in [\![\dot{\pi}_{A_1 \cup A_2}(g_1 \dot{\bowtie}_E g_2)]\!]_{\mathfrak{I}}$, and let $(\mathsf{r}_1 : p_1), \dots, (\mathsf{r}_{k_1} : p_{k_1}) \in [\![g_1]\!]_{\mathfrak{I}}$ and $(\mathsf{s}_1 : q_1), \dots, (\mathsf{s}_{k_2} : q_{k_2}) \in [\![g_2]\!]_{\mathfrak{I}}$ be all tuple-count pairs such that, for every i_1, $1 \le i_1 \le k_1$, $\mathsf{r}_{i_1} \equiv_{A_1} \mathsf{t}$, and, for every i_2, $1 \le i_2 \le k_2$, $\mathsf{s}_{i_2} \equiv_{A_2} \mathsf{t}$. Let $p = p_1 + \dots + p_{k_1}$ and $q = q_1 + \dots + q_{k_2}$. By construction, we have that, for every i_1 and i_2, $1 \le i_1 \le k_1$ and $1 \le i_2 \le k_2$, $(\mathsf{r}_{i_1} \cdot \mathsf{s}_{i_2} : p_{i_1} \cdot q_{i_2}) \in [\![g_1 \dot{\bowtie}_E g_2]\!]_{\mathfrak{I}}$. Hence, $(\mathsf{t}|_{A_1} : p) \in [\![\dot{\pi}_{A_1}(g_1)]\!]_{\mathfrak{I}}$, $(\mathsf{t}|_{A_2} : q) \in [\![\dot{\pi}_{A_2}(g_2)]\!]_{\mathfrak{I}}$, and $n = p \cdot q$. By $g_1 \stackrel{\sim}{=}_{A_1} h_1$ and $g_2 \stackrel{\sim}{=}_{A_2} h_2$, we have $(\mathsf{t}|_{A_1} : p') \in [\![\dot{\pi}_{A_1}(h_1)]\!]_{\mathfrak{I}}$ and $(\mathsf{t}|_{A_2} : q') \in [\![\dot{\pi}_{A_2}(h_2)]\!]_{\mathfrak{I}}$. Let $(\mathsf{r}'_1 : p'_1), \dots, (\mathsf{r}'_{l_1} : p'_{l_1}) \in [\![h_1]\!]_{\mathfrak{I}}$ and $(\mathsf{s}'_1 : q'_1), \dots, (\mathsf{s}'_{l_2} : q'_{l_2}) \in [\![h_2]\!]_{\mathfrak{I}}$ be all tuple-count pairs such that, for every j_1, $1 \le j_1 \le l_1$, $\mathsf{r}'_{j_1} \equiv_{A_1} \mathsf{t}$, and, for every j_2, $1 \le j_2 \le l_2$, $\mathsf{s}'_{j_2} \equiv_{A_2} \mathsf{t}$. By construction, $p' = p'_1 + \dots + p'_{l_1}$ and $q' = q'_1 + \dots + q'_{l_2}$ and, for every j_1 and j_2, $1 \le j_1 \le l_1$ and $1 \le j_2 \le l_2$, $\mathsf{r}'_{j_1} \cdot \mathsf{s}'_{j_2}$ satisfies E and $(\mathsf{r}'_{j_1} \cdot \mathsf{s}'_{j_2} : p'_{j_1} \cdot q'_{j_2}) \in [\![h_1 \dot{\bowtie}_E h_2]\!]_{\mathfrak{I}}$. We conclude $(\mathsf{t} : p' \cdot q') \in [\![\dot{\pi}_{A_1 \cup A_2}(h_1 \dot{\bowtie}_E h_2)]\!]_{\mathfrak{I}}$ and $n' = p' \cdot q'$.

Next, we prove *(ii)* by bootstrapping the proof of *(i)*. Observe that, by $g_1 \doteq_{A_1} h_1$ and $g_2 \doteq_{A_2} h_2$, we have that $p = p'$ and $q = q'$. Hence, $n = n'$ and $(\mathsf{t} : n) \in [\![\dot{\pi}_{A_1 \cup A_2}(g_1 \dot{\bowtie}_E g_2)]\!]_{\mathfrak{I}}$ if and only if $(\mathsf{t} : n) \in [\![\dot{\pi}_{A_1 \cup A_2}(h_1 \dot{\bowtie}_E h_2)]\!]_{\mathfrak{I}}$.

The other statements can be proven in analogous ways. □

The rules of Theorem 2 can be specialized to semi-joins only:

Corollary 1. *Let* g_1, g_2, h_1, *and* h_2 *be expressions which satisfy the conditions of Theorem 2 and let* $\hat{=}$ *be either* $\doteq$ *or* $\stackrel{\sim}{=}$. *If* $g_1 \hat{=}_{A_1} h_1$ *and* $g_2 \stackrel{\sim}{=}_{A_2} h_2$, *then* $g_1 \dot{\ltimes}_E g_2 \hat{=}_{A_1} h_1 \dot{\ltimes}_E h_2$.

4.2 Deduplication

The second optimization goal we have set ourselves is to eliminate the need for removing duplicates. This is possible if we can push down deduplication operators to a level where subexpressions are guaranteed to evaluate to set relations, in which case deduplication becomes redundant.

The rewrite rules relevant for pushing down deduplication are the following:

Proposition 5. *Let* e, e_1, e_2 *be expressions over* $\mathfrak{D}$. *We have*

(i) $\dot{\delta}(\dot{\sigma}_E(e)) \doteq \dot{\sigma}_E(\dot{\delta}(e))$;
(ii) $\dot{\delta}(\dot{\pi}_B(e) \doteq \dot{\pi}_B(\dot{\delta}(e))$ *if* B *is a key of* e;
(iii) $\dot{\delta}(\dot{\rho}_f(e)) \doteq \dot{\rho}_f(\dot{\delta}(e))$;
(iv) $\dot{\delta}(\dot{\delta}(e)) \doteq \dot{\delta}(e)$;
(v) $\dot{\delta}(e_1 \cup e_2) \doteq \dot{\delta}(e_1) \dot{\sqcup} \dot{\delta}(e_2)$;
(vi) $\dot{\delta}(e_1 \dot{\cap} e_2) \doteq \dot{\delta}(e_1) \dot{\cap} \dot{\delta}(e_2)$;
(vii) $\dot{\delta}(e_1 \dot{\bowtie}_E e_2) \doteq \dot{\delta}(e_1) \dot{\bowtie}_E \dot{\delta}(e_2)$;
(viii) $\dot{\delta}(e_1 \dot{\ltimes}_E e_2) \doteq \dot{\delta}(e_1) \dot{\ltimes}_E e_2$;
(ix) $\dot{\delta}(e_1 \dot{\sqcup} e_2) \doteq \dot{\delta}(e_1) \dot{\sqcup} \dot{\delta}(e_2)$; *and*
(x) $\dot{\delta}(\iota_f(e)) \doteq \iota_f(\dot{\delta}(e))$.

One cannot push down deduplication through difference, however, as $\dot{\delta}(e_1 \dot{-} e_2) \doteq \dot{\delta}(e_1) \dot{-} \dot{\delta}(e_2)$ does *not* hold in general.

If, in the end, deduplication operates on expressions that evaluate to set relations, it can be eliminated altogether:

Proposition 6. *Let e be an expression over $\mathfrak{D}$. If e is a set relation, then $\dot{\delta}(e) \doteq e$.*

4.3 Other Rewrite Rules

To use the rewrite rules of Proposition 1, Theorem 1, Theorem 2, Proposition 2, and Proposition 5 in the most general way possible, we also need to know how the other operators interact with strong-B-equivalence and weak-B-equivalence. For the unary operators, we have the following:

Proposition 7. *Let g and h be expressions over $\mathfrak{D}$ with $\mathrm{A} \subseteq \mathcal{S}(g;\mathfrak{D}) \cap \mathcal{S}(h;\mathfrak{D})$. Let $\hat{=}$ be either $\doteq$ or $\dot{\simeq}$. If $g \mathrel{\hat{=}_{\mathrm{A}}} h$, then we have*

(i) $\dot{\sigma}_E(g) \mathrel{\hat{=}_{\mathrm{A}}} \dot{\sigma}_E(h)$ *if E is a set of equalities with* $\mathrm{attrs}(E) \subseteq \mathrm{A}$;
(ii) $\dot{\pi}_{\mathrm{B}_g}(g) \mathrel{\hat{=}_{\mathrm{A}\cap\mathrm{B}}} \dot{\pi}_{\mathrm{B}_h}(h)$ *if $\mathrm{B}_g \subseteq \mathcal{S}(g;\mathfrak{D})$, $\mathrm{B}_h \subseteq \mathcal{S}(h;\mathfrak{D})$, and $\mathrm{B} = \mathrm{B}_g \cap \mathrm{B}_h$;*
(iii) $\dot{\rho}_{f_g}(g) \mathrel{\hat{=}_{\{f_g(a) \mid a \in \mathrm{A}\}}} \dot{\rho}_{f_h}(h)$ *if $f_g : \mathcal{S}(g;\mathfrak{D}) \to \mathrm{B}_g$ and $f_h : \mathcal{S}(h;\mathfrak{D}) \to \mathrm{B}_h$ are bijections with, for all $a \in \mathrm{A}$, $f_g(a) = f_h(a)$;*
(iv) $\dot{\delta}(g) \mathrel{\hat{=}_{\mathrm{A}}} \dot{\delta}(h)$; *and*
(v) $\iota_f(g) \mathrel{\hat{=}_{\mathrm{A}\cup\mathrm{B}}} \iota_f(h)$ *if $\mathrm{B} \subseteq (\mathfrak{N} - (\mathcal{S}(g;\mathfrak{D}) \cup \mathcal{S}(h;\mathfrak{D})))$ and f is a set of assignment-pairs such that, for each $b \in \mathrm{B}$, there exists $(b := x) \in f$ with $x \in (\mathrm{A} \cup \mathfrak{U})$.*

For the binary operators, we observe that the θ-join and θ-semi-join operators are completely covered by Theorem 2 and Corollary 1.

For the union and max-union operators, we have the following:

Proposition 8. *Let g_1, g_2, h_1, and h_2 be expressions over $\mathfrak{D}$ and let A be a set of attributes with $\mathrm{A}_g = \mathcal{S}(g_1;\mathfrak{D}) = \mathcal{S}(g_2;\mathfrak{D})$, $\mathrm{A}_h = \mathcal{S}(h_1;\mathfrak{D}) = \mathcal{S}(h_2;\mathfrak{D})$, and $\mathrm{A} \subseteq (\mathrm{A}_g \cap \mathrm{A}_h)$. Let $\hat{=}$ be either $\doteq$ or $\dot{\simeq}$. If $g_1 \mathrel{\hat{=}_{\mathrm{A}}} h_1$ and $g_2 \mathrel{\hat{=}_{\mathrm{A}}} h_2$, then we have $g_1 \mathbin{\dot{\cup}} g_2 \mathrel{\hat{=}_{\mathrm{A}}} h_1 \mathbin{\dot{\cup}} h_2$ and $g_1 \mathbin{\dot{\sqcup}} g_2 \mathrel{\dot{\simeq}_{\mathrm{A}}} h_1 \mathbin{\dot{\sqcup}} h_2$. If, in addition, $\mathrm{A} = \mathrm{A}_g = \mathrm{A}_h$, then we also have $g_1 \mathbin{\dot{\sqcup}} g_2 \mathrel{\doteq_{\mathrm{A}}} h_1 \mathbin{\dot{\sqcup}} h_2$.*

Finally, for the operators intersection and difference, we propose the following straightforward rewrite rules:

Proposition 9. *Let g_1, g_2, h_1 and h_2 be expressions over $\mathfrak{D}$ with $\mathcal{S}(g_1;\mathfrak{D}) = \mathcal{S}(g_2;\mathfrak{D})$ and $\mathcal{S}(h_1;\mathfrak{D}) = \mathcal{S}(h_2;\mathfrak{D})$. Let $\hat{=}$ be either $\doteq$ or $\dot{\simeq}$. If $g_1 \mathrel{\hat{=}} h_1$ and $g_2 \mathrel{\hat{=}} h_2$, then we have $g_1 \mathbin{\dot{\cap}} g_2 \mathrel{\hat{=}} h_1 \mathbin{\dot{\cap}} h_2$; and if $g_1 \doteq h_1$ and $g_2 \doteq h_2$, then $g_1 \dot{-} g_2 \doteq h_1 \dot{-} h_2$.*

The above rewrite rules for max-union, intersection, and difference are very restrictive. Next, we illustrate that we cannot simply relax these restrictive rewrite rules to more general strong-B-equivalence or weak-B-equivalence rewrite rules:

Example 8. Let $\text{A} = \{m, n\}$, let U, V, V', and W be four relation names, and let $\mathfrak{I}$ be the database instance mapping these four relation names to the following multiset relations over A:

$$[\![U]\!]_{\mathfrak{I}} = \{(\{m \mapsto \texttt{u}, n \mapsto \texttt{v}\} : 1), (\{m \mapsto \texttt{u}, n \mapsto \texttt{w}\} : 1)\};$$
$$[\![V]\!]_{\mathfrak{I}} = \{(\{m \mapsto \texttt{u}, n \mapsto \texttt{v}\} : 2)\}; \qquad [\![V']\!]_{\mathfrak{I}} = \{(\{m \mapsto \texttt{u}, n \mapsto \texttt{v}\} : 3)\};$$
$$[\![W]\!]_{\mathfrak{I}} = \{(\{m \mapsto \texttt{u}, n \mapsto \texttt{w}\} : 2)\}.$$

We have $V \doteqdot V'$ and, due to $[\![\dot{\pi}_m(U)]\!]_{\mathfrak{I}} = [\![\dot{\pi}_m(V)]\!]_{\mathfrak{I}} = [\![\dot{\pi}_m(W)]\!]_{\mathfrak{I}} = \{(\{m \mapsto \texttt{u}\} : 2)\}$, we have $U \doteq_m V \doteq_m W$. We also have

$$[\![V \mathbin{\dot{-}} V']\!]_{\mathfrak{I}} = \emptyset; \qquad [\![V' \mathbin{\dot{-}} V]\!]_{\mathfrak{I}} = \{(\{m \mapsto \texttt{u}, n \mapsto \texttt{v}\} : 1)\};$$
$$[\![V \mathbin{\dot{\sqcup}} V]\!]_{\mathfrak{I}} = [\![V]\!]_{\mathfrak{I}} \qquad [\![V \mathbin{\dot{\sqcup}} W]\!]_{\mathfrak{I}} = [\![V]\!]_{\mathfrak{I}} \cup [\![W]\!]_{\mathfrak{I}};$$
$$[\![V \mathbin{\dot{\cap}} U]\!]_{\mathfrak{I}} = \{(\{m \mapsto \texttt{u}, n \mapsto \texttt{v}\} : 1)\}; \quad [\![V \mathbin{\dot{\cap}} W]\!]_{\mathfrak{I}} = \emptyset;$$
$$[\![V \mathbin{\dot{-}} U]\!]_{\mathfrak{I}} = \{(\{m \mapsto \texttt{u}, n \mapsto \texttt{v}\} : 1)\}; \quad [\![V \mathbin{\dot{-}} W]\!]_{\mathfrak{I}} = \{(\{m \mapsto \texttt{u}, n \mapsto \texttt{v}\} : 2)\}.$$

Hence, $V \mathbin{\dot{-}} V' \not\doteqdot V' \mathbin{\dot{-}} V$, $V \mathbin{\dot{\sqcup}} V \not\doteq_m V \mathbin{\dot{\sqcup}} W$, $V \mathbin{\dot{\cap}} U \not\doteq_m V \mathbin{\dot{\cap}} W$, $V \mathbin{\dot{\cap}} U \not\doteq_m V \mathbin{\dot{\cap}} W$, and $V \mathbin{\dot{-}} U \not\doteq_m V \mathbin{\dot{-}} W$. Let $\mathfrak{I}'$ be the database instance obtained from $\mathfrak{I}$ by changing all the counts in $[\![U]\!]_{\mathfrak{I}}$ to 2. Then,

$$[\![V \mathbin{\dot{-}} U]\!]_{\mathfrak{I}'} = \emptyset; \qquad [\![V \mathbin{\dot{-}} W]\!]_{\mathfrak{I}'} = \{(\{m \mapsto \texttt{u}, n \mapsto \texttt{v}\} : 2)\},$$

and, hence, we also have $V \mathbin{\dot{-}} U \not\doteq_m V \mathbin{\dot{-}} W$. We must conclude that we cannot hope for meaningful rewrite rules for max-union, intersection, and difference if the conditions of Propositions 8 and 9 are not satisfied.

Besides the rewrite rules we introduced, the usual strong-equivalent (multiset) relational algebra rewrite rules can, of course, also be used in the setting of non-strong-equivalent rewriting. This includes rules for pushing down selections and projections [5,12] and the usual associativity and distributivity rules for union, intersection, difference, and joins [2].

5 Deriving Structural Query Information

Some of the rewrite rules of Proposition 1, Theorem 2, and Proposition 6 can only be applied if it is known that some subexpression is a set relation or has certain keys. Therefore, we introduce rules to derive this information.

We use the well-known functional dependencies as a framework to reason about keys. A *functional dependency over* A is of the form $\text{B} \rightarrow \text{C}$, with $\text{B}, \text{C} \subseteq \text{A}$. Let $\text{B} \rightarrow \text{C}$ be a functional dependency over A. A relation $\mathscr{R}$ over A satisfies $\text{B} \rightarrow \text{C}$ if, for every pair of tuples $\mathsf{r}, \mathsf{s} \in \mathscr{R}$ with $\mathsf{r} \equiv_{\text{B}} \mathsf{s}$, we have $\mathsf{r} \equiv_{\text{C}} \mathsf{s}$. A multiset relation $\langle \mathscr{R}; \tau \rangle$ over A satisfies $\text{B} \rightarrow \text{C}$ if $\mathscr{R}$ satisfies $\text{B} \rightarrow \text{C}$. Let D be a set of functional dependencies over A. The *closure of* D, which we denote by $+(D)$, is the set of all functional dependencies over A that are logically implied by D [1]. By $\text{fds}(e; \mathfrak{D})$, we denote the functional dependencies we derive from expression

e over $\mathfrak{D}$. We define $\text{fds}(e;\mathfrak{D})$ inductively. The base cases are relation names $R \in \mathbf{N}$, for which we have $\text{fds}(R;\mathfrak{D}) = +(\{\text{K} \to \mathbf{A}(R) \mid \text{K} \in \mathbf{K}(R)\})$. For the other operations, we have

$$\begin{aligned}
\text{fds}(\dot{\sigma}_E(e);\mathfrak{D}) &= +(\{\text{A} \to \text{B} \mid \exists \text{C}\ \text{B} \subseteq \det(\text{C};E) \wedge (\text{A} \to \text{C}) \in \text{fds}(e;\mathfrak{D})\});\\
\text{fds}(\dot{\pi}_{\text{C}}(e);\mathfrak{D}) &= \{\text{A} \to \text{B} \cap \text{C} \mid \text{A} \subseteq \text{C} \wedge (\text{A} \to \text{B}) \in \text{fds}(e;\mathfrak{D})\};\\
\text{fds}(\dot{\rho}_f(e);\mathfrak{D}) &= \{f(\text{A}) \to f(\text{B}) \mid (\text{A} \to \text{B}) \in \text{fds}(e;\mathfrak{D})\};\\
\text{fds}(\dot{\delta}(e);\mathfrak{D}) &= \text{fds}(e;\mathfrak{D});\\
\text{fds}(e_1 \mathbin{\dot{\cup}} e_2;\mathfrak{D}) &= +(\emptyset);\\
\text{fds}(e_1 \mathbin{\dot{\cap}} e_2;\mathfrak{D}) &= +((\text{fds}(e_1;\mathfrak{D}) \cup \text{fds}(e_2;\mathfrak{D})));\\
\text{fds}(e_1 \mathbin{\dot{-}} e_2;\mathfrak{D}) &= \text{fds}(e_1;\mathfrak{D});\\
\text{fds}(e_1 \mathbin{\dot{\bowtie}} e_2;\mathfrak{D}) &= +((\text{fds}(e_1;\mathfrak{D}) \cup \text{fds}(e_2;\mathfrak{D})));\\
\text{fds}(e_1 \mathbin{\dot{\bowtie}_E} e_2;\mathfrak{D}) &= \text{fds}(\dot{\sigma}_E(e_1 \mathbin{\dot{\bowtie}} e_2);\mathfrak{D});\\
\text{fds}(e_1 \mathbin{\dot{\ltimes}_E} e_2;\mathfrak{D}) &= \text{fds}(\dot{\pi}_{\mathcal{S}(e_1;\mathfrak{D})}(e_1 \mathbin{\dot{\bowtie}_E} e_2);\mathfrak{D})\\
\text{fds}(e_1 \mathbin{\dot{\sqcup}} e_2;\mathfrak{D}) &= +(\emptyset);\\
\text{fds}(\iota_f(e);\mathfrak{D}) &= +(\{(\{x\} \cap \mathcal{S}(e;\mathfrak{D}) \to \{b\}) \mid (b := x) \in f\} \cup \text{fds}(e;\mathfrak{D})).
\end{aligned}$$

We define $\text{keys}(e;\mathfrak{D}) = \{\text{A} \mid (\text{A} \to \mathcal{S}(e;\mathfrak{D})) \in \text{fds}(e;\mathfrak{D})\}$ to be the keys we derive from expression e over $\mathfrak{D}$. Next, we define the predicate $\text{set}(e;\mathfrak{D})$ which is `true` if we can derive that expression e over $\mathfrak{D}$ always evaluates to a set relation. We define $\text{set}(e;\mathfrak{D})$ inductively. The base cases are relation names $R \in \mathbf{N}$, for which we have $\text{set}(R;\mathfrak{D}) = \mathbf{S}(R)$. For the other operations, we have

$$\begin{aligned}
\text{set}(\dot{\sigma}_E(e);\mathfrak{D}) &= \text{set}(e;\mathfrak{D}); & \text{set}(\dot{\pi}_{\text{B}}(e);\mathfrak{D}) &= \text{set}(e;\mathfrak{D}) \wedge \text{B} \in \text{keys}(e;\mathfrak{D});\\
\text{set}(\dot{\rho}_f(e);\mathfrak{D}) &= \text{set}(e;\mathfrak{D}); & \text{set}(\dot{\delta}(e);\mathfrak{D}) &= \texttt{true};\\
\text{set}(e_1 \mathbin{\dot{\cup}} e_2;\mathfrak{D}) &= \texttt{false}; & \text{set}(e_1 \mathbin{\dot{\cap}} e_2;\mathfrak{D}) &= \text{set}(e_1;\mathfrak{D}) \vee \text{set}(e_2;\mathfrak{D});\\
\text{set}(e_1 \mathbin{\dot{-}} e_2;\mathfrak{D}) &= \text{set}(e_1;\mathfrak{D}); & \text{set}(e_1 \mathbin{\dot{\bowtie}_E} e_2;\mathfrak{D}) &= \text{set}(e_1;\mathfrak{D}) \wedge \text{set}(e_2;\mathfrak{D});\\
\text{set}(e_1 \mathbin{\dot{\ltimes}_E} e_2;\mathfrak{D}) &= \text{set}(e_1;\mathfrak{D}); & \text{set}(e_1 \mathbin{\dot{\sqcup}} e_2;\mathfrak{D}) &= \text{set}(e_1;\mathfrak{D}) \wedge \text{set}(e_2;\mathfrak{D});\\
\text{set}(\iota_f(e);\mathfrak{D}) &= \text{set}(e;\mathfrak{D}).
\end{aligned}$$

The derivation rules for $\text{fds}(e;\mathfrak{D})$ and $\text{set}(e;\mathfrak{D})$ are not complete: it is not guaranteed that $\text{fds}(e;\mathfrak{D})$ contains *all* functional dependencies that must hold in $[\![e]\!]_{\mathfrak{I}}$, for every database instance $\mathfrak{I}$ over $\mathfrak{D}$. Likewise, it is not guaranteed that $\text{set}(e;\mathfrak{D}) = \texttt{false}$ implies that e is *not* a set relation.

Example 9. Let $\mathtt{u}_1, \mathtt{u}_2 \in \mathfrak{U}$ with $\mathtt{u}_1 \neq \mathtt{u}_2$ and let e be an expression over $\mathfrak{D}$. Consider the expressions $e_1 = \dot{\sigma}_{a=\mathtt{u}_1, a=\mathtt{u}_2}(e)$ and $e_2 = e - e$. For every database instance $\mathfrak{I}$ over $\mathfrak{D}$, we have $[\![e_1]\!]_{\mathfrak{I}} = [\![e_2]\!]_{\mathfrak{I}} = \emptyset$. Hence, every functional dependency $\text{A} \to \text{B}$ with $\text{A}, \text{B} \subseteq \mathcal{S}(e;\mathfrak{D})$ holds on $[\![e_1]\!]_{\mathfrak{I}}$ and on $[\![e_2]\!]_{\mathfrak{I}}$. In addition, both $[\![e_1]\!]_{\mathfrak{I}}$ and $[\![e_2]\!]_{\mathfrak{I}}$ are set relations. If $\text{fds}(e;\mathfrak{D}) = \emptyset$, then we derive that $\text{fds}(e_1;\mathfrak{D}) = +(\{\emptyset \to \{a\}\})$ and $\text{fds}(e_2;\mathfrak{D}) = +(\emptyset)$. Observe that $\text{fds}(e_1;\mathfrak{D})$ contains all functional dependencies over $\mathcal{S}(e_1;\mathfrak{D})$ if and only if $\mathcal{S}(e;\mathfrak{D}) = \{a\}$ and that $\text{fds}(e_2;\mathfrak{D})$ never contains all functional dependencies over $\mathcal{S}(e_1;\mathfrak{D})$. We also derive that $\text{set}(e_1; \mathfrak{D}) = \text{set}(e_2; \mathfrak{D}) = \texttt{false}$, despite e_1 and e_2 evaluating to set relations.

Although the above derivation rules are not complete, they are sound:

Theorem 3. *Let e be an expression over $\mathfrak{D}$. The derivation rules for* $\text{fds}(e; \mathfrak{D})$ *and* $\text{set}(e; \mathfrak{D})$ *are sound: if $\mathfrak{I}$ is a database instance over $\mathfrak{D}$, then*

(i) *$[\![e]\!]_{\mathfrak{I}}$ satisfies every functional dependency in* $\text{fds}(e; \mathfrak{D})$*; and*
(ii) *if* $\text{set}(e; \mathfrak{D}) = \texttt{true}$*, then $[\![e]\!]_{\mathfrak{I}}$ is a set relation.*

We introduce the following notions. Let $\langle \mathscr{R}_1; \tau_1 \rangle$ and $\langle \mathscr{R}_2; \tau_2 \rangle$ be multiset relations over A. We say that $\langle \mathscr{R}_1; \tau_1 \rangle$ *is a weak subset of* $\langle \mathscr{R}_2; \tau_2 \rangle$, denoted by $\langle \mathscr{R}_1; \tau_1 \rangle \mathrel{\tilde{\subseteq}} \langle \mathscr{R}_2; \tau_2 \rangle$, if $\mathscr{R}_1 \subseteq \mathscr{R}_2$. We say that $\langle \mathscr{R}_1; \tau_1 \rangle$ *is a strong subset of* $\langle \mathscr{R}_2; \tau_2 \rangle$, denoted by $\langle \mathscr{R}_1; \tau_1 \rangle \mathrel{\dot{\subseteq}} \langle \mathscr{R}_2; \tau_2 \rangle$, if $(\mathtt{t} : n) \in \langle \mathscr{R}_1; \tau_1 \rangle$ implies $(\mathtt{t} : m) \in \langle \mathscr{R}_2; \tau_2 \rangle$ with $n \leq m$.[8] Lemma 2 lists the main properties of these notions needed to prove Theorem 3.

Lemma 2. *Let $\langle \mathscr{R}_1; \tau_1 \rangle$ and $\langle \mathscr{R}_2; \tau_2 \rangle$ be multiset relations over A. We have*

(i) *$\langle \mathscr{R}_1; \tau_1 \rangle \mathrel{\tilde{\subseteq}} \langle \mathscr{R}_2; \tau_2 \rangle$ if $\langle \mathscr{R}_1; \tau_1 \rangle \mathrel{\dot{\subseteq}} \langle \mathscr{R}_2; \tau_2 \rangle$;*
(ii) *$\langle \mathscr{R}_1; \tau_1 \rangle$ satisfies $\mathrm{B} \to \mathrm{C}$ if $\langle \mathscr{R}_1; \tau_1 \rangle \mathrel{\tilde{\subseteq}} \langle \mathscr{R}_2; \tau_2 \rangle$ and $\langle \mathscr{R}_2; \tau_2 \rangle$ satisfies functional dependency $\mathrm{B} \to \mathrm{C}$; and*
(iii) *$\langle \mathscr{R}_1; \tau_1 \rangle$ is a set relation if $\langle \mathscr{R}_1; \tau_1 \rangle \mathrel{\dot{\subseteq}} \langle \mathscr{R}_2; \tau_2 \rangle$ and $\langle \mathscr{R}_2; \tau_2 \rangle$ is a set relation.*

6 Rewriting the Example Queries

To illustrate the techniques introduced in this paper, we provide a detailed rewriting of the queries exhibited in Example 3. We start by considering the queries $\dot{\delta}(\dot{\pi}_{age}(e))$ and $\dot{\delta}(\dot{\pi}_{cname,type}(e))$ with $e = \dot{\sigma}_{type=\texttt{food}}(Customer \bowtie Bought \bowtie Product)$. As a first step, we use well-known push-down rules for selection [12], and we get $e \doteq Customer \bowtie Bought \bowtie_{type=\texttt{food}} Product$.

Consider the query $e' = Bought \bowtie_{type=\texttt{food}} Product$, subquery of the above query. We have $\{pname\} \in \text{keys}(Product; \mathfrak{D})$ and $Product$ is a set relation. Hence, we can apply Theorem 2.*(iv)* and Theorem 2.*(vi)* with $f = \{type := \texttt{food}\}$, and we obtain $e' \doteq_{\{cname,pname\}} Bought \ltimes_{type=\texttt{food}} Product$ and $e' \doteq_{\{cname,pname,type\}} i_f(Bought \ltimes_{type=\texttt{food}} Product)$. Let $e'' = Bought \ltimes_{type=\texttt{food}} Product$. For the rewriting of $\dot{\delta}(\dot{\pi}_{age}(e))$, we use Proposition 1.*(iii)* and Proposition 1.*(v)* and infer $e' \mathrel{\tilde{\doteq}}_{\{cname\}} e''$. Using Theorem 2.*(iii)* and transitivity, we infer $e \mathrel{\dot{=}}_{\{cname,age\}} Customer \ltimes e''$, and, using Proposition 1.*(v)*, we infer $e \mathrel{\tilde{\doteq}}_{\{age\}} Customer \ltimes e''$. Finally, we can apply Proposition 1.*(iv)* and Proposition 1.*(vi)* to conclude $\dot{\delta}(\dot{\pi}_{age}(e)) \doteq \dot{\delta}(\dot{\pi}_{age}(Customer \ltimes e''))$. Hence, $\dot{\delta}(\dot{\pi}_{age}(e)) \doteq \dot{\delta}(\dot{\pi}_{age}(Customer \ltimes (Bought \ltimes_{type=\texttt{food}} Product)))$, which is the query resulting from optimizing $\dot{\delta}(\dot{\pi}_{age}(e))$ in Example 3.

---

[8] Notice that $\langle \mathscr{R}_1; \tau_1 \rangle \mathrel{\tilde{\subseteq}} \langle \mathscr{R}_2; \tau_2 \rangle$ does not imply that $\langle \mathscr{R}_1; \tau_1 \rangle$ is fully included in $\langle \mathscr{R}_2; \tau_2 \rangle$: there can be tuples $\mathtt{t} \in \mathscr{R}_1$ for which $\tau_1(\mathtt{t}) > \tau_2(\mathtt{t})$.

For the rewriting of $\dot{\delta}(\dot{\pi}_{cname,type}(e))$, we directly use Proposition 1.*(iii)* to obtain $e' \doteq_{\{cname,type\}} i_f(e'')$. We use Theorem 2.*(ii)* and transitivity to obtain $e \doteq_{\{cname,type\}} Customer \mathbin{\dot{\bowtie}} i_f(e'')$. Using Proposition 2.*(vi)* and commutativity of θ-join, we conclude $Customer \mathbin{\dot{\bowtie}} i_f(e'') \doteq_{\{cname,type\}} i_f(Customer \mathbin{\dot{\bowtie}} e'')$. Next, consider the query $Customer \mathbin{\dot{\bowtie}} e''$, subquery of the query $i_f(Customer \mathbin{\dot{\bowtie}} e'')$. Using Theorem 2.*(iii)*, we infer $Customer \mathbin{\dot{\bowtie}} e'' \mathrel{\tilde{\doteq}}_{\{cname\}} Customer \mathbin{\dot{\ltimes}} e''$ and we apply Proposition 7.*(v)* with f on both sides to obtain $i_f(Customer \mathbin{\dot{\bowtie}} e'') \mathrel{\tilde{\doteq}}_{\{cname,type\}} i_f(Customer \mathbin{\dot{\ltimes}} e'')$. Using transitivity, we get $e \mathrel{\tilde{\doteq}}_{\{cname,type\}} i_f(Customer \mathbin{\dot{\ltimes}} e'')$ and we apply Proposition 7.*(ii)* with $\mathrm{B} = \{cname, type\}$ to obtain $\dot{\pi}_{\mathrm{B}}(e) \mathrel{\tilde{\doteq}}_{\{cname,type\}} \dot{\pi}_{\mathrm{B}}(i_f(Customer \mathbin{\dot{\ltimes}} e''))$. Next, we apply Proposition 2.*(ii)* on $\dot{\pi}_{\mathrm{B}}(i_f(Customer \mathbin{\dot{\ltimes}} e''))$ and transitivity to obtain $\dot{\pi}_{\mathrm{B}}(e) \mathrel{\tilde{\doteq}}_{\{cname,type\}} i_f(\dot{\pi}_{cname}(Customer \mathbin{\dot{\ltimes}} e'')$. Then we apply Proposition 7.*(iv)* on both sides and we use Propositions 1.*(iv)* and 1.*(vii)* to obtain $\dot{\delta}(\dot{\pi}_{\mathrm{B}}(e)) \doteq \dot{\delta}(i_f(\dot{\pi}_{cname}(Customer \mathbin{\dot{\ltimes}} e'')))$. We observe that *Customer* is a set relation. Hence, $i_f(\dot{\pi}_{cname}(Customer \mathbin{\dot{\ltimes}} e''))$ is also set relation. Finally, we can now apply Proposition 6 and transitivity to conclude $\dot{\delta}(\dot{\pi}_{cname,type}(e)) \doteq i_f(\dot{\pi}_{cname}(Customer \mathbin{\dot{\ltimes}} e''))$ and, hence, $\dot{\delta}(\dot{\pi}_{cname,type}(e)) \doteq i_{type:=\texttt{food}}(\dot{\pi}_{cname}(Customer \mathbin{\dot{\ltimes}} (Bought \mathbin{\dot{\ltimes}}_{type=\texttt{food}} Product)))$, which is the query resulting from optimizing $\dot{\delta}(\dot{\pi}_{cname,type}(e))$ in Example 3.

7 Conclusion

In this work, we provide a formal framework for optimizing SQL queries using query rewriting rules aimed at optimizing query evaluation for relational algebra queries using multiset semantics. The main goals of our rewrite rules are the automatic elimination of costly join steps, in favor of semi-join steps, and the automatic elimination of deduplication steps. We believe that our rewrite rules can be applied to many practical queries on which traditional techniques fall short. Hence, we believe that our results provide a promising strengthening of traditional query rewriting and optimization techniques. Based upon the ideas of our work, there are several clear directions for future work.

We have primarily studied the automatic rewriting of queries using joins into queries using semi-joins instead. In the setting of SQL, this optimization is usually obtained by rewriting joins into `WHERE ... IN ...` clauses. The anti-semi-join plays a similar role in performing `WHERE ... NOT IN ...` clauses. As such, it is only natural to ask whether our framework can be extended to also automatically rewrite towards anti-semi-join operators. A careful investigation is needed to fully incorporate anti-semi-joins in our framework, however.

The multiset relational algebra we studied does not cover all features provided by SQL. Among the missing features are aggregation and recursive queries (via `WITH RECURSIVE`), and both are candidates for further study. With respect to recursive queries, we observe that in the setting of graph query languages, usage of transitive closure to express reachability can automatically be rewritten to very fast fixpoint queries [6,7]. Similar optimizations also apply to simple `WITH RECURSIVE` queries, but it remains open whether a general technique exists to optimize such queries.

Acknowledgement. This material is based upon work supported by the National Science Foundation under Grant No. #1606557.

References

1. Abiteboul, S., Hull, R., Vianu, V. (eds.): Foundations of Databases, 1st edn. Addison-Wesley Publishing Company, Boston (1995)
2. Albert, J.: Algebraic properties of bag data types. In: Proceedings of the 17th International Conference on Very Large Data Base, pp. 211–219. VLDB 1991, Morgan Kaufmann Publishers Inc. (1991)
3. Bernstein, P.A., Chiu, D.M.W.: Using semi-joins to solve relational queries. J. ACM **28**(1), 25–40 (1981). https://doi.org/10.1145/322234.322238
4. Dayal, U., Goodman, N., Katz, R.H.: An extended relational algebra with control over duplicate elimination. In: Proceedings of the 1st ACM SIGACT-SIGMOD Symposium on Principles of Database Systems, pp. 117–123. PODS 1982, ACM (1982). https://doi.org/10.1145/588111.588132
5. Grefen, P.W.P.J., de By, R.A.: A multi-set extended relational algebra: a formal approach to a practical issue. In: Proceedings of 1994 IEEE 10th International Conference on Data Engineering, pp. 80–88. IEEE (1994). https://doi.org/10.1109/ICDE.1994.283002
6. Hellings, J., Pilachowski, C.L., Van Gucht, D., Gyssens, M., Wu, Y.: From relation algebra to semi-join algebra: an approach for graph query optimization. In: Proceedings of The 16th International Symposium on Database Programming Languages. ACM (2017). https://doi.org/10.1145/3122831.3122833
7. Hellings, J., Pilachowski, C.L., Van Gucht, D., Gyssens, M., Wu, Y.: From relation algebra to semi-join algebra: an approach to graph query optimization. Comput. J. **64**(5), 789–811 (2020). https://doi.org/10.1093/comjnl/bxaa031
8. International Organization for Standardization: ISO/IEC 9075–1: Information technology - database languages - SQL (2011)
9. Klausner, A., Goodman, N.: Multirelations: semantice and languages. In: Proceedings of the 11th International Conference on Very Large Data Bases, pp. 251–258. VLDB 1985, VLDB Endowment (1985)
10. Lamperti, G., Melchiori, M., Zanella, M.: On multisets in database systems. In: Calude, C.S., Paun, G., Rozenberg, G., Salomaa, A. (eds.) WMC 2000. LNCS, vol. 2235, pp. 147–215. Springer, Heidelberg (2001). https://doi.org/10.1007/3-540-45523-X_9
11. Paulley, G.N.: Exploiting Functional Dependence in Query Optimization. Ph.D. thesis, University of Waterloo (2000)
12. Ullman, J.D.: Principles of Database and Knowledge-Base Systems: Volume II: The New Technologies. W.H. Freeman & Co, San Francisco (1990)
13. Yannakakis, M.: Algorithms for acyclic database schemes. In: Proceedings of the Seventh International Conference on Very Large Data Bases, vol. 7, pp. 82–94. VLDB 1981, VLDB Endowment (1981)

Properties of System W and Its Relationships to Other Inductive Inference Operators

Jonas Haldimann and Christoph Beierle(✉)

Faculty of Mathematics and Computer Science,
FernUniversität in Hagen, 58084 Hagen, Germany
{jonas.haldimann,christoph.beierle}@fernuni-hagen.de

Abstract. System W is a recently introduced inference method for conditional belief bases with some notable properties like capturing system Z and thus rational closure and, in contrast to system Z, fully satisfying syntax splitting. This paper further investigates properties of system W. We show how system W behaves with respect to postulates put forward for nonmonotonic reasoning like rational monotony, weak rational monotony, or semi-monotony. We develop tailored postulates ensuring syntax splitting for any inference operator based on a strict partial order on worlds. By showing that system W satisfies these axioms, we obtain an alternative and more general proof that system W satisfies syntax splitting. We explore how syntax splitting affects the strict partial order underlying system W and exploit this for answering certain queries without having to determine the complete strict partial order. Furthermore, we investigate the relationships among system W and other inference methods, showing that, for instance, lexicographic inference extends both system W and c-inference, and leading to a full map of interrelationships among various inductive inference operators.

1 Introduction

For answering the question of what should be entailed by a conditional belief base [21] many approaches have been proposed. Established examples are p-entailment [19] which coincides with system P [1], system Z [9,24] which coincides with rational closure [21], or lexicographic inference [22]. Two further model based approaches are reasoning with a single c-representation [12,13] and c-inference [2] which is defined with respect to all c-representations of a belief base.

This paper addresses the inference operator system W [17,18] that has already been shown to have interesting properties like capturing and extending the inference relations induced by system Z and c-inference or avoiding the drowning problem [6]. In a recent short paper, it has been pointed out that system W also satisfies syntax splitting [10], a concept originally introduced by Parikh [23] for postulates for the revision of belief sets and later extended to other frameworks [15,25] and to inductive inference from conditional belief bases [14].

I. Varzinczak (Ed.): FoIKS 2022, LNCS 13388, pp. 206–225, 2022.
https://doi.org/10.1007/978-3-031-11321-5_12

The main objective of this paper is to advance the investigation of properties of system W. We introduce the notion of strict partial order-based (SPO-based) inductive inference operators of which system W is an instance. We adapt the syntax splitting postulates from [14] to SPO-based inductive inference operators. By showing that system W fulfils the resulting postulates we extend our work in [10] and obtain a full formal proof that system W complies to syntax splitting. Note that the highly desirable property of syntax splitting is indeed a distinguishing feature: Among the inference methods investigated in [14], only reasoning with c-representations satisfies syntax splitting. Our results make system W, besides c-inference, another inference operator to fully comply with syntax splitting; in addition, system W also extends rational closure [21] which is not the case for c-inference. Furthermore, we demonstrate how splitting properties may be exploited in certain situations for simplifying query answering with system W even if the syntax splitting postulates are not applicable.

While it is known that system W extends rational closure, we establish the properties of system W regarding rationality and monotony related postulates suggested for nonmonotonic reasoning. Finally, we elaborate the relationship of system W to other inference relations, leading to a full map of interrelationships among inductive inference operators.

In summary, the main contributions of this paper are

- adapting splitting postulates to SPO-based inductive inference operators,
- proving that system W fulfils these syntax splitting postulates,
- investigating rationality and monotony related postulates for system W,
- showing how syntax splitting can be exploited to simplify query answering for system W even in cases where syntax splitting postulates are not applicable,
- establishing a map of relations among system W and other inference methods.

In Sect. 2, we recall the background on conditional logic. In Sect. 3 we introduce SPO-based inductive inference operators and corresponding syntax splitting postulates. We recall system W and investigate its properties in Sect. 4. Section 5 shows that system W fulfils the syntax splitting postulates. Section 6 deals with the effect of syntax splittings on the induced SPO and how this can be helpful for answering certain queries. In Sect. 7 we elaborate the relationships among inductive inference operators, before we conclude in Sect. 8.

2 Reasoning with Conditional Logic

A *(propositional) signature* is a finite set Σ of propositional variables. For a signature Σ, we denote the propositional language over Σ by $\mathcal{L}_\Sigma$. Usually, we denote elements of signatures with lowercase letters $a, b, c, \ldots$ and formulas with uppercase letters $A, B, C, \ldots$. We may denote a conjunction $A \wedge B$ by AB and a negation $\neg A$ by $\overline{A}$ for brevity of notation. The set of interpretations over a signature Σ is denoted as Ω_Σ. Interpretations are also called *worlds* and Ω_Σ is called the *universe*. An interpretation $\omega \in \Omega_\Sigma$ is a *model* of a formula $A \in \mathcal{L}_\Sigma$ if A holds in ω. This is denoted as $\omega \models A$. The set of models of a formula (over a signature Σ) is denoted as $Mod_\Sigma(A) = \{\omega \in \Omega_\Sigma \mid \omega \models A\}$ or sometimes as Ω_A. The Σ in $Mod_\Sigma(A)$ can be omitted if the signature is clear from the context. A formula A *entails* a formula B if $Mod_\Sigma(A) \subseteq Mod_\Sigma(B)$. By slight abuse of notation we sometimes interpret worlds as the corresponding complete conjunction of all elements in the signature in either positive or negated form.

Worlds over (sub-)signatures can be merged or marginalized. Let Σ be a signature with disjoint sub-signatures Σ_1, Σ_2 such that $\Sigma = \Sigma_1 \cup \Sigma_2$. Let $\omega_1 \in \Omega_{\Sigma_1}$ and $\omega_2 \in \Omega_{\Sigma_2}$. Then $(\omega_1 \cdot \omega_2)$ denotes the world from Ω_Σ that assigns the truth values for variables in Σ_1 as ω_1 and truth values for variables in Σ_2 as ω_2. For $\omega \in \Omega_\Sigma$, the world from Ω_{Σ_1} that assigns the truth values for variables in Σ_1 as ω is denoted as $\omega_{|\Sigma_1}$.

A *conditional* $(B|A)$ connects two formulas A, B and represents the rule "If A then usually B", where A is called the *antecedent* and B the *consequent* of the conditional. The conditional language over a signature Σ is denoted as $(\mathcal{L}|\mathcal{L})_\Sigma = \{(B|A) \mid A, B \in \mathcal{L}_\Sigma\}$. A finite set of conditionals is called a *conditional belief base*. We use a three-valued semantics of conditionals in this paper [8]. For a world ω a conditional $(B|A)$ is either *verified* by ω if $\omega \models AB$, *falsified* by ω if $\omega \models A\overline{B}$, or *not applicable* to ω if $\omega \models \overline{A}$. Popular models for conditional belief bases are ranking functions (also called ordinal conditional functions, OCF) [27] and total preorders (TPO) on Ω_Σ [7].

3 Syntax Splitting and SPO-Based Inductive Inference

Reasoning with conditionals is often modelled by inference relations. An *inference relation* is a binary relation $\mid\!\sim$ on formulas over an underlying signature Σ with the intuition that $A \mid\!\sim B$ means that A (plausibly) entails B. (Non-monotonic) inference is closely related to conditionals: an inference relation $\mid\!\sim$ can also be seen as a set of conditionals $\{(B|A) \mid A, B \in \mathcal{L}_\Sigma, A \mid\!\sim B\}$. The following definition formalizes the inductive completion of a belief base to an inference relation.

Definition 1 (inductive inference operator [14]). *An* inductive inference operator *is a mapping $C : \Delta \mapsto \mid\!\sim_\Delta$ that maps each belief base to an inference relation such that direct inference (DI) and trivial vacuity (TV) are fulfilled, i.e.,*

(DI) *if $(B|A) \in \Delta$ then $A \mid\!\sim_\Delta B$ and*
(TV) *if $\Delta = \emptyset$ and $A \mid\!\sim_\Delta B$ then $A \models B$.*

Examples for inductive inference operators are p-entailment [19], system Z [24], model-based reasoning with respect to a c-representation [12,13], or various forms of c-inference taking sets of c-representations into account [2,3].

The concept of syntax splitting was originally developed by Parikh [23] describing that a belief set contains independent information over different parts of the signature. He proposed a postulate (P) stating that for a belief set with a syntax splitting the revision with a formula relevant to only one such part should only affect the information about that part of the signature. The notion of syntax splitting was later extended to other representations of beliefs such as ranking functions [15] and belief bases [14]. Furthermore, in [14] properties for inductive inference operators, that govern the behaviour for belief bases with syntax splitting, were formulated. To formulate syntax splitting postulates for inductive inference operators, we need a notion of syntax splitting for belief bases.

Definition 2 (syntax splitting for belief bases (adapted from [14])). *Let Δ be a belief base over a signature Σ. A partitioning $\{\Sigma_1, \dots, \Sigma_n\}$ of Σ is a* syntax splitting *for Δ if there is a partitioning $\{\Delta_1, \dots, \Delta_n\}$ of Δ such that $\Delta_i \subseteq (\mathcal{L}|\mathcal{L})_{\Sigma_i}$ for every $i = 1, \dots, n$.*

In this paper, we focus on syntax splittings $\{\Sigma_1, \Sigma_2\}$ of Δ with two parts and corresponding partition $\{\Delta_1, \Delta_2\}$, denoted as $\Delta = \Delta_1 \bigcup\limits_{\Sigma_1,\Sigma_2} \Delta_2$. Results for belief bases with syntax splittings in more than two parts can be obtained by iteratively applying the postulates and constructions presented here.

Example 1 (Δ_{ve}). Consider the signature $\Sigma = \{m, b, e, t, g\}$ for modelling aspects about vehicles with the intended meaning (m) being a *m*otorized vehicle, (b) being a *b*ike, (e) having an *e*lectric motor, (t) having *t*wo wheels, and (g) requiring *g*asoline. The belief base

$$\Delta_{ve} = \{(m|e), (g|m), (\overline{g}|me), (t|b)\}$$

over Σ states that vehicles with an electric motor are usually motorized vehicles, motorized vehicles usually require gasoline, motorized vehicles with an electric motor usually do not require gasoline, and bikes usually have two wheels. This belief base has a syntax splitting $\Delta_{ve} = \Delta_1 \bigcup\limits_{\Sigma_1,\Sigma_2} \Delta_2$ with $\Sigma_1 = \{m, e, g\}$, $\Sigma_2 = \{b, t\}$ and

$$\Delta_1 = \{(m|e), (g|m), (\overline{g}|me)\}, \ \Delta_2 = \{(t|b)\}.$$

For belief bases with syntax splitting, the postulate (Rel) describes that conditionals corresponding to one part of the syntax splitting do not have any influence on inferences that only use the other part of the syntax splitting, i.e., that only conditionals from the considered part of the syntax splitting are relevant.

(Rel) [14] An inductive inference operator $C : \Delta \mapsto {\mid\!\sim}_\Delta$ satisfies **(Rel)** if for any $\Delta = \Delta_1 \bigcup\limits_{\Sigma_1,\Sigma_2} \Delta_2$, and for any $A, B \in \mathcal{L}_{\Sigma_i}$ for $i = 1, 2$ we have that

$$A {\mid\!\sim}_\Delta B \quad \text{iff} \quad A {\mid\!\sim}_{\Delta_i} B. \tag{1}$$

The postulate (Ind) describes that inferences should not be affected by beliefs in formulas over other sub-signatures in the splitting, i.e., inferences using only atoms from one part of the syntax splitting should be drawn independently of beliefs about other parts of the splitting.

(Ind) [14] An inductive inference operator $C : \Delta \mapsto \mathrel{|\!\sim}_\Delta$ satisfies **(Ind)** if for any $\Delta = \Delta_1 \bigcup_{\Sigma_1,\Sigma_2} \Delta_2$, and for any $A, B \in \mathcal{L}_{\Sigma_i}$, $D \in \mathcal{L}_{\Sigma_j}$ for $i, j \in \{1,2\}, i \neq j$ such that D is consistent, we have

$$A \mathrel{|\!\sim}_\Delta B \quad \text{iff} \quad AD \mathrel{|\!\sim}_\Delta B. \tag{2}$$

Syntax splitting is the combination of (Rel) and (Ind):

(SynSplit) [14] An inductive inference operator satisfies **(SynSplit)** if it satisfies (Rel) and (Ind).

The effect of the postulates is illustrated by the following example.

Example 2. Consider the belief base $\Delta_{ve} = \Delta_1 \bigcup_{\Sigma_1,\Sigma_2} \Delta_2$ from Example 1. We can deduce $b \mathrel{|\!\sim}_{\Delta_{ve}} t$ with any inductive inference operator $C : \Delta \mapsto \mathrel{|\!\sim}_\Delta$ because of (DI). The postulate (Rel) requires that $b \mathrel{|\!\sim}_{\Delta_2} t$ should hold because the formulas b and t contain only variables from Σ_2. The postulate (Ind) requires that $be \mathrel{|\!\sim}_\Delta t$ should hold because e contains only variables from Σ_1.

Note that in Example 2 we cannot deduce either $be \mathrel{|\!\sim}^z_{\Delta_{ve}} t$ with system Z or $be \mathrel{|\!\sim}^P_{\Delta_{ve}} t$ with system P; the additional information e from an independent part of the signature prevents the deduction of t. Therefore, both p-entailment and system Z do not fulfil (SynSplit).

In [14], the properties (Rel), (Ind), and (SynSplit) are lifted to total preorders and ranking functions since both TPOs and OCFs induce inference relations. But there are also nonmonotonic inference methods induced by a strict partial order on worlds that cannot be expressed by a TPO or an OCF, e.g., structural inference [16] or system W [18]. Therefore, in the following we will develop syntax splitting postulates tailored to inference relations induced by strict partial orders.

Strict partial orders (SPOs) are irreflexive, transitive, and antisymmetric binary relations. Analogously to inductive inference operator for TPOs, a SPO-based inductive inference operators uses a SPO on worlds as intermediate step. To define SPO-based inductive inference operators, we lift SPOs on worlds to formulas first.

Definition 3. *Let $\prec$ be a strict partial order on Ω_Σ and $A, B \in \mathcal{L}_\Sigma$. Then*

$$A \prec B \quad \textit{iff} \quad \textit{for every } \omega' \in \Omega_B \textit{ there is an } \omega \in \Omega_A \textit{ such that } \omega \prec \omega'. \tag{3}$$

Note that we can use (3) to define when a strict partial order $\prec$ on worlds is a model of a conditional $(B|A)$ by setting $\prec \models (B|A)$ iff $AB \prec A\overline{B}$, i.e., iff the verification AB of $(B|A)$ is strictly smaller in $\prec$ than the falsification $A\overline{B}$ of $(B|A)$; furthermore $\prec \models \Delta$ if $\prec \models (B|A)$ for every $(B|A) \in \Delta$. Using this concept, every strict partial order $\prec$ induces an inference relation $\mathrel{|\!\sim}_\prec$ by

$$A \mathrel{|\!\sim}_\prec B \quad \text{iff} \quad AB \prec A\overline{B}. \tag{4}$$

Definition 4 (SPO-based inductive inference operator). *A* SPO-based inductive inference operator *is a mapping* $C^{spo} : \Delta \mapsto \prec_\Delta$ *that maps a belief base to a SPO* $\prec_\Delta$ *such that* $\prec_\Delta \models \Delta$ *and* $\prec_\emptyset = \emptyset$*. The induced inference* $\mid\!\sim_\Delta$ *is obtained from* $\prec_\Delta$ *as in* (4).

The syntax splitting postulates can now be formulated for the case of SPO-based inductive inference operators.

(Rel^{spo}) A SPO-based inductive inference operator $C^{spo} : \Delta \mapsto \prec_\Delta$ satisfies **(Rel^{spo})** if for $\Delta = \Delta_1 \bigcup_{\Sigma_1,\Sigma_2} \Delta_2$ and for $A, B \in \mathcal{L}_{\Sigma_i}$ for $i = 1, 2$, we have

$$A \prec_\Delta B \quad \text{iff} \quad A \prec_{\Delta_i} B. \tag{5}$$

(Ind^{spo}) A SPO-based inductive inference operator $C^{spo} : \Delta \mapsto \prec_\Delta$ satisfies **(Ind^{spo})** if for any $\Delta = \Delta_1 \bigcup_{\Sigma_1,\Sigma_2} \Delta_2$, and for any $A, B \in \mathcal{L}_{\Sigma_i}$, $D \in \mathcal{L}_{\Sigma_j}$ for $i, j \in \{1, 2\}, i \neq j$, such that D is consistent, we have

$$A \prec_\Delta B \quad \text{iff} \quad AD \prec_\Delta BD. \tag{6}$$

(SynSplit^{spo}) A SPO-based inductive inference operator $C^{spo} : \Delta \mapsto \prec_\Delta$ satisfies **(SynSplit^{spo})** if it satisfies (Rel^{spo}) and (Ind^{spo}).

The new postulates for SPO-based inductive inference operators cover the initial postulates (Rel) and (Ind).

Proposition 1. *If a SPO-based inductive inference operator satisfies (Rel^{spo}), then it satisfies (Rel).*

Proof. Let $\Delta = \Delta_1 \bigcup_{\Sigma_1,\Sigma_2} \Delta_2$. Let $i \in \{1, 2\}$ and $A, B \in \mathcal{L}_{\Sigma_i}$. As $AB, A\overline{B} \in \mathcal{L}_{\Sigma_i}$, we have $A \mid\!\sim_\Delta B$ iff $AB \prec_\Delta A\overline{B}$ iff $AB \prec_{\Delta_i} A\overline{B}$ iff $A \mid\!\sim_{\Delta_i} B$. □

Proposition 2. *If an inductive inference operator for strict partial orders satisfies (Ind^{spo}), then it satisfies (Ind).*

Proof. Let $\Delta = \Delta_1 \bigcup_{\Sigma_1,\Sigma_2} \Delta_2$. Let $i, j \in \{1, 2\}, i \neq j$ and $A, B \in \mathcal{L}_{\Sigma_i}$, $D \in \mathcal{L}_{\Sigma_j}$ such that D is consistent. As $AB, A\overline{B} \in \mathcal{L}_{\Sigma_i}$, we have $A \mid\!\sim_\Delta B$ iff $AB \prec_\Delta A\overline{B}$ iff $ABD \prec_\Delta A\overline{B}D$ iff $AD \mid\!\sim_\Delta B$. □

Hence, if a SPO-based inductive inference operator C satisfies both (Rel^{spo}) and (Ind^{spo}), and thus also (SynSplit^{spo}), then it also satisfies (SynSplit).

Having the postulates for SPO-based inference at hand enables easier and more succinct proofs for showing that an inductive inference operator for SPOs satisfies syntax splitting. In the following, we will exploit this for showing that system W fully complies with syntax splitting.

4 System W and the Preferred Structure on Worlds

System W is an inductive inference operator [18] that takes into account the inclusion maximal tolerance partition of a belief base Δ, which is also used for the definition of system Z [24].

Definition 5 (inclusion maximal tolerance partition [24]**).** *A conditional* $(B|A)$ *is* tolerated *by* $\Delta = \{(B_i|A_i) \mid i = 1,\dots,n\}$ *if there exists a world* $\omega \in \Omega_\Sigma$ *such that* ω *verifies* $(B|A)$ *and* ω *does not falsify any conditional in* Δ*, i.e.,* $\omega \models AB$ *and* $\omega \models \bigwedge_{i=1}^{n}(\overline{A_i} \vee B_i)$*. The inclusion maximal* tolerance partition $OP(\Delta) = (\Delta^0,\dots,\Delta^k)$ *of a consistent belief base* Δ *is the ordered partition of* Δ *where each* Δ^i *is the inclusion maximal subset of* $\bigcup_{j=i}^{n} \Delta^j$ *that is tolerated by* $\bigcup_{j=i}^{n} \Delta^j$.

It is well-known that $OP(\Delta)$ exists iff Δ is consistent; moreover, because the Δ^i are chosen inclusion-maximal, the tolerance partitioning is unique [24].

In addition to $OP(\Delta)$, system W also takes into account the structural information which conditionals are falsified. System W is based on a binary relation called *preferred structure on worlds* $<^{\mathsf{w}}_{\Delta}$ over Ω_Σ induced by every consistent belief base Δ.

Definition 6 (ξ^j, ξ, preferred structure $<^{\mathsf{w}}_{\Delta}$ on worlds [18]**).** *Consider a consistent belief base* $\Delta = \{r_i = (B_i|A_i) \mid i \in \{1,\dots,n\}\}$ *with the tolerance partition* $OP(\Delta) = (\Delta^0,\dots,\Delta^k)$*. For* $j = 0,\dots,k$*, the functions* ξ^j *and* ξ *are the* functions mapping worlds to the set of falsified conditionals *from the set* Δ^j *in the tolerance partition and from* Δ*, respectively, given by*

$$\xi^j(\omega) := \{r_i \in \Delta^j \mid \omega \models A_i\overline{B_i}\}, \tag{7}$$

$$\xi(\omega) := \{r_i \in \Delta \mid \omega \models A_i\overline{B_i}\}. \tag{8}$$

The preferred structure on worlds *is given by the binary relation* $<^{\mathsf{w}}_{\Delta} \subseteq \Omega \times \Omega$ *defined by, for any* $\omega, \omega' \in \Omega$*,*

$$\omega <^{\mathsf{w}}_{\Delta} \omega' \text{ iff there exists an } m \in \{0,\dots,k\} \text{ such that}$$

$$\xi^i(\omega) = \xi^i(\omega') \quad \forall i \in \{m+1,\dots,k\} \text{ and} \tag{9}$$

$$\xi^m(\omega) \subsetneqq \xi^m(\omega'). \tag{10}$$

Thus, $\omega <^{\mathsf{w}}_{\Delta} \omega'$ if and only if ω falsifies strictly fewer conditionals than ω' in the partition with the biggest index m where the conditionals falsified by ω and ω' differ. Note, that $<^{\mathsf{w}}_{\Delta}$ is a strict partial order [18, Lemma 3].

Definition 7 (system W, $\mid\!\sim^{\mathsf{w}}_{\Delta}$ [18]**).** *Let* Δ *be a belief base and* A, B *be formulas. Then* B *is a* system W inference from A (in the context of Δ)*, denoted* $A \mid\!\sim^{\mathsf{w}}_{\Delta} B$*, if for every* $\omega' \in \Omega_{A\overline{B}}$ *there is an* $\omega \in \Omega_{AB}$ *such that* $\omega <^{\mathsf{w}}_{\Delta} \omega'$*.*

Thus, employing Definition 4, since $<^{\mathsf{w}}_{\Delta}$ is a strict partial order, system W is a SPO-based inductive inference operator $C^{\mathsf{w}} : \Delta \mapsto <^{\mathsf{w}}_{\Delta}$.

System W fulfils system P, extends system Z [24] and c-inference [2], and enjoys further desirable properties for nonmonotonic reasoning like avoiding the drowning problem [6,18]. We illustrate system W with an example.

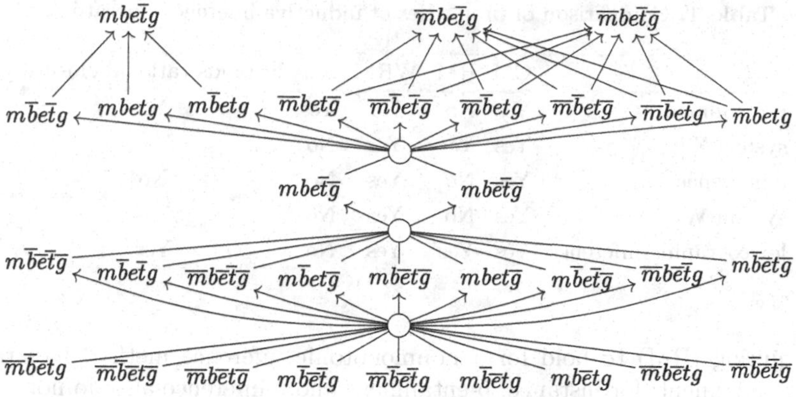

Fig. 1. The preferred structure on worlds induced by the belief base Δ_{ve} from Example 3. Edges that can be obtained from transitivity are omitted.

Example 3. Consider the belief base $\Delta_{ve} = \{(m|e), (g|m), (\overline{g}|me), (t|b)\}$ over the signature $\Sigma = \{m, b, e, t, g\}$ from Example 1. The inclusion maximal tolerance partition of Δ is $OP(\Delta_{ve}) = (\Delta^0, \Delta^1)$ with $\Delta^0 = \{(g|m), (t|b)\}$ and $\Delta^1 = \{(m|e), (\overline{g}|me)\}$. The resulting preferred structure $<^{\mathsf{w}}_{\Delta_{ve}}$ is given in Fig. 1. Using $<^{\mathsf{w}}_{\Delta_{ve}}$ we can verify that for each world ω' with $\omega' \models be\overline{t}$ there is a world ω with $\omega \models bet$ such that $\omega <^{\mathsf{w}}_{\Delta_{ve}} \omega'$. Therefore, with system W we can infer $be \mid\!\sim^{\mathsf{w}}_{\Delta_{ve}} t$, complying with (SynSplit) (see Example 2).

As system W is a form of nonmonotonic reasoning, it does not fulfil monotony. For instance, in Example 3 we have $m \mid\!\sim^{\mathsf{w}}_{\Delta_{ve}} g$ but $me \not\mid\!\sim^{\mathsf{w}}_{\Delta_{ve}} g$.

Several weaker notions of monotony have been introduced for nonmonotonic reasoning. As system W fulfils system P, system W fulfils cautious monotony:

(CM) If $A \mid\!\sim B$ and $A \mid\!\sim C$ then $AB \mid\!\sim C$.

Three further monotony related properties are rational monotony (RM), weak rational monotony (WRM), and semi-monotony (SM) [9,26]:

(RM) If $A \mid\!\sim B$ and $A \not\mid\!\sim \overline{C}$ then $AC \mid\!\sim B$.
(WRM) If $\top \mid\!\sim B$ and $\top \not\mid\!\sim \overline{A}$ then $A \mid\!\sim B$.
(SM) If $\Delta \subseteq \Delta'$ and $A \mid\!\sim_{\Delta} B$ then $A \mid\!\sim_{\Delta'} B$.

First, we observe that system W does not satisfy rational monotony, in contrast to, e.g., system Z or lexicographic inference [22].

Proposition 3. *System W does not fulfil rational monotony (RM).*

Proof. Let $\Sigma = \{a, b, c\}$. Consider the belief base $\Delta = \{(\overline{a}c \vee \overline{b}c|\top), (\overline{a} \vee \overline{b}|\top), (\overline{a}|\top), (\overline{a}\overline{b}|\top)\}$ and the formulas $A = \neg(\overline{a}\overline{b}c)$, $B = \overline{a}\overline{b}\overline{c} \vee abc$, and $C = \neg(\overline{a}c)$. We have $A \mid\!\sim^{\mathsf{w}}_{\Delta} B$ and $A \not\mid\!\sim^{\mathsf{w}}_{\Delta} \overline{C}$ but not $AC \mid\!\sim^{\mathsf{w}}_{\Delta} B$. □

Table 1. Comparison of properties of inductive inference operators.

	CM	RM	WRM	SM	Extends rational closure
p-entailment	Yes	No	No	Yes	No
system Z	Yes	Yes	Yes	No	Yes
c-inference	Yes	No	Yes	No	No
system W	Yes	No	Yes	No	Yes
lexicographic inference	Yes	Yes	Yes	No	Yes

Requiring (RM) to hold for a nonmonotonic inference method is a rather strict requirement; for instance, p-entailment and c-inference also do not satisfy (RM). On the other hand, extending rational closure [21] is a desirable property of nonmonotonic inference [22], and it has been shown that system W extends rational closure [18, Proposition 13]. While system W does not satisfy (RM) it does satisfy the weaker notion (WRM).

Proposition 4. *System W fulfils weak rational monotony (WRM).*

Proof. Let Σ be a signature, Δ a consistent belief base and $\mid\!\sim^{\mathsf{w}}_{\Delta}$ the instance of a system W inference over Σ induced by Δ with the preferred order $<^{\mathsf{w}}_{\Delta}$. Let $A, B \in \mathcal{L}_\Sigma$ such that $\top \mid\!\sim^{\mathsf{w}}_{\Delta} B$ and $\top \not\mid\!\sim^{\mathsf{w}}_{\Delta} \overline{A}$.

Let $\Omega_{nf} \subseteq \Omega_\Sigma$ be the set of all worlds that falsify none of the conditionals in Δ. As Δ is consistent, Ω_{nf} is not empty. The worlds in Ω_{nf} are lower than other worlds with respect to $<^{\mathsf{w}}_{\Delta}$: for every $\omega' \in \Omega_{nf}, \omega \in \Omega_\Sigma \setminus \Omega_{nf}$ we have $\omega' <^{\mathsf{w}}_{\Delta} \omega$. For worlds $\omega, \omega' \in \Omega_{nf}$ neither $\omega <^{\mathsf{w}}_{\Delta} \omega'$ nor $\omega' <^{\mathsf{w}}_{\Delta} \omega$ holds. Because $\top \mid\!\sim^{\mathsf{w}}_{\Delta} B$, every world in Ω_{nf} models B, i.e. $\Omega_{nf} \subseteq Mod(B)$. Because $\top \not\mid\!\sim^{\mathsf{w}}_{\Delta} \overline{A}$, there is at least one world $\omega^A \in \Omega_{nf}$ such that $\omega \models A$.

Let ω be a model of $A\overline{B}$. $\Omega_{nf} \subseteq Mod(B)$ (see above) entails that $\omega \notin \Omega_{nf}$. Because $\omega^A \in \Omega_{nf}$ and $\omega \notin \Omega_{nf}$ we know that $\omega^A <^{\mathsf{w}}_{\Delta} \omega$. Furthermore, we have $\omega^A \models AB$. Therefore, $A \mid\!\sim^{\mathsf{w}}_{\Delta} B$. In summary, system W fulfils (WRM). □

While p-entailment fulfils (SM), system W, like many other inference methods, does not satisfy (SM).

Proposition 5. *System W does not fulfil semi-monotony (SM).*

Proof. Let $\Sigma = \{a, b\}$ and consider the belief bases $\Delta = \{(b|\top)\}$ and $\Delta' = \{(b|\top), (\overline{b}|a)\}$. We have $a \mid\!\sim^{\mathsf{w}}_{\Delta} b$ but $a \not\mid\!\sim^{\mathsf{w}}_{\Delta'} b$. □

Table 1 summarizes some properties of system W and of other inductive inference operators for comparison. System W behaves similar to c-inference with respect to the properties (CM), (RM), (WRM), and (SM), but in contrast to c-inference, it extends rational closure [21].

5 Syntax Splitting and System W

In this section, we evaluate system W with respect to syntax splitting. As system W is an SPO-based inductive inference operator mapping a belief base Δ to the preferred structure $<^{\mathsf{w}}_{\Delta}$ we can apply the postulates (Rel^{spo}), (Ind^{spo}), and (SynSplit^{spo}). For proving that system W fulfils these postulates, we will show four lemmas on the properties of $<^{\mathsf{w}}_{\Delta}$ in the presence of a syntax splitting $\Delta = \Delta_1 \bigcup_{\Sigma_1,\Sigma_2} \Delta_2$. While these lemmas are stated in the short paper [10], but without proofs, we will now present full proofs for these lemmas.

Note that we consider the belief bases Δ_1, Δ_2 as belief bases over $\Sigma = \Sigma_1 \cup \Sigma_2$ in this section. Thus, in particular $<^{\mathsf{w}}_{\Delta_1}$ and $<^{\mathsf{w}}_{\Delta_2}$ are relations on Ω_Σ and the inference relations induced by Δ_1, Δ_2 are calculated with respect to Σ.

The following Lemma 1 shows how a syntax splitting on a belief base carries over to the corresponding inclusion maximal tolerance partitioning.

Lemma 1. *Let $\Delta = \Delta_1 \bigcup_{\Sigma_1,\Sigma_2} \Delta_2$ be a consistent belief base with syntax splitting. Let $OP(\Delta) = (\Delta^0, \ldots, \Delta^k)$ and let $OP(\Delta_i) = (\Delta_i^0, \ldots, \Delta_i^{l_i})$ for $i = 1, 2$.*

1. *For $i = 1, 2$ and $j = 0, \ldots, l_i$ it is $\Delta_i^j = \Delta^j \cap \Delta_i$ and thus especially $\Delta_i^j \subseteq \Delta^j$.*
2. $\max\{l_1, l_2\} = k$
3. *If $l_1 \leqslant l_2$, we have* $\Delta^j = \begin{cases} \Delta_1^j \cup \Delta_2^j & \text{for } j = 1, \ldots, l_1 \\ \Delta_2^j & \text{for } j = l_1 + 1, \ldots, k \end{cases}$

Proof. **Ad 1.:** W.l.o.g. consider $i = 1$. We will show that $\Delta_1^0 = \Delta^0 \cap \Delta_1$ for $\Delta_1 \neq \emptyset$. This implies $\Delta_1^j = \Delta^j \cap \Delta_1$ for $j = 1, \ldots, l_1$ as the tolerance partitioning can be constructed by recursively selecting the conditionals tolerated by all other conditionals in the belief base.

Every conditional $r \in \Delta_1$ is tolerated by all conditionals in $(\mathcal{L}|\mathcal{L})_{\Sigma_2}$ that are not contradictory. Especially, r is tolerated by all conditionals in Δ_2. Therefore, r is tolerated by all conditionals in Δ iff it is tolerated by all conditional in Δ_1.

If $q \in \Delta_1^0$, then it is in Δ_1 and tolerated by every conditional in Δ_1. Hence, q is also in Δ and tolerated by every conditional in Δ (see above). Thus, $q \in \Delta^0 \cap \Delta_1$. If $p \in \Delta^0 \cap \Delta_1$, then it is in Δ_1 and it is tolerated by every conditional in Δ. Therefore, q is also tolerated by Δ_1. Thus, $q \in \Delta_1^0$. Together, we have $\Delta_1^0 = \Delta^0 \cap \Delta_1$.

Ad 2: As $\Delta_i \subseteq \Delta$ for $i = 1, 2$ we have $l_1, l_2 \leqslant k$. As Δ_k is not empty, it contains at least one conditional r. This r is either in Δ_1 or in Δ_2. Let $i \in \{1, 2\}$ be such that $r \in \Delta_i$. Then there is some m such that $r \in \Delta_i^m$. With (1.) we have $\Delta_i^m \subseteq \Delta^m$. As r can only be in one set of the tolerance partitioning of Δ we have $m = k$. Therefore, the tolerance partitioning of Δ_i has at least k elements, i.e., $l_i = k$.

Ad 3: First, consider the case that $j \leqslant l_1$ and therefore also $j \leqslant l_2$. As $\Delta_i^j \subseteq \Delta^j$ for $i = 1, 2$ we get $\Delta^j \supseteq \Delta_1^j \cup \Delta_2^j$. Every $r \in \Delta^j$ is either in Δ_1 or in Δ_2. W.l.o.g. assume $r \in \Delta_1$. Then there is some m such that $r \in \Delta_1^m$. With (1.)

we have $\Delta_1^m \subseteq \Delta^m$. As r can only be in one set of the tolerance partitioning of Δ we have $m = j$. Therefore, $\Delta^j \subseteq \Delta_1^j \cup \Delta_2^j$ and thus $\Delta^j = \Delta_1^j \cup \Delta_2^j$.

Now consider $l_1 < j \leqslant k$. From (1.) we get $\Delta^j \supseteq \Delta_2^j$. Analogous to the first case, we know that every conditional $r \in \Delta^j$ is either in Δ_1^j or Δ_2^j. Because $j > l_1$ there is no Δ_1^j in the tolerance partitioning of Δ_1, and thus we have $r \in \Delta_2^j$. Therefore, $\Delta^j \subseteq \Delta_2^j$ and thus $\Delta^j = \Delta_2^j$. □

If we have $\omega <^{\mathsf{w}}_{\Delta} \omega'$, then there is some conditional r that falsifies ω' but not ω and thus causes the $\subsetneqq$ relation in (9) in Definition 6. If $\Delta = \Delta_1 \bigcup_{\Sigma_1,\Sigma_2} \Delta_2$, this r is either in Δ_1 or in Δ_2. Lemma 2 states that the relation $\omega <^{\mathsf{w}}_{\Delta} \omega'$ can also be obtained using only Δ_1 or only Δ_2.

Lemma 2. *Let $\Delta = \Delta_1 \bigcup_{\Sigma_1,\Sigma_2} \Delta_2$ and let $\omega, \omega' \in \Omega$. If $\omega <^{\mathsf{w}}_{\Delta} \omega'$, then $\omega <^{\mathsf{w}}_{\Delta_1} \omega'$ or $\omega <^{\mathsf{w}}_{\Delta_2} \omega'$.*

Proof. Let $OP(\Delta) = (\Delta^0, \ldots, \Delta^k)$ and let ξ, ξ^i be the functions as in Definition 6. Let $\omega, \omega' \in \Omega$ be worlds with $\omega <^{\mathsf{w}}_{\Delta} \omega'$. By definition of $<^{\mathsf{w}}_{\Delta}$ there is an $m \in \{0, \ldots, k\}$ such that $\xi^i(\omega) = \xi^i(\omega')$ for every $i = m+1, \ldots, k$ and $\xi^m(\omega) \subsetneqq \xi^m(\omega')$. Therefore, there is an $r \in \Delta$ such that $r \in \xi^m(\omega')$ and $r \notin \xi^m(\omega)$. The conditional r is either in Δ_1 or Δ_2. Assume that $r \in \Delta_x$ with x being either 1 or 2. Let $OP(\Delta_x) = (\Delta_x^0, \ldots, \Delta_x^l)$ be the tolerance partition of Δ_x. Let ξ_x, ξ_x^i for $i = 0, \ldots, k$ be the functions mapping worlds to the set of falsified conditionals for Δ_x. Because $\Delta_x^i \subseteq \Delta^i$ (see Lemma 1) and $\xi^i(\omega) = \xi^i(\omega')$ for every $i = m+1, \ldots, k$ we have $\xi_x^i(\omega) = \xi_x^i(\omega')$ for every $i = m+1, \ldots, l$. And because $r \in \xi_x^m(\omega')$ and $r \notin \xi_x^m(\omega)$ we have $\xi_x^m(\omega) \subsetneqq \xi_x^m(\omega')$. Therefore, $\omega <^{\mathsf{w}}_{\Delta_x} \omega'$.

Hence, we have $\omega <^{\mathsf{w}}_{\Delta_1} \omega'$ or $\omega <^{\mathsf{w}}_{\Delta_2} \omega'$. □

Note, that both $\omega <^{\mathsf{w}}_{\Delta_1} \omega'$ and $\omega <^{\mathsf{w}}_{\Delta_2} \omega'$ might be true.

Example 4. Consider the belief base $\Delta_{ve} = \Delta_1 \bigcup_{\Sigma_1,\Sigma_2} \Delta_2$ from Examples 1 and 3 together with the worlds $\omega = mb\overline{e}tg$ and $\omega' = mbe\overline{t}\overline{g}$. We have $\omega <^{\mathsf{w}}_{\Delta_{ve}} \omega'$. It also holds that $\omega <^{\mathsf{w}}_{\Delta_1} \omega'$ and $\omega <^{\mathsf{w}}_{\Delta_2} \omega'$.

Lemma 3 considers the reverse direction of Lemma 2 and shows a situation where we can infer $\omega <^{\mathsf{w}}_{\Delta} \omega'$ from $\omega <^{\mathsf{w}}_{\Delta_1} \omega'$ for a belief base with syntax splitting.

Lemma 3. *Let $\Delta = \Delta_1 \bigcup_{\Sigma_1,\Sigma_2} \Delta_2$ and let $\omega, \omega' \in \Omega$. If $\omega <^{\mathsf{w}}_{\Delta_1} \omega'$ and $\omega_{|\Sigma_2} = \omega'_{|\Sigma_2}$, then $\omega <^{\mathsf{w}}_{\Delta} \omega'$.*

Proof. Let $\omega, \omega' \in \Omega$ with $\omega <^{\mathsf{w}}_{\Delta_1} \omega'$ and $\omega_{|\Sigma_2} = \omega'_{|\Sigma_2}$. Let $OP(\Delta) = (\Delta^0, \ldots, \Delta^k)$ and let ξ, ξ^i be the functions as in Definition 6. Let $OP(\Delta_1) = (\Delta_1^0, \ldots, \Delta_1^l)$ be the tolerance partition of Δ_1. Let ξ_x, ξ_x^i for $i = 0, \ldots, k$ be the functions mapping worlds to the set of falsified conditionals for Δ_x. By definition of $<^{\mathsf{w}}_{\Delta_1}$ there is an $m \in \{0, \ldots, l\}$ such that $\xi_1^i(\omega) = \xi_1^i(\omega')$ for every

$i = m+1, \ldots, l$ and $\xi_1^m(\omega) \subsetneqq \xi_1^m(\omega')$. With Lemma 1 we have $\xi_x^j(\omega^*) = \{r \in \Delta_x^j \mid \omega^* \text{ falsifies } r\} = \{r \in \Delta_x \cap \Delta^j \mid \omega^* \text{ falsifies } r\} = \xi^j \cap \Delta_x$ for every world ω^*, $x = 1, 2$ and $j = 0, \ldots, l$. This implies $\xi^i(\omega) \cap \Delta_1 = \xi_1^i(\omega) = \xi_1^i(\omega') = \xi^i(\omega') \cap \Delta_1$ for every $i = m+1, \ldots, k$ and $\xi^m(\omega) \cap \Delta_1 = \xi_1^m(\omega) \subsetneqq \xi_1^m(\omega) = \xi^m(\omega') \cap \Delta_1$.

Because $\omega_{|\Sigma_2} = \omega'_{|\Sigma_2}$, we have $\xi^i(\omega) \cap \Delta_2 = \xi^i(\omega') \cap \Delta_2$ for every $i = 0, \ldots, k$. Hence, we have $\xi^i(\omega) = \xi^i(\omega')$ for every $i = m+1, \ldots, k$ and $\xi^m(\omega) \subsetneqq \xi^m(\omega')$ and therefore $\omega <^{\mathsf{w}}_{\Delta} \omega'$. □

The next Lemma 4 captures that the variable assignment for variables that do not occur in the belief set has no influence on the position of a world in the resulting preferential structure on worlds.

Lemma 4. *Let* $\Delta = \Delta_1 \bigcup_{\Sigma_1, \Sigma_2} \Delta_2$ *and* $\omega^a, \omega^b, \omega' \in \Omega_\Sigma$ *with* $\omega^a{}_{|\Sigma_1} = \omega^b{}_{|\Sigma_1}$. *Then we have* $\omega^a <^{\mathsf{w}}_{\Delta_1} \omega'$ *iff* $\omega^b <^{\mathsf{w}}_{\Delta_1} \omega'$.

Proof. Let $\omega^a, \omega^b, \omega'$ be as introduced above. As $\omega^a{}_{|\Sigma_1} = \omega^b{}_{|\Sigma_1}$ and $\Delta_1 \subseteq (\mathcal{L}|\mathcal{L})_{\Sigma_1}$ the worlds ω^a and ω^b falsify the same conditionals in Δ_1. Whether $\omega^X <^{\mathsf{w}}_{\Delta_1} \omega'$ holds or does not depends entirely on the conditionals in Δ_1 falsified by ω^X and ω' for $X \in \{a, b\}$. Because ω^a and ω^b behave alike with respect to falsification of conditionals in Δ_1, we have that $\omega^a <^{\mathsf{w}}_{\Delta_1} \omega'$ iff $\omega^b <^{\mathsf{w}}_{\Delta_1} \omega'$. □

Using these lemmas, we show that system W fulfils both (Rel^{spo}) and (Ind^{spo}).

Proposition 6. *System W fulfils (Rel^{spo}).*

Proof. Let $\Delta = \Delta_1 \bigcup_{\Sigma_1, \Sigma_2} \Delta_2$ be a belief base with syntax splitting and let $A, B \in \mathcal{L}_{\Sigma_1}$ be propositional formulas. W.l.o.g. we need to show that

$$A <^{\mathsf{w}}_{\Delta} B \text{ if and only if } A <^{\mathsf{w}}_{\Delta_1} B. \tag{11}$$

Direction $\Rightarrow$ *of* (11): Assume that $A <^{\mathsf{w}}_{\Delta} B$. We need to show that $A <^{\mathsf{w}}_{\Delta_1} B$. Let ω' be any world in Ω_B. Now choose $\omega'_{min} \in \Omega$ such that

1. $\omega'_{min} \leqslant^{\mathsf{w}}_{\Delta} \omega'$,
2. $\omega'_{|\Sigma_1} = \omega'_{min|\Sigma_1}$, and
3. there is no world ω'_{min2} with $\omega'_{min2} < \omega'_{min}$ that fulfils (1.) and (2.).

Such an ω'_{min} exists because ω' fulfils properties (1.) and (2.), $<^{\mathsf{w}}_{\Delta}$ is irreflexive and transitive, and there are only finitely many worlds in Ω.

Because of (2.) and because $\omega' \models B$ we have that $\omega'_{min} \models B$. Because $A <^{\mathsf{w}}_{\Delta} B$, there is a world ω such that $\omega \models A$ and $\omega <^{\mathsf{w}}_{\Delta} \omega'_{min}$. Lemma 2 yields that $\omega <^{\mathsf{w}}_{\Delta_1} \omega'_{min}$ or $\omega <^{\mathsf{w}}_{\Delta_2} \omega'_{min}$.

The case $\omega <^{\mathsf{w}}_{\Delta_2} \omega'_{min}$ is not possible: Assuming $\omega <^{\mathsf{w}}_{\Delta_2} \omega'_{min}$, it follows that $\omega'_{min2} = (\omega'_{min|\Sigma_1} \cdot \omega_{|\Sigma_2}) <^{\mathsf{w}}_{\Delta_2} \omega'_{min}$ with Lemma 4. With Lemma 3 it follows that $\omega'_{min2} <^{\mathsf{w}}_{\Delta} \omega'_{min}$. This contradicts (3.). Hence, $\omega <^{\mathsf{w}}_{\Delta_1} \omega'_{min}$. Because of (2.)

and Lemma 4 it follows that $\omega <^{\mathsf{w}}_{\Delta_1} \omega'$. As we can find an ω such that $\omega <^{\mathsf{w}}_{\Delta_1} \omega'$ and $\omega \models A$ for every $\omega' \models B$ we have that $A <^{\mathsf{w}}_{\Delta_1} B$.

Direction $\Leftarrow$ *of* (11): Assume that $A <^{\mathsf{w}}_{\Delta_1} B$. We need to show that $A <^{\mathsf{w}}_{\Delta} B$. Let ω' be any world in Ω_B. Because $A <^{\mathsf{w}}_{\Delta_1} B$, there is a world ω^* such that $\omega^* \models A$ and $\omega^* <^{\mathsf{w}}_{\Delta_1} \omega'$. Let $\omega = (\omega^*_{|\Sigma_1} \cdot \omega'_{|\Sigma_2})$. Because $\omega^* \models A$ we have that $\omega \models A$. Furthermore, with Lemma 4 it follows that $\omega <^{\mathsf{w}}_{\Delta_1} \omega'$ and thus $\omega <^{\mathsf{w}}_{\Delta} \omega'$ with Lemma 3. As we can construct ω such that $\omega <^{\mathsf{w}}_{\Delta} \omega'$ and $\omega \models A$ for every $\omega' \models B$ we have that $A <^{\mathsf{w}}_{\Delta} B$. □

Proposition 7. *System W fulfils (Ind^{spo}).*

Proof. Let $\Delta = \Delta_1 \bigcup_{\Sigma_1,\Sigma_2} \Delta_2$ be a belief base with syntax splitting. W.l.o.g. let $A, B \in \mathcal{L}_{\Sigma_1}$ and $C \in \mathcal{L}_{\Sigma_2}$ be propositional formulas such that C is consistent. We need to show that

$$A <^{\mathsf{w}}_{\Delta} B \text{ if and only if } AC <^{\mathsf{w}}_{\Delta} BC. \tag{12}$$

Direction $\Rightarrow$ *of* (12): Assume that $A <^{\mathsf{w}}_{\Delta} B$. We need to show that $AC <^{\mathsf{w}}_{\Delta} BC$. Let ω' be any world in Ω_{BC}. Now choose $\omega'_{min} \in \Omega$ such that

1. $\omega'_{min} \leqslant^{\mathsf{w}}_{\Delta} \omega'$,
2. $\omega'_{|\Sigma_1} = \omega'_{min|\Sigma_1}$, and
3. there is no world $\omega'_{min2} < \omega'_{min}$ that fulfils (1.) and (2.).

Such an ω'_{min} exists because ω' fulfils properties (1.) and (2.), $<^{\mathsf{w}}_{\Delta}$ is irreflexive and transitive, and there are only finitely many worlds in Ω. Because of (2.) and because $\omega' \models BC$ we have that $\omega'_{min} \models B$. Because $A <^{\mathsf{w}}_{\Delta} B$, there is a world ω^* such that $\omega^* \models A$ and $\omega^* <^{\mathsf{w}}_{\Delta} \omega'_{min}$. Lemma 2 yields that either $\omega^* <^{\mathsf{w}}_{\Delta_1} \omega'_{min}$ or $\omega^* <^{\mathsf{w}}_{\Delta_2} \omega'_{min}$.

The case $\omega^* <^{\mathsf{w}}_{\Delta_2} \omega'_{min}$ is not possible: Assuming $\omega^* <^{\mathsf{w}}_{\Delta_2} \omega'_{min}$, it follows that $\omega'_{min2} = (\omega'_{min|\Sigma_1} \cdot \omega^*_{|\Sigma_2}) <^{\mathsf{w}}_{\Delta_2} \omega'_{min}$ with Lemma 4. With Lemma 3 it follows that $\omega'_{min2} <^{\mathsf{w}}_{\Delta} \omega'_{min}$. This contradicts (3.). Hence, $\omega^* <^{\mathsf{w}}_{\Delta_1} \omega'_{min}$. Let $\omega = (\omega^*_{|\Sigma_1} \cdot \omega'_{|\Sigma_2})$. Because $\omega^* \models A$ we have that $\omega \models A$. Because $\omega' \models C$ we have that $\omega \models C$. Because of (2.) and Lemma 4 it follows that $\omega <^{\mathsf{w}}_{\Delta_1} \omega'$ and thus with Lemma 3 it follows that $\omega <^{\mathsf{w}}_{\Delta} \omega'$. As we can construct an ω for every ω' we have that $AC \mid\!\sim^{\mathsf{w}}_{\Delta} BC$.

Direction $\Leftarrow$ *of* (12): Assume that $AC <^{\mathsf{w}}_{\Delta} BC$. We need to show that $A <^{\mathsf{w}}_{\Delta} B$. Let ω' be any world in Ω_B. Now choose $\omega'_{min} \in \Omega$ such that

1. $\omega'_{min} \models C$
2. $\omega'_{|\Sigma_1} = \omega'_{min|\Sigma_1}$, and
3. there is no world $\omega^{**\prime} < \omega'_{min}$ that fulfils (1.) and (2.).

Such an ω'_{min} exists because C is consistent, $<^{\mathsf{w}}_{\Delta}$ is irreflexive and transitive, and there are only finitely many worlds in Ω. Because of (2.) and because $\omega' \models B$ we have that $\omega'_{min} \models B$. Because of (1.) we have that $\omega'_{min} \models C$.

Because $AC <^{\mathsf{w}}_{\Delta} BC$, there is a world ω^* such that $\omega^* \models AC$ and $\omega^* <^{\mathsf{w}}_{\Delta} \omega'_{min}$. Lemma 2 yields that either $\omega^* <^{\mathsf{w}}_{\Delta_1} \omega'_{min}$ or $\omega^* <^{\mathsf{w}}_{\Delta_2} \omega'_{min}$.

The case $\omega^* <^{\mathsf{w}}_{\Delta_2} \omega'_{min}$ is not possible: Assuming $\omega^* <^{\mathsf{w}}_{\Delta_2} \omega'_{min}$, it follows that $\omega'_{min2} = (\omega'_{min|\Sigma_1} \cdot \omega^*{}_{|\Sigma_2}) <^{\mathsf{w}}_{\Delta_2} \omega'_{min}$ with Lemma 4. With Lemma 3 it follows that $\omega'_{min2} <^{\mathsf{w}}_{\Delta} \omega'_{min}$. This contradicts (3.). Hence, $\omega^* <^{\mathsf{w}}_{\Delta_1} \omega'_{min}$. Let $\omega = (\omega^*{}_{|\Sigma_1} \cdot \omega'{}_{|\Sigma_2})$. Because $\omega^* \models A$ we have that $\omega \models A$. Because of (2.) and Lemma 4 it follows that $\omega <^{\mathsf{w}}_{\Delta_1} \omega'$ and thus with Lemma 3 $\omega <^{\mathsf{w}}_{\Delta} \omega'$.

As we can construct an ω for every ω' we have that $A <^{\mathsf{w}}_{\Delta} B$. □

From the fulfilment of the syntax splitting postulates for SPO-based inductive inference operators we can conclude that system W also fulfils the general syntax splitting postulates.

Proposition 8. *System W fulfils (Rel).*

Proof. This follows from Proposition 6 and Proposition 1. □

Proposition 9. *System W fulfils (Ind).*

Proof. This follows from Proposition 7 and Proposition 2. □

Combining Propositions 8 and 9 yields that system W fulfils (SynSplit).

Proposition 10. *System W fulfils (SynSplit).*

Thus, by employing the SPO-based syntax splitting postulates, we have established a complete formal proof that system W fulfils (SynSplit).

6 Syntax Splitting and the Preferred Structure on Worlds

Let us assume that we have a belief base $\Delta = \Delta_1 \bigcup_{\Sigma_1,\Sigma_2} \Delta_2$ with syntax splitting. To answer a query "does A entail B" for $A, B \in (\mathcal{L}|\mathcal{L})_{\Sigma_1}$ (or $A, B \in (\mathcal{L}|\mathcal{L})_{\Sigma_2}$) we can employ the syntax splitting and consider only Δ_1 (or Δ_2) to answer that query. But if A, B contain variables from both Σ_1 and Σ_2, or if we want to calculate the preferred structure over Δ, (SynSplit) is not applicable. In this section, we show how a syntax splitting affects the preferred structure on worlds over the whole belief base.

Definition 8 **($hdl(\omega, \omega', \Delta)$).** *Let $OP(\Delta) = (\Delta^0, \ldots, \Delta^k)$ and let ξ, ξ^i be as in Definition 6. We define* the number of the highest differentiating layer

$$hdl(\omega, \omega', \Delta) = \max\{i \in \mathbb{N} \mid \xi^i(\omega) \neq \xi^i(\omega')\},$$

i.e., $\xi^i(\omega) \neq \xi^i(\omega')$ and $\xi^j(\omega) = \xi^j(\omega')$ for $i = hdl(\omega, \omega', \Delta)$ and $j > i$. If ω, ω' falsify the same conditionals in Δ, we define $hdl(\omega, \omega', \Delta) = -1$.

Example 5. Consider the belief base Δ_{ve} from Example 1 with $OP(\Delta_{ve}) = (\Delta^0, \Delta^1)$ with $\Delta^0 = \{(g|m), (t|b)\}$ for $\Delta^1 = \{(m|e), (\overline{g}|me)\}$ (see Example 3). For $\omega = mbetg$ and $\omega' = mbet\overline{g}$ we have $hdl(\omega, \omega', \Delta_{ve}) = 1$ because ω falsifies $(\overline{g}|me)$ from Δ^1 and ω' does not. The verification behaviour of ω, ω' regarding Δ^0 does not make a difference.

For $\omega'' = \overline{m}\overline{b}e\overline{t}g$ and $\omega''' = \overline{m}be\overline{t}g$ we have $hdl(\omega'', \omega''', \Delta_{ve}) = 0$ because $\xi^1(\omega'') = \xi^1(\omega''') = \{(m|e)\}$ and $\xi^0(\omega'') \neq \xi^0(\omega''')$.

The following proposition states that for a belief based Δ with syntax splitting we can calculate the preferred structure on worlds separately on the parts of the syntax splitting for deciding whether $\omega <^{\mathsf{w}}_{\Delta} \omega'$ holds for a pair of worlds ω, ω'. The number $hdl(\omega, \omega', \Delta)$ indicates how the preferred structures induced by each of the sub-bases should be combined. Sometimes it is only necessary to consider one of the parts in the syntax splitting, making answering queries easier.

Proposition 11. *Let* $\Delta = \Delta_1 \bigcup_{\Sigma_1, \Sigma_2} \Delta_2$ *and* $\omega, \omega' \in \Omega_\Sigma$. *Then* $\omega <^{\mathsf{w}}_{\Delta} \omega'$ *iff either*

- $hdl(\omega, \omega', \Delta_1) > hdl(\omega, \omega', \Delta_2)$ *and* $\omega_{|\Sigma_1} <^{\mathsf{w}}_{\Delta_1} \omega'_{|\Sigma_1}$ *or*
- $hdl(\omega, \omega', \Delta_1) < hdl(\omega, \omega', \Delta_2)$ *and* $\omega_{|\Sigma_2} <^{\mathsf{w}}_{\Delta_2} \omega'_{|\Sigma_2}$ *or*
- $hdl(\omega, \omega', \Delta_1) = hdl(\omega, \omega', \Delta_2)$ *and* $\omega_{|\Sigma_1} <^{\mathsf{w}}_{\Delta_1} \omega'_{|\Sigma_1}$ *and* $\omega_{|\Sigma_2} <^{\mathsf{w}}_{\Delta_2} \omega'_{|\Sigma_2}$.

Proof. Let $\Delta = \Delta_1 \bigcup_{\Sigma_1, \Sigma_2} \Delta_2$ and $\omega, \omega' \in \Omega_\Sigma$. We will prove both directions of the "iff" seperately.

Direction $\Rightarrow$: Assume $\omega <^{\mathsf{w}}_{\Delta} \omega'$. We can distinguish three cases for the numbers $k = hdl(\omega, \omega', \Delta_1)$ and $l = hdl(\omega, \omega', \Delta_2)$.

Case 1: $hdl(\omega, \omega', \Delta_1) > hdl(\omega, \omega', \Delta_2)$ In this case, we have $\xi^k_1(\omega) \neq \xi^k_1(\omega')$ and $\xi^k_2(\omega) = \xi^k_2(\omega')$. Additionally, we have $\xi^i_1(\omega) = \xi^i_1(\omega')$ and $\xi^i_2(\omega) = \xi^i_2(\omega')$ for $i > k$. With Lemma 1 it follows that $\xi^i(\omega) = \xi^i(\omega')$ for $i > k$ and $\xi^k(\omega) \neq \xi^k(\omega')$. As $\omega <^{\mathsf{w}}_{\Delta} \omega'$ we know that $\xi^k(\omega) \subsetneqq \xi^k(\omega')$ and therefore $\xi^k_1(\omega) \subsetneqq \xi^k_1(\omega')$ because of $\xi^k_2(\omega) = \xi^k_2(\omega')$ and Lemma 1. As the falsification of conditionals in Δ_1 over Σ_1 does not depend on the assignments of truth values for Σ_1, we have that $\xi^k_1(\omega_{|\Sigma_1}) \subsetneqq \xi^k_1(\omega'_{|\Sigma_1})$ and therefore $\omega_{|\Sigma_1} <^{\mathsf{w}}_{\Delta_1} \omega'_{|\Sigma_1}$.

Case 2: $hdl(\omega, \omega', \Delta_1) < hdl(\omega, \omega', \Delta_2)$ Analogous to Case 1 we conclude $\omega_{|\Sigma_2} <^{\mathsf{w}}_{\Delta_2} \omega'_{|\Sigma_2}$.

Case 3: $hdl(\omega, \omega', \Delta_1) = hdl(\omega, \omega', \Delta_2)$ In this case, we have $\xi^k_1(\omega) \neq \xi^k_1(\omega')$ and $\xi^k_2(\omega) \neq \xi^k_2(\omega')$. Additionally, we have $\xi^i_1(\omega) = \xi^i_1(\omega')$ and $\xi^i_2(\omega) = \xi^i_2(\omega')$ for $i > k$. With Lemma 1 it follows that $\xi^i(\omega) = \xi^i(\omega')$ for $i > k$ and $\xi^k(\omega) \neq \xi^k(\omega')$. As $\omega <^{\mathsf{w}}_{\Delta} \omega'$ we know that $\xi^k(\omega) \subsetneqq \xi^k(\omega')$. Therefore, we have $\xi^k_1(\omega) \subsetneqq \xi^k_1(\omega')$ and $\xi^k_2(\omega) \subsetneqq \xi^k_2(\omega')$. As the falsification of conditionals in Δ_1 over Σ_1 does not depend on the assignments of truth values for Σ_1, we have that $\xi^k_1(\omega_{|\Sigma_1}) \subsetneqq \xi^k_1(\omega'_{|\Sigma_1})$ and $\xi^k_2(\omega_{|\Sigma_2}) \subsetneqq \xi^k_2(\omega'_{|\Sigma_2})$. Hence, $\omega_{|\Sigma_1} <^{\mathsf{w}}_{\Delta_1} \omega'_{|\Sigma_1}$ and $\omega_{|\Sigma_2} <^{\mathsf{w}}_{\Delta_2} \omega'_{|\Sigma_2}$.

Direction $\Leftarrow$: Assume one of the three items in the proposition are true.

Case 1: $hdl(\omega, \omega', \Delta_1) > hdl(\omega, \omega', \Delta_2)$ and $\omega_{|\Sigma_1} <^{\mathsf{w}}_{\Delta_1} \omega'_{|\Sigma_1}$ We have $\xi^k_1(\omega) \neq \xi^k_1(\omega')$ and $\xi^k_2(\omega) = \xi^k_2(\omega')$. Additionally, we have $\xi^i_1(\omega) = \xi^i_1(\omega')$ and

$\xi_2^i(\omega) = \xi_2^i(\omega')$ for $i > k$. Because of $\omega_{|\Sigma_1} <^{\mathsf{w}}_{\Delta_1} \omega'_{|\Sigma_1}$, we have $\xi_1^k(\omega) \subsetneqq \xi_1^k(\omega')$. With Lemma 1 it follows that $\xi^i(\omega) = \xi^i(\omega')$ for $i > k$ and $\xi^k(\omega) \subsetneqq \xi^k(\omega')$. Therefore $\omega <^{\mathsf{w}}_{\Delta} \omega'$.

Case 2: $hdl(\omega, \omega', \Delta_1) < hdl(\omega, \omega', \Delta_2)$ and $\omega_{|\Sigma_2} <^{\mathsf{w}}_{\Delta_2} \omega'_{|\Sigma_2}$ Analogous to Case 1 we have $\omega <^{\mathsf{w}}_{\Delta} \omega'$.

Case 3: $hdl(\omega, \omega', \Delta_1) = hdl(\omega, \omega', \Delta_2)$ and $\omega_{1|\Sigma_1} <^{\mathsf{w}}_{\Delta_1} \omega_{2|\Sigma_1}$ and $\omega_{|\Sigma_2} <^{\mathsf{w}}_{\Delta_2} \omega'_{|\Sigma_2}$ We have $\xi_1^i(\omega) = \xi_1^i(\omega')$ and $\xi_2^i(\omega) = \xi_2^i(\omega')$ for $i > k$. Because of $\omega_{|\Sigma_1} <^{\mathsf{w}}_{\Delta_1} \omega'_{|\Sigma_1}$, we have $\xi_1^k(\omega) \subsetneqq \xi_1^k(\omega')$. Because of $\omega_{|\Sigma_2} <^{\mathsf{w}}_{\Delta_2} \omega'_{|\Sigma_2}$, we have $\xi_2^k(\omega) \subsetneqq \xi_2^k(\omega')$. With Lemma 1 it follows that $\xi^i(\omega) = \xi^i(\omega')$ for $i > k$ and $\xi^k(\omega) \subsetneqq \xi^k(\omega')$. Therefore, $\omega <^{\mathsf{w}}_{\Delta} \omega'$. □

Proposition 11 can be used to decide whether $\omega <^{\mathsf{w}}_{\Delta} \omega'$ holds in presence of a syntax splitting $\Delta = \Delta_1 \bigcup_{\Sigma_1, \Sigma_2} \Delta_2$ considering Δ_1 and Δ_2 separately.

Example 6. Consider again the belief base $\Delta_{ve} = \Delta_1 \bigcup_{\Sigma_1, \Sigma_2} \Delta_2$ from Example 1 and Example 3. To check whether $\omega <^{\mathsf{w}}_{\Delta_{ve}} \omega'$ for $\omega = mbe\overline{t}\overline{g}$ and $\omega' = mbetg$ holds, we first calculate $h_1 = hdl(\omega, \omega', \Delta_1) = 1$ and $h_2 = hdl(\omega, \omega', \Delta_2) = 0$. As $h_1 > h_2$ we only have to check whether $\omega <^{\mathsf{w}}_{\Delta_1} \omega'$ holds. As this is the case, we can conclude that $\omega <^{\mathsf{w}}_{\Delta_{ve}} \omega'$.

Applications of this approach include the calculation of $<^{\mathsf{w}}_{\Delta}$ or parts of this relation. Deciding whether $\omega <^{\mathsf{w}}_{\Delta} \omega'$ holds also coincides with deciding whether Δ entails the base conditional $(\omega|\omega \vee \omega')$ with system W. More generally, deciding whether a conditional $(B|A)$ is entailed by a certain belief base requires only comparing models of the antecedent A in $<^{\mathsf{w}}_{\Delta}$. If A has only few models, it is not necessary to calculate the complete preferred structure $<^{\mathsf{w}}_{\Delta}$ but only a small part of it to decide if $(B|A)$ holds. Here, Proposition 11 might be of use as well for simplifying inferences.

7 Relations Among Inductive Inference Operators

Some of the relations of system W to other inductive inference operators have already been established. In [18] it is shown that system W properly captures and strictly extends p-entailment, system Z, and c-inference. We will now investigate and establish the relationship between system W and lexicographic inference.

Lexicographic inference [22] is an inductive inference operator that, similar to system W, is based on an ordering on worlds which is obtained by considering the falsified conditionals in each partition of the inclusion maximal tolerance partitioning of the belief base. But unlike system W, lexicographic inference only considers the number of falsified conditionals in each part of the partition instead of checking for a subsumption of the sets of falsified conditionals. The following definition of lexicographic inference is adapted from [22]; in it we employ the notation introduced in Definition 6 and use $\min_{\prec} S$ to denote the minima in the set S with respect to the ordering $\prec$.

Definition 9 ($<^{lex}_{\Delta}$**, lexicographic inference).** *The* lexicographic ordering *on vectors in* $\mathbb{N}^n$ *is defined by* $(v_1, \dots, v_n) <^{lex} (w_1, \dots, w_n)$ *iff there is a* $k \in \{1, \dots, n\}$ *such that* $v_k < w_k$ *and* $v_j = w_j$ *for* $j = k+1, \dots, n$.

The binary relation $\leqslant^{lex}_{\Delta} \subseteq \Omega \times \Omega$ *on worlds induced by a belief base* Δ *with* $|OP(\Delta)| = n$ *is defined by, for any* $\omega, \omega' \in \Omega$,

$$\omega \leqslant^{lex}_{\Delta} \omega' \quad \text{if} \quad (|\xi^1_{\Delta}(\omega)|, \dots, |\xi^n_{\Delta}(\omega)|) \leqslant^{lex} (|\xi^1_{\Delta}(\omega')|, \dots, |\xi^n_{\Delta}(\omega')|).$$

For formulas F, G, A, B, lexicographic inference $\mid\!\sim^{lex}_{\Delta}$ *is induced by* $<^{lex}_{\Delta}$:

$$F <^{lex}_{\Delta} G \quad \text{iff} \quad \min_{<^{lex}_{\Delta}}\{\omega \in \Omega \mid \omega \models F\} <^{lex}_{\Delta} \min_{<^{lex}_{\Delta}}\{\omega \in \Omega \mid \omega \models G\}$$

$$A \mid\!\sim^{lex}_{\Delta} B \quad \text{iff} \quad AB <^{lex}_{\Delta} A\overline{B}$$

As the definitions for lexicographic inference and system W are similar, we want to compare these inference operators in more detail. First, we observe that the condition for $\omega <^{lex}_{\Delta} \omega'$ is stronger than the condition for $\omega <^{\mathsf{w}}_{\Delta} \omega'$.

Proposition 12. *Let* Δ *be a belief base and* ω, ω' *be worlds. Then* $\omega <^{\mathsf{w}}_{\Delta} \omega'$ *implies* $\omega <^{lex}_{\Delta} \omega'$.

Proof. If $\omega <^{\mathsf{w}}_{\Delta} \omega'$ then there is a k such that $\xi^k(\omega) \subsetneqq \xi^k(\omega')$ and $\xi^i(\omega) = \xi^i(\omega')$ for $i > k$. This implies $|\xi^k(\omega)| < |\xi^k(\omega')|$ and $|\xi^i(\omega)| = |\xi^i(\omega')|$ for $i > k$. Hence, $\omega <^{lex}_{\Delta} \omega'$. □

Using Proposition 12, we can show that every system W entailment is also an entailment for lexicographic inference.

Proposition 13. *Lexicographic inference captures system W, i.e., for a belief base* Δ *and formulas* A, B *it holds that if* $A \mid\!\sim^{\mathsf{w}}_{\Delta} B$ *then* $A \mid\!\sim^{lex}_{\Delta} B$.

Proof. Let $A \mid\!\sim^{\mathsf{w}}_{\Delta} B$. For every $\omega' \in Mod(A\overline{B})$, by the definition of system W there is another world $\omega \in Mod(AB)$ such that $\omega <^{\mathsf{w}}_{\Delta} \omega'$. This implies that for every $\omega' \in Mod(A\overline{B})$ there is a world $\omega \in Mod(AB)$ such that $\omega <^{lex}_{\Delta} \omega'$. Hence, $AB <^{lex}_{\Delta} A\overline{B}$ and therefore $A \mid\!\sim^{lex}_{\Delta} B$. □

Moreover, some lexicographic inferences are not licensed by system W.

Proposition 14. *There are belief bases* Δ *such that* $\mid\!\sim^{\mathsf{w}}_{\Delta} \subsetneqq \mid\!\sim^{lex}_{\Delta}$.

Proof. Consider Δ_{ve} from Example 1 as well as the worlds $\omega = mbetg$ and $\omega' = \overline{m}be\overline{t}g$. We have that $(|\xi^0_{\Delta}(\omega)|, |\xi^1_{\Delta}(\omega)|) = (0, 1)$ and $(|\xi^0_{\Delta}(\omega')|, |\xi^1_{\Delta}(\omega')|) = (1, 1)$ (see Example 3 for the ordered partition of Δ). Hence, $\omega <^{lex}_{\Delta} \omega'$ and therefore $\mid\!\sim^{lex}_{\Delta} \models r$ with $r = (\omega|\omega \vee \omega')$. But $\omega \not<^{\mathsf{w}}_{\Delta} \omega'$, and therefore $\mid\!\sim^{\mathsf{w}}_{\Delta} \not\models r$. □

These propositions allow us to establish further relationships among the inductive inference operators.

Proposition 15. *Lexicographic inference captures and strictly extends c-inference, i.e., every c-inference is also a lexicographic inference but there are lexicographic inferences that are not c-inferences.*

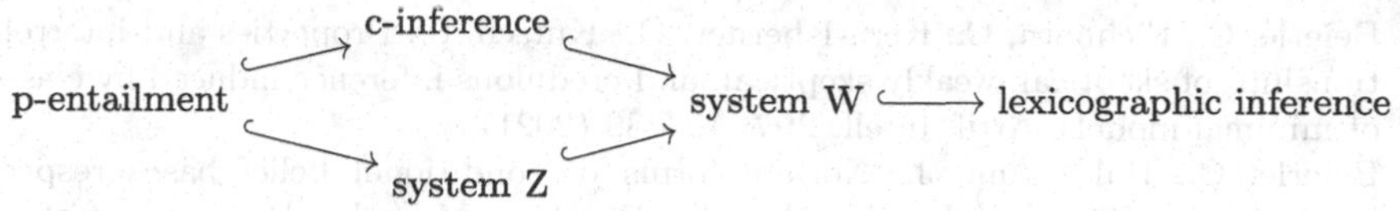

Fig. 2. Overview over relationships between different inference operators. An arrow $I_1 \hookrightarrow I_2$ indicates that inductive inference operator I_1 is captured by I_2 and that I_1 is strictly extended by I_2 for some belief bases.

Proof. Because system W captures and strictly extends c-inference [18, Proposition 11], this follows from Propositions 13 and 14. □

In a similar way, as system W captures system Z, Propositions 13 and 14 imply that lexicographic inference, which has very recently be shown to satisfy syntax splitting [11], captures and strictly extends system Z and thus rational closure. An overview over the relations between the inference relations mentioned in this paper can be found in Fig. 2.

8 Conclusions and Further Work

In this paper, we investigated the properties of the inductive inference operator system W. We show that system W fulfils (WRM) but not (RM) and or (SM). We generalize results on syntax splitting and system W, introduce postulates for SPO-based inductive inference operators, and prove that these postulates are fulfilled by system W. We demonstrate that the influence of syntax splitting on the preferred structure on worlds underlying system W may simplify query answering. Furthermore, by establishing the relationship among system W and other inductive inference operators we obtain a full map of capturing relations for established inductive inference operators. In our current work, we are enlarging this map by comparing system W to further inference approaches for conditional belief bases. We are currently also extending the reasoning plattform InfOCF-Web [20] by a web-based implementation of system W, and we work on more efficient algorithms for reasoning with system W. Furthermore, we investigate normal forms of conditional belief bases respecting system W inferences [4,5].

Acknowledgement. We thank the anonymous reviewers for their detailed and helpful comments. This work was supported by the Deutsche Forschungsgemeinschaft (DFG, German Research Foundation), grant BE 1700/10-1 awarded to Christoph Beierle as part of the priority program "Intentional Forgetting in Organizations" (SPP 1921).

References

1. Adams, E.: The logic of conditionals. Inquiry **8**(1–4), 166–197 (1965)
2. Beierle, C., Eichhorn, C., Kern-Isberner, G., Kutsch, S.: Properties of skeptical c-inference for conditional knowledge bases and its realization as a constraint satisfaction problem. Ann. Math. Artif. Intell. **83**(3–4), 247–275 (2018)

3. Beierle, C., Eichhorn, C., Kern-Isberner, G., Kutsch, S.: Properties and interrelationships of skeptical, weakly skeptical, and credulous inference induced by classes of minimal models. Artif. Intell. **297**, 103489 (2021)
4. Beierle, C., Haldimann, J.: Normal forms of conditional belief bases respecting inductive inference. In: Keshtkar, F., Franklin, M. (eds.) Proceedings of the Thirty-Fifth International Florida Artificial Intelligence Research Society Conference (FLAIRS), Hutchinson Island, Florida, USA, 15–18 May 2022 (2022)
5. Beierle, C., Haldimann, J.: Normal forms of conditional knowledge bases respecting system p-entailments and signature renamings. Ann. Math. Artif. Intell. **90**(2), 149–179 (2022)
6. Benferhat, S., Cayrol, C., Dubois, D., Lang, J., Prade, H.: Inconsistency management and prioritized syntax-based entailment. In: Proceedings of IJCAI 1993, vol. 1, pp. 640–647. Morgan Kaufmann Publishers, San Francisco (1993)
7. Darwiche, A., Pearl, J.: On the logic of iterated belief revision. Artif. Intell. **89**(1–2), 1–29 (1997)
8. de Finetti, B.: La prévision, ses lois logiques et ses sources subjectives. Ann. Inst. H. Poincaré **7**(1), 1–68 (1937). Engl. transl. Theory of Probability, J. Wiley & Sons, 1974
9. Goldszmidt, M., Pearl, J.: Qualitative probabilities for default reasoning, belief revision, and causal modeling. Artif. Intell. **84**(1–2), 57–112 (1996)
10. Haldimann, J., Beierle, C.: Inference with system W satisfies syntax splitting. In: Principles of Knowledge Representation and Reasoning: Proceedings of the 19th International Conference, KR 2022, Haifa, Israel, 31 July–5 August 2022 (2022, to appear)
11. Heyninck, J., Kern-Isberner, G., Meyer, T.: Lexicographic entailment, syntax splitting and the drowning problem. In: 31st International Joint Conference on Artificial Intelligence, IJCAI 2022. ijcai.org (2022, to appear)
12. Kern-Isberner, G.: Conditionals in Nonmonotonic Reasoning and Belief Revision. LNCS, vol. 2087. Springer, Heidelberg (2001). https://doi.org/10.1007/3-540-44600-1_3
13. Kern-Isberner, G.: A thorough axiomatization of a principle of conditional preservation in belief revision. Ann. Math. Artif. Intell. **40**(1–2), 127–164 (2004)
14. Kern-Isberner, G., Beierle, C., Brewka, G.: Syntax splitting = relevance + independence: New postulates for nonmonotonic reasoning from conditional belief bases. In: Calvanese, D., Erdem, E., Thielscher, M. (eds.) Principles of Knowledge Representation and Reasoning: Proceedings of the 17th International Conference, KR 2020, pp. 560–571. IJCAI Organization (2020)
15. Kern-Isberner, G., Brewka, G.: Strong syntax splitting for iterated belief revision. In: IJCAI-2017, pp. 1131–1137 (2017)
16. Kern-Isberner, G., Eichhorn, C.: Structural inference from conditional knowledge bases. Stud. Logica. **102**(4), 751–769 (2014)
17. Komo, C., Beierle, C.: Nonmonotonic inferences with qualitative conditionals based on preferred structures on worlds. In: Schmid, U., Klügl, F., Wolter, D. (eds.) KI 2020. LNCS (LNAI), vol. 12325, pp. 102–115. Springer, Cham (2020). https://doi.org/10.1007/978-3-030-58285-2_8
18. Komo, C., Beierle, C.: Nonmonotonic reasoning from conditional knowledge bases with system W. Ann. Math. Artif. Intell. **90**(1), 107–144 (2021). https://doi.org/10.1007/s10472-021-09777-9
19. Kraus, S., Lehmann, D., Magidor, M.: Nonmonotonic reasoning, preferential models and cumulative logics. Artif. Intell. **44**(1–2), 167–207 (1990)

20. Kutsch, S., Beierle, C.: InfOCF-Web: an online tool for nonmonotonic reasoning with conditionals and ranking functions. In: Zhou, Z. (ed.) Proceedings of the Thirtieth International Joint Conference on Artificial Intelligence, IJCAI 2021, Virtual Event/Montreal, Canada, 19–27 August 2021, pp. 4996–4999. ijcai.org (2021)
21. Lehmann, D., Magidor, M.: What does a conditional knowledge base entail? Artif. Intell. **55**, 1–60 (1992)
22. Lehmann, D.: Another perspective on default reasoning. Ann. Math. Artif. Intell. **15**(1), 61–82 (1995). https://doi.org/10.1007/BF01535841
23. Parikh, R.: Beliefs, belief revision, and splitting languages. Logic Lang. Comput. **2**, 266–278 (1999)
24. Pearl, J.: System Z: a natural ordering of defaults with tractable applications to nonmonotonic reasoning. In: Proceedings of TARK 1990, pp. 121–135. Morgan Kaufmann (1990)
25. Peppas, P., Williams, M.A., Chopra, S., Foo, N.Y.: Relevance in belief revision. Artif. Intell. **229**((1–2)), 126–138 (2015)
26. Reiter, R.: A logic for default reasoning. Artif. Intell. **13**, 81–132 (1980)
27. Spohn, W.: Ordinal conditional functions: a dynamic theory of epistemic states. In: Harper, W., Skyrms, B. (eds.) Causation in Decision, Belief Change, and Statistics, II, pp. 105–134. Kluwer Academic Publishers (1988)

Towards the Evaluation of Action Reversibility in STRIPS Using Domain Generators

Tobias Schwartz(✉), Jan H. Boockmann, and Leon Martin

University of Bamberg, Bamberg, Germany
{tobias.schwartz,jan.boockmann,leon.martin}@uni-bamberg.de

Abstract. Robustness is a major prerequisite for using AI systems in real world applications. In the context of AI planning, the reversibility of actions, i.e., the possibility to undo the effects of an action using a reverse plan, is one promising direction to achieve robust plans. Plans only made of reversible actions are resilient against goal changes during plan execution. This paper presents a naive implementation of a non-deterministic theoretical algorithm for determining action reversibility in STRIPS planning. However, evaluating action reversibility systems turns out to be a difficult challenge, as standard planning benchmarks are hardly applicable. We observed that manually crafted domains and in particular those obtained from domain generators easily contain bias. Based on an existing domain generator, we propose two slight variations that exhibit a completely different search tree characteristics. We use these domain generators to evaluate our implementation in close comparison to an existing ASP implementation and show that different generators indeed favor different implementations. Thus, a variety of domain generators is a necessary foundation for the evaluation of action reversibility systems.

1 Introduction

A classical planning problem [8] is typically solved upon finding a sequence of actions, i.e., a plan, that leads from the predefined initial state to a desired goal state. In this context, the notion of *action reversibility* has extensively been studied recently [2,3,5,6,11] and is used to assess whether the effects of applying an action can be undone using a *reverse plan*. Although this property is less important in a static execution environment, it is crucial in a dynamic execution environment, e.g., as found in autonomous spacecraft control [14] or cloud management [13] domains, where the desired goal state might change during the execution of a precomputed plan. While a plan containing non-reversible actions might lead to dead ends upon a goal change introduced by the environment, a plan with only reversible actions is resilient against this change. A similar issue arises in a closed world setting when using online planning strategies since these are susceptible to dead ends [4]. Consequently, implementing action reversibility algorithms in current AI planning tools can be useful.

I. Varzinczak (Ed.): FoIKS 2022, LNCS 13388, pp. 226–236, 2022.
https://doi.org/10.1007/978-3-031-11321-5_13

The works in [3,11] introduced a framework for action reversibility that generalizes different reversibility notions studied in the past [5,6]. Besides a complexity analysis, which inherits the PSPACE completeness of basic STRIPS planning [1], the work in [11] proposes a non-deterministic algorithm for computing their generalized notion of uniform φ-reversibility, but leaves an implementation and evaluation of their algorithm to future work. In a follow-up paper [2], the authors provide an Answer Set Programming (ASP) based implementation, evaluate it using STRIPS domains obtained from a domain generator, and motivate a comparison with a procedural implementation for future work.

In this paper, we follow their calls to action and evaluate the performance of our prototypical breadth-first search (BFS) and depth-first search (DFS) implementation of their theoretical algorithm, before comparing the results to their ASP implementation. We show that the domain generator considered so far does not suffice for evaluating other action reversibility algorithms in STRIPS without bias. To elaborate, we propose two additional domain generators leading to domains with multiple reverse plans and with potential dead ends. Our results indicate that a rigorous analysis of the bias contained in generated domains is necessary to properly evaluate implementations of action reversibility systems.

2 Background and Related Work

We follow the idea of [7] and use the same naming conventions as [11] to define a STRIPS planning problem as a tuple $\Pi = (\mathcal{F}, \mathcal{A}, s_0, \mathcal{G})$ consisting of facts $\mathcal{F}$, actions $\mathcal{A}$, an initial state $s_0 \subseteq \mathcal{F}$ and a goal specification $\mathcal{G} \subseteq \mathcal{F}$. In this context, the fact set $\mathcal{F}$ contains atomic statements about the world, and a state is a subset of facts $s \subseteq \mathcal{F}$. Each action is a tuple $a = \langle pre(a), add(a), del(a) \rangle$, where $pre(a) \subseteq \mathcal{F}$ denotes the preconditions of action a, $add(a) \subseteq \mathcal{F}$ and $del(a) \subseteq \mathcal{F}$ denote the positive add effects and negative delete effects, respectively. We assume that actions are well-formed, i.e., $add(a) \cap del(a) = \emptyset$ and $pre(a) \cap add(a) = \emptyset$. Action a in state s is applicable iff $pre(a) \subseteq s$. Applying an action a in state s, given that a is applicable with respect to s, yields the state $a[s] = (s \setminus del(a)) \cup add(a)$. An action sequence $\pi = \langle a_1, \ldots, a_n \rangle$ is applicable in state s_0 iff there is a sequence of states $\langle s_1, \ldots, s_n \rangle$ with $0 \leq i \leq n$ such that a_i is applicable in s_{i-1} and $a_i[s_{i-1}] = s_i$. Applying an action sequence π in state s_0 yields $\pi[s_0] = s_n$. $|\pi|$ denotes the length of an action sequence π.

Intuitively, an action a can be considered reversible if there exists a reverse plan that revokes all changes introduced by a. In the literature, different notions of reversibility have been studied: the work of [5] introduces the notion of undoability, which defines an action a to be undoable iff there exists a reverse plan for every state s reachable from the initial state s_0 in a STRIPS planning problem Π. Their notion of *universal* undoability lifts this reachability restriction and holds if there exists a reverse plan for any state $s \in 2^{\mathcal{F}}$. The work of [6] operates on a restricted notion of reversibility and considers an action a reversible iff there exists a reverse plan that is independent of the state s in which a was applied.

The works in [3,11] provide a generalized framework that distinguishes several notions of reversibility where an action a is

1. φ-reversible iff a is S-reversible in the set of models S of the propositional formulas φ over $\mathcal{F}$. An action a is S-reversible iff there exists a reverse plan $\pi = \langle a_1, \ldots, a_n \rangle \in \mathcal{A}^n$ for every state $s \in S$ where a is applicable in $a[s]$ such that $\pi[a[s]] = s$;
2. reversible in Π iff a is $\mathcal{R}_\Pi$-reversible with respect to a STRIPS planning problem Π, where $\mathcal{R}_\Pi$ denotes the set of states reachable from the initial state s_0 of Π;
3. (universally) reversible iff a is $2^{\mathcal{F}}$-reversible; and
4. uniformly φ-reversible iff a is φ-reversible using the same reverse plan π.

As noted in [11], the second notion coincides with the work of [5], while a uniform restriction of the third notion coincides with the concept studied by [6]. This uniform restriction requires that the computed reverse plan is applicable disregarding the state in which action a was applied. Note that such a restriction is of practical interest because reverse plans can thereby be computed upfront.

Algorithm 1: Uniform φ-reversibility of an action a (adopted from [11]).

Input : A set of actions $\mathcal{A}$, an action $a \in A$
Output: A formula φ, a reverse plan π

$F^+ = (pre(a) \setminus del(a)) \cup add(a)$
$F^- = del(a)$
$F^0 = \emptyset$
$\pi = \langle\rangle$
while $pre(a) \not\subseteq F^+ \vee F^0 \cap F^- \neq \emptyset$ **do**
  non-deterministically choose $a' \in \mathcal{A}$ such that $pre(a') \cap F^- = \emptyset$
  if a' does not exist **then return** $\bot, \langle\rangle$
  $F^0 = F^0 \cup (pre(a') \setminus F^+)$
  $F^+ = (F^+ \setminus del(a')) \cup add(a')$
  $F^- = (F^- \setminus add(a')) \cup del(a')$
  $\pi = \pi \cdot a'$
$\varphi = \bigwedge_{l \in F^+ \cup F^0} l \wedge \bigwedge_{l \in F^-} \neg l$
return φ, π

For reference, Algorithm 1 depicts the non-deterministic algorithm proposed by [11] to compute a reverse plan π for an action a and an associated formula φ, such that a is at least uniformly φ-reversible. Note that F^+/F^- represent sets of facts that are true/false within a state, while F^0 represents the necessary preconditions. Since uniform reversibility is state independent, the algorithm only requires a set of actions $\mathcal{A}$ and the action a to be reversed as input. Upon termination, it provides a reverse plan π and a formula φ yielding states in which the reverse plan is applicable as output.

Listing 1.1. Single path domain generator (adopted from [2]).

```
(define (domain single-path-<i>)
(:requirements :strips)
(:predicates (f0) ... (f<i>))

(:action del-all
:precondition (and (f0) ... (f<i>))
:effect (and (not (f0)) ... (not (f<i>))))

(:action add-f0
:effect (f0))

...

(:action add-f<i>
:precondition (f<i-1>)
:effect (f<i>)))
```

3 Domains for Evaluating Action Reversibility

In contrast to traditional planning problems, the complexity of action reversibility primarily depends on the domain and not on particular instances. Hence, the characteristics of the domain are key for a proper evaluation. This section discusses why existing planning benchmarks, such as the IPC domains, and the domain generator proposed by [2], are not sufficient for an in-depth evaluation of action reversibility systems. We subsequently propose two concrete domain generators that yield search trees containing multiple paths suitable as reverse plans and potential dead ends.

3.1 Existing Benchmarks for Planning

The International Planning Competition (IPC) provides a large number of domains and accompanying instances encoded in the Planning Domain Definition Language (PDDL) [9] that enable an empirical comparison of planning systems. Recall that Algorithm 1 computes a uniform notion of reversibility and thereby only relies on the set of actions $\mathcal{A}$. Contrary to traditional planning where in particular PDDL instances are of interest, the evaluation of action reversibility systems solely depends on the PDDL domains. The work in [5] examined the STRIPS domains from IPC'98 to IPC'14 in their experiments and observed that most actions in these domains are not reversible. They have further noted that the reverse plans for the remaining reversible actions have very small reverse plans, i.e., often of length 1 where the effects of an action a can be completely undone in a single step by applying a so-called *inverse action* $\overline{a}$ [10]. For example in the well-known Blocksworld domain, the action `pickup` is the inverse action of action `putdown` and vice versa.

Listing 1.2. Our multiple paths domain generator.

```
1   (define (domain multiple−paths−<i >)
2   (:requirements :strips)
3   (:predicates (f0) ... (f<i >))
4
5   (:action del−all
6   :precondition (and (f0) ... (f<i >))
7   :effect (and (not (f0)) ... (not (f<i >))))
8
9   (:action add−f0
10  :effect (f0))
11
12  ...
13
14  (:action add−f<i >
15  :precondition (f<i−1>)
16  :effect (and (f<i >) (not (f0)) ... (not (f<i−1>)))))
```

Note that the existence of small reverse plans in STRIPS planning domains is not necessarily surprising since the complexity of a planning problem primarily arises from the PDDL instances, i.e., the concrete initial state and goal state. This is also the case for domains studied in the field of real-time planning [4] where domains typically contain a few actions only and the complexity of the modeled system is encoded in the states. In contrast, complexity of the action reversibility problem primarily arises from the action to be reversed in the context of other available actions in the respective domain. Accordingly, the IPC domains in particular are not well-suited for a sophisticated evaluation of action reversibility systems, due to the domains containing very small reverse plans.

3.2 Single Path Domain

To overcome the limitations of the IPC domains for evaluating action reversibility, the authors of [2] use a handcrafted domain generator. This generator, as shown in Listing 1.1, receives a single integer i as input and generates the PDDL domain `single-path-<i>` where action `del-all` is to be reversed. A reverse plan must reestablish the precondition, i.e., the predicates `f0` to `f<i>`. These predicates can be regained using the actions (`add-f<i>`) that each require the add effect from their preceding action `f<i-1>` to regain predicate `f<i>`. Hence, each action `add-f<i>` has to be called exactly once in ascending order to reverse the effect of action `del-all`. With respect to the PDDL domain, this property is primarily encoded by the effect of the `add-f<i>` action, i.e., :`effect` (`(f<i>)`) (see line 16 in Listing 1.1). The length of the reverse plan for action `del-all` grows linear with respect to input i, precisely $|\pi| = i$.

The search trees produced by this domain comprise exactly a single reverse plan containing the actions `add-f<i>` in ascending order. Note that this is not a problem for the ASP implementation [2], which follows a guess-and-check pattern. Their approach first guesses action sequences of fixed length, but disregards their applicability. Whether a sequence of actions is executable is checked in the subsequent step. Hence, the performance of their approach, disregarding the impact of the ASP encoding and ASP solver, is expected to primarily depend on the length of the reverse plan and not on the search tree characteristics.

Thereby, a single path domain is only suitable for the evaluation of approaches that follow a similar pattern. However, this is not the case for search algorithms, which consider the applicability of actions during the construction and traversal of the search tree. For those algorithms, the length of the reverse plan and the characteristics of the search tree are expected to have an impact on their performance. Accordingly, the single path domain generator is not sufficient to evaluate arbitrary action reversibility systems.

3.3 Multiple Path Domain

We propose to extend the aforementioned single path domain such that multiple paths, each representing a valid reverse plan, exist. This is achieved by making permutations of the prior reverse plan also feasible. To do so, our multiple paths domain generator, as shown in Listing 1.2, adjusts the effects of actions `add-f<i>`. Instead of only adding predicate `f<i>`, these actions now delete all prior predicates `f0` to `f<i-1>`. As result, these have to be regained by subsequently executing all previous actions. This change modifies the characteristics of the search tree and also lifts the length of the reverse plan from linear to quadratic with respect to the input i, precisely $|\pi| = \sum_{k=1}^{i} k$.

Obviously, computing a reverse plan of quadratic length with respect to i requires significantly more time than one of linear length. However, since all applicable paths resemble valid reverse plans, the performance of a DFS approach should not differ much from the single path domain given that both reverse plans have the same length. By contrast, the performance of a BFS approach is expected to be much worse due to the increased size of the search tree. The guess-and-check pattern employed by the aforementioned ASP implementation should not be affected significantly by the difference in the search tree, but rather by the increased length of the reverse plan.

3.4 Dead End Domains

Both domains discussed so far have the property that any applicable path in the search tree always leads to a valid reverse plan, which strongly favors a DFS approach. In real world planning problems, executing an arbitrary action does not necessarily provide progress towards the goal. In hindsight, an action might have been unnecessary for reaching the desired goal state or even led to a dead end, i.e., a state where no further actions are applicable. From a search perspective, this requires the need to backtrack to a previous configuration and follow a different path instead. Consequently, a DFS approach can in this context be drastically slower, because it follows potentially long paths that eventually reach a dead end. In contrast, dead end paths longer than the reverse plan should hardly affect a BFS disregarding the larger search tree.

Listing 1.3. Our dead ends domain generator.

```
1  (define (domain dead-ends-<i>)
2  (:requirements :strips)
3  (:predicates (f0) ... (f<i>) (token))
4
5  (:action del-all
6  :precondition (and  (f0) ... (f<i>))
7  :effect (and  (not (f0)) ... (not (f<i>)) (token)))
8
9  (:action consume
10 :precondition (token)
11 :effect (not (token)))
12
13 (:action add-f0
14 :precondition (token)
15 :effect (f0))
16
17 ...
18
19 (:action add-f<i>
20 :precondition (f<i-1>) (token)
21 :effect (and (f<i>) (not (f0)) ... (not (f<i-1>)))))
```

PDDL domains containing dead ends can be constructed similarly to the single and multiple path domain generators. Recall that the starting state upon computing reverse plans for an action a is the state $a[s]$, i.e., the state after applying a. This state is characterized by $(pre(a) \setminus del(a)) \cup add(a)$, which resembles the initial value of F^+ in Algorithm 1. Further note that this starting state is empty for the single and multiple path domain generator, because $pre(del\text{-}all) = del(del\text{-}all) \wedge add(del\text{-}all) = \emptyset$ holds. While the derived search trees for these domains do not contain dead ends, this observation can be generalized even further: If there exists a reverse plan for action a and $pre(a) = del(a) \wedge add(a) = \emptyset$ then the search tree for computing action reversibility does not contain a dead end. The proof follows from the necessary existence of actions that do not contain preconditions and can be used to produce predicates, e.g., the actions add-f<i>. A reverse plan would not be possible in the absence of such actions.

Dead end domains can be constructed with the help of a token predicate, which is required but cannot be produced by other actions, and adding it to the add effects of the to be reversed action. Our proposed dead ends domain generator is shown in Listing 1.3 and accordingly adds the token predicate to the add effect of action del-all and adds it to the precondition of all other actions. The new action consume irreversibly removes this token predicate upon execution, thereby resulting in a dead end, because no action is applicable anymore. Such a "malicious" action can potentially always be executed during the search.

4   Experiments

In what follows, we briefly describe our implementation of Algorithm 1, and primarily discuss the findings of our conducted performance evaluation for the

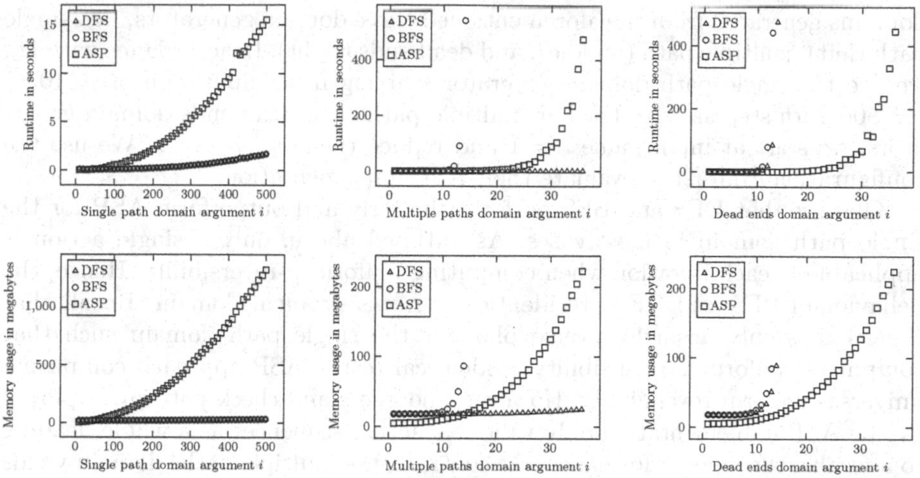

Fig. 1. Runtime (top) and memory usage (bottom) of our procedural BFS and DFS and the ASP implementation [2] for the single path (left), multiple paths (middle), and dead ends (right) domains with a timeout of 10 min.

DFS, BFS, and ASP implementation with respect to the aforementioned single path, multiple path, and dead ends domain generators. Our implementation, the domain generators, and the results of our experiments are available online at https://github.com/TobiasSchwartz/strips-reversibility-benchmarks.

4.1 Implementation

Our implementation of Algorithm 1 is written in Python and uses PDDL as input format. Non-deterministic choice of the next action a' in line 6, which requires a suitable selection strategy in a deterministic implementation, is implemented as follows: We successively construct a search tree where paths correspond to action sequences and explore it either in a BFS or DFS manner. Note that we do not employ heuristics for action selection on purpose to exclude the possibility of overfitting with respect to the small number of domains used in the evaluation. We consider two nodes in our search tree equivalent if they share the same values for F^+, F^-, and F^0, but ignore the value of π. As an optimization, we use this notion to introduce a cycle detection and do not consider a candidate action if its application yields an equivalent node, which has already or is to be traversed.

4.2 Results and Discussion

We conducted our experiments using a desktop PC with a stock AMD Ryzen™ 5 3600 and 16 GB of RAM. Figure 1 depicts the runtime (top) and memory usage (bottom) for the three implementations, i.e., our DFS and BFS implementation, and the simple encoding for the ASP implementation of [2]. We use domains generated from the aforementioned three domain generators, i.e., single path (left), multiple path (middle), and dead ends (right). In accordance with [2], we use the single path domain generator starting from input value $i = 10$ to $i = 500$ with step size 10. For our multiple paths and dead ends domain generators, we start at input value $i = 1$ and reduce the step size to 1. We use the configuration from [2] to evaluate their ASP implementation.

Observe that DFS and BFS perform similarly and outperform ASP for the single path domain in most cases. As outlined above, only a single action is applicable in each iteration when computing uniform φ-reversibility. Hence, the behavior of BFS and DFS are identical for this problem domain. Recall that there exists only a single reverse plan for the single path domain such that computing uniform φ-reversibility is identical to the ASP approach computing universal uniform reversibility. However, the guess-and-check pattern employed by the ASP implementation only validates action sequences as a whole, leading to a weaker runtime performance. By contrast, the multiple paths domain yields a large number of possible reverse plans. Hence, as expected, the DFS approach performs best for these domains, because every admissible action sequence eventually yields an applicable reverse plan. However, this poses a problem for the BFS approach due to the large number of potential states, which is also reflected by the increased memory usage. Albeit the naive use of a guess-and-check pattern, the ASP approach outperforms the BFS approach. We assume that this observation can be ascribed to the fact that our BFS implementation is not optimized with respect to performance and memory usage in contrast to the ASP solver underlying the ASP approach. Finally, in the dead ends domain we again find a large number of potential states. But, in this domain, not every action sequence necessarily has a valid reverse plan. Accordingly, both DFS and BFS require more time and space than the ASP approach.

In summary, our results show that minor changes to the domain generation and thus the resulting domains can have a major impact on the performance of different action reversibility algorithms. In particular, evaluating search based algorithms, e.g., the aforementioned DFS and BFS, using existing domain generators can easily lead to a biased result.

5 Conclusion

In this paper, we implemented the notion of uniform φ-reversibility based on the non-deterministic algorithm proposed by [11]. With this we follow their call to provide an actual implementation for said theoretical algorithm. Our implementation considers a breadth-first search and depth-first search variant. Following the suggestion of [2], we evaluate our procedural implementation in close comparison to their Answer Set Programming implementation. Evaluating action reversibility, however, turns out to be a difficult challenge on its own. In contrast to traditional planning problems, the complexity of action reversibility primarily depends on the planning domain and not on particular planning instances. However, the majority of existing planning benchmark problems, e.g., from the IPC, are complex at instance level, whereas actions in the domain are often, if reversible at all, trivially reversible using inverse actions. The domain generator introduced by [2] generates STRIPS domains of increasing complexity. We show that these contain a strong bias favoring other search strategies over their ASP approach. We have designed two domain generators that produce domains containing multiple reverse plans and containing dead ends. All generators, albeit similar in construction, produce domains that drastically favor different approaches. We thereby highlight the challenge of constructing and using domain generators with respect to evaluating action reversibility systems.

Regarding future work, we consider the design of further domain generators worth exploring in order to broaden the variety of generable domains, i.e., different problem classes, and thereby improve the validity of obtained performance results. Understanding the characteristics of domain generators allow to identify the most promising approach for a given problem instance paving the way for automated algorithm selection [12]. Finally, a solid benchmark of different domain generators enables to assess whether future algorithm improvements do result in a better performance in general or only for particular problem classes.

Acknowledgements. We would like to thank the anonymous reviewers for their insightful feedback. This work has been partially supported by BMBF funding for the project Dependable Intelligent Software Lab. Financial support is gratefully acknowledged.

References

1. Bylander, T.: The computational complexity of propositional STRIPS planning. Artif. Intell. **69**(1–2), 165–204 (1994)
2. Chrpa, L., Faber, W., Fiser, D., Morak, M.: Determining action reversibility in STRIPS using answer set programming. In: Workshop Proceedings Co-located with ICLP 2020. CEUR Workshop Proceedings, vol. 2678. CEUR-WS.org (2020). http://ceur-ws.org/Vol-2678/paper2.pdf
3. Chrpa, L., Faber, W., Morak, M.: Universal and uniform action reversibility. In: KR 2021, pp. 651–654 (2021). https://doi.org/10.24963/kr.2021/63
4. Cserna, B., Doyle, W.J., Ramsdell, J.S., Ruml, W.: Avoiding dead ends in real-time heuristic search. In: AAAI 2018, pp. 1306–1313. AAAI Press (2018). https://www.aaai.org/ocs/index.php/AAAI/AAAI18/paper/view/17405
5. Daum, J., Torralba, Á., Hoffmann, J., Haslum, P., Weber, I.: Practical undoability checking via contingent planning. In: ICAPS 2016, pp. 106–114. AAAI Press (2016). http://www.aaai.org/ocs/index.php/ICAPS/ICAPS16/paper/view/13091
6. Eiter, T., Erdem, E., Faber, W.: Undoing the effects of action sequences. J. Appl. Log. **6**(3), 380–415 (2008)
7. Ghallab, M., Nau, D.S., Traverso, P.: Automated Planning - Theory and Practice. Elsevier, San Francisco (2004)
8. Ghallab, M., Nau, D.S., Traverso, P.: Automated Planning and Acting. Cambridge University Press, Cambridge (2016)
9. Haslum, P., Lipovetzky, N., Magazzeni, D., Muise, C.: An Introduction to the Planning Domain Definition Language. Synthesis Lectures on Artificial Intelligence and Machine Learning. Morgan & Claypool Publishers (2019)

10. Koehler, J., Hoffmann, J.: On reasonable and forced goal orderings and their use in an agenda-driven planning algorithm. J. Artif. Intell. Res. **12**, 338–386 (2000)
11. Morak, M., Chrpa, L., Faber, W., Fiser, D.: On the reversibility of actions in planning. In: KR 2020, pp. 652–661 (2020)
12. Rice, J.R.: The algorithm selection problem. Adv. Comput. **15**, 65–118 (1976). https://doi.org/10.1016/S0065-2458(08)60520-3
13. Weber, I., Wada, H., Fekete, A.D., Liu, A., Bass, L.: Automatic undo for cloud management via AI planning. In: HotDep 2012. USENIX Association (2012)
14. Williams, B.C., Nayak, P.P.: A reactive planner for a model-based executive. In: IJCAI 1997, pp. 1178–1185. Morgan Kaufmann (1997). http://ijcai.org/Proceedings/97-2/Papers/056.pdf

Author Index

Printed in the United States
by Baker & Taylor Publisher Services